High Performance Materials in Aerospace

High Performance Materials in Aerospace

Edited by

Harvey M. Flower

Professor of Materials Science
Imperial College of Science, Technology and Medicine
London, UK

CHAPMAN & HALL

London · Glasgow · Weinheim · New York · Tokyo · Melbourne · Madras

Published by Chapman & Hall, 2–6 Boundary Row, London SE1 8HN, UK

Chapman & Hall, 2–6 Boundary Row, London SE1 8HN, UK

Blackie Academic & Professional, Wester Cleddens Road, Bishopbriggs, Glasgow G64 2NZ, UK

Chapman & Hall GmbH, Pappelallee 3, 69469 Weinheim, Germany

Chapman & Hall USA, One Penn Plaza, 41st Floor, New York NY 10119, USA

Chapman & Hall Japan, ITP-Japan, Kyowa Building, 3F, 2–2–1 Hirakawacho, Chiyoda-ku, Tokyo 102, Japan

Chapman & Hall Australia, Thomas Nelson Australia, 102 Dodds Street, South Melbourne, Victoria 3205, Australia

Chapman & Hall India, R. Seshadri, 32 Second Main Road, CIT East, Madras 600 035, India

First edition 1995

Typset in 10/12 Times by Pure Tech Corporation, Pondicherry, India
Printed in Great Britain at the University Press, Cambridge.

ISBN 0 412 53350 2

A catalogue record for this book is available from the British Library

Library of Congress Catalog Card Number: 94–68243

∞ Printed on acid-free text paper, manufactured in accordance with ANSI/NISO Z39.48–1992 (Permanence of Paper).

Contents

Contributors

S. Andrews
HIP Ltd, Chesterfield, UK

D. P. Davies
Materials Laboratory, Westland Helicopters Ltd, Yeovil, UK

D. M. Dawson
Rolls-Royce plc, Derby, UK

D. Driver
Centre for Adhesive Technology, Abingdon Hall, Cambridge, UK

H. M. Flower
Department of Materials, Imperial College of Science, Technology and Medicine, London, UK

P. J. Gregson
Department of Engineering Materials, University of Southampton, UK

H. Jones
Department of Materials Engineering, University of Sheffield, UK

M. McLean
Department of Materials, Imperial College of Science, Technology and Medicine, London, UK

N. Marks
Westland Helicopters Ltd, Yeovil, UK

P. G. Partridge
Interface Analysis Centre, University of Bristol, UK

C. J. Peel
Materials and Structures Department, Defence Research Agency, Farnborough, UK

B. A. Rickinson
HIP Ltd, Chesterfield, UK

D. Stephen
IEP Structures Ltd, Holbrook works, Halfway, Sheffield, UK

J. C. Williams
Engineering Materials and Technology Laboratories, GE Aircraft Engines, Cincinnati, USA

A. Wisbey
Materials and Structures Department, Defence Research Agency, Farnborough, UK

Preface

Aerospace presents an extremely challenging environment for structural materials and the development of new, or improved, materials: processes for material and for component production are the subject of continuous research activity. It is in the nature of high performance materials that the steps of material and of component production should not be considered in isolation from one another. Indeed, in some cases, the very process of material production may also incorporate part or all of the component production itself and, at the very least, will influence the choice of material/component production method to be employed. However, the developments currently taking place are to be discovered largely within the confines of specialist conferences or books each dedicated to perhaps a single element of the overall process.

In this book contributors, experts drawn from both academia and the aerospace industry, have joined together to combine their individual knowledge to examine high performance aerospace materials in terms of their production, structure, properties and applications. The central interrelationships between the development of structure through the production route and between structure and the properties exhibited in the final component are considered. It is hoped that the book will be of interest to students of aeronautical engineering and of materials science, together with those working within the aerospace industry.

Harvey M. Flower
Imperial College

1

Design requirements for aerospace structural materials

C.J. Peel and P.J. Gregson

1.1 INTRODUCTION

Structural materials used in the aerospace industry require a certain balance in physical and mechanical properties to enable their safe and efficient use. For the purposes of this chapter, these properties can be arbitrarily separated into those that affect the structural performance of the application *ab initio* and those limiting its service performance. Both types of property have an influence on the cost of the structure, either in its purchase or in ownership, and both types are therefore important in making the best selection of material for a specific application. Additionally, the development of an improved aerospace material will be seen to be dependent upon an enhanced balance of properties and rarely upon the improvement of a single characteristic. It is the purpose of this chapter to attempt to identify those features of a material that are required to optimize efficient structural application.

It should be realized that a commercial transport aircraft may be required to fly for a service life of as much as 60 000 hours spread over perhaps thirty years and encompassing 20 000 flights and that in this lifetime it will taxi on runways for a distance in excess of 100 000 miles. During this lifetime it will experience adverse climatic conditions, repetitive thermal cycles and, of course, hundreds of thousands of minor fatigue cycles imposed upon the major ground-to-air excursions. To a large extent the duty cycles imposed upon an aeroengine mirror those suffered by the airframe except, of course, that the thermal cycle is a more dominant element. Military aircraft will be required to survive perhaps one-tenth of the life of a civil transport with a service life of typically 6000 hours. While load levels induced by acceleration in manoeuvre are

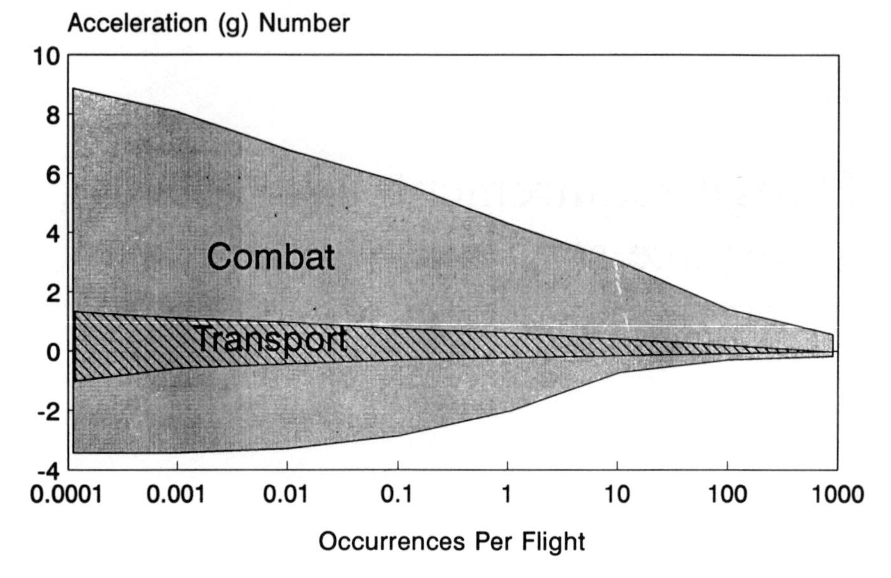

Figure 1.1 Typical acceleration-induced loadings for transport and combat aircraft.

high in military aircraft in comparison with civil transports, where gust loadings assume greater importance (Figure 1.1), the maximum levels of stress allowed in both types of structure tend to be similar. This is because the levels in stress generated per unit acceleration (or unit *g*) are designed to be much higher in transport aircraft than in their combat counterparts and consequently many design aspects and problems are common to both types of aircraft.

It is noteworthy that, although aircraft structures have been produced using high strength materials such as titanium alloy and steel, the overwhelming proportion of aircraft structures is made primarily of aluminium alloy. The proportions of aircraft structures produced in competing materials for aircraft newly built in 1960, 1980 and projected for 1995 are identified in Figure 1.2. Integration of these relative proportions over the last five decades would reveal the overwhelming predominance of the lightweight aluminium alloys but there are, however, significant future prospects claimed for the enhancement of the relative contributions of polymeric composite materials (Figure 1.3).

It is also important to recognize that the aeroengine introduces the requirement for efficient performance at much elevated temperature in relatively confined spaces and hence currently forces the use of high density, high temperature materials. Typical maximum application temperatures for existing and projected materials are shown in Figure 1.4.

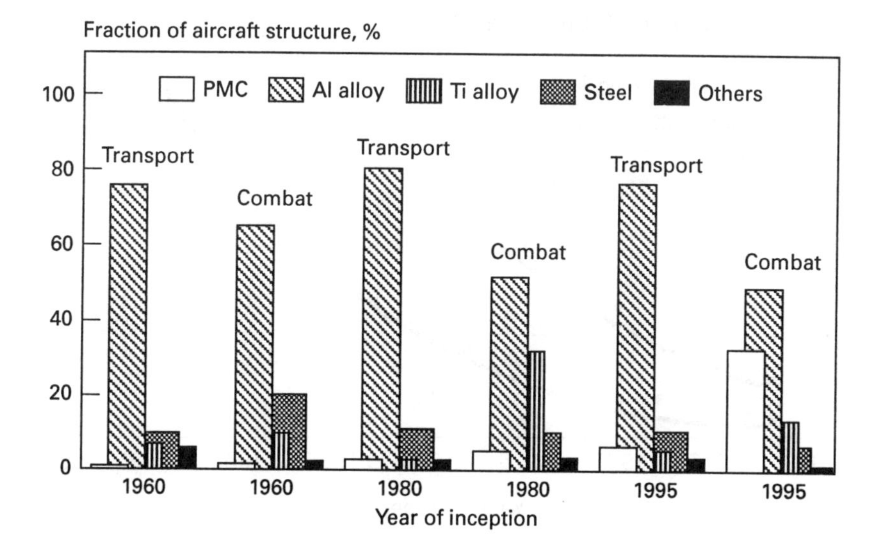

Figure 1.2 Distribution of materials in structural application, including landing gear.

1.2 PROPERTIES THAT AFFECT STRUCTURAL EFFICIENCY *AB INITIO*

Certain physical and mechanical properties of materials have a predominant effect on the efficiency of an aerospace structural design throughout a specified service envelope. This is because the efficiency of an airframe or aeroengine is dependent upon structural mass. In general terms the key properties identified are those of density, strength, stiffness, and a combination of properties here loosely categorized as damage tolerance. In aerospace terms these design properties may also need to be considered at ambient, reduced and, for engines and the airframes of supersonic transport aircraft, elevated temperatures.

To a certain extent the relative importance of these selected properties on design efficiency can be judged from the results of design studies published by Ekvall *et al.* [2] in which the impact of unit property changes on structural mass of selected airframes can be seen to emphasize the value of reduced material density (Figure 1.5). It is argued in this reference and related papers that improvements in the strength or stiffness of a material are relatively inefficient in comparison with a reduction in density for several critical reasons. For example, different parts of any one component or structural unit may be critical in terms of strength or stiffness or damage tolerance such that a unilateral improvement in a singular mechanical property may only affect a portion

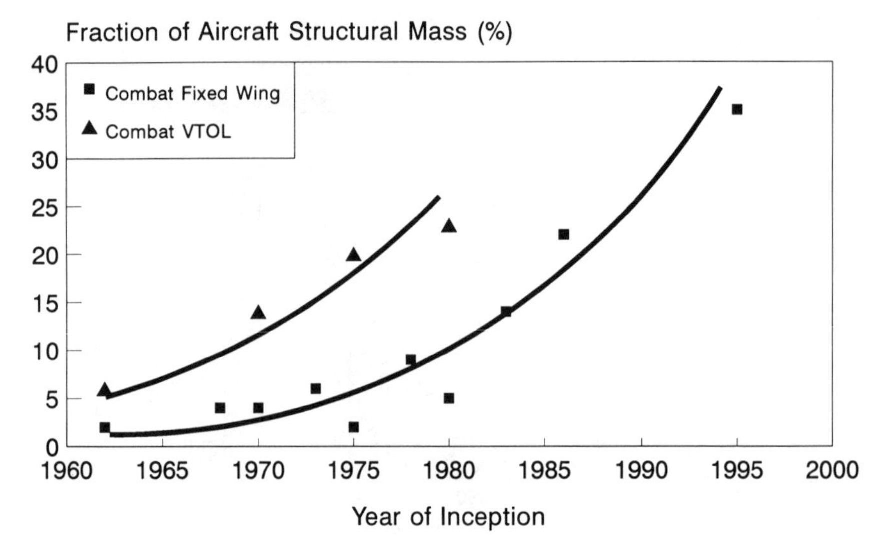

Figure 1.3 Polymeric composites in combat aircraft – primary structure excluding radomes etc. [1]

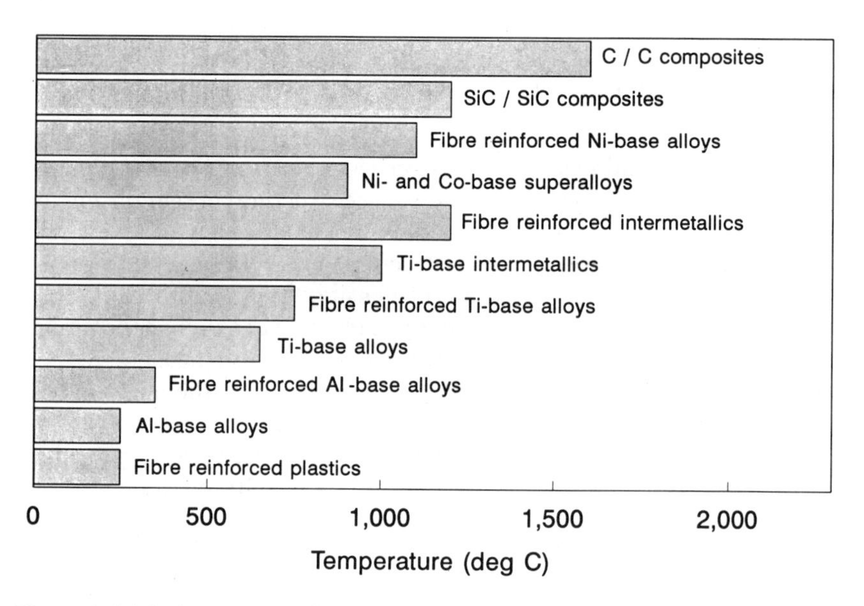

Figure 1.4 Maximum operating temperatures both existing and projected aerospace materials.

of the structure while a density reduction will affect all the components made of the selected material. This may limit the weight savings that may be achieved by an increase in strength, for example, to perhaps half of

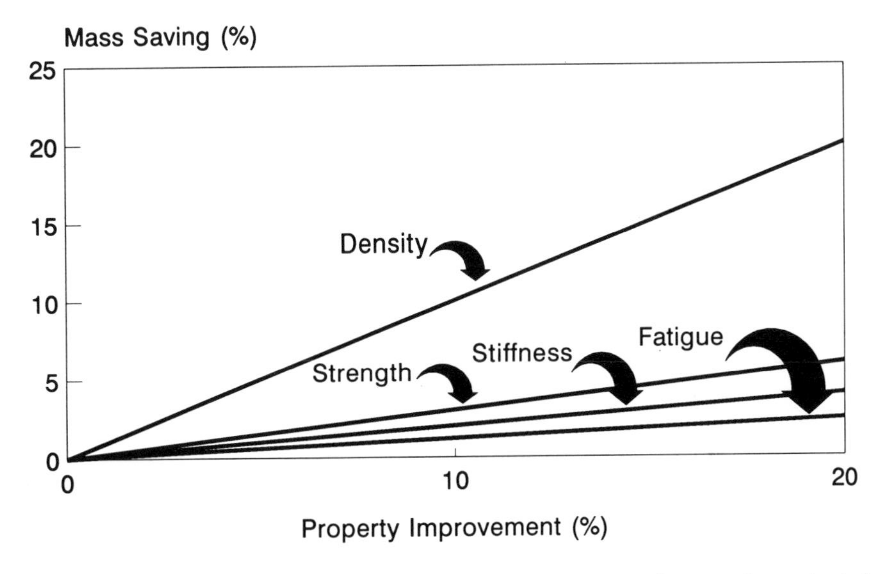

Figure 1.5 Effect of property improvement on structural mass for selected combat aircraft [2].

the structure. That is, mass reductions can be described in a somewhat simplistic manner, where:

$$m_n/m_o = \sum f_i r_i$$

m_n and m_o respectively define the new and old masses of a selected structure comprised of new and old materials. The influence of individual mechanical properties on mass is described in terms of the sum of the products of a property ratio, r_i, and the fraction, f_i of the structural mass critical in that property. Table 1.1 contains values of f_i typical of a modern combat aircraft and describes the property ratios. It can be seen that the ratio of densities appears in each property ratio, illustrating the overall benefit of density reduction.

It is quite possible that any component of an airframe will be marginally critical in several properties simultaneously. For example, an increase in stiffness may require a matching increase in strength and an increase in strength may require a matching increase in fatigue strength or damage tolerance. That is, an improvement in one property of a material may result in an increase in sensitivity to another. For example, increased working stress levels selected to exploit extra strength may result in premature fatigue failures unless a concomitant improvement in fatigue strength has also been achieved. To emphasize this difficulty it should be noted that damage mechanisms affecting metallic structures such as fatigue and creep are extremely stress dependent. However, mass

Table 1.1 Mass fraction critical in terms of property ratios

Design mode	*Property ratio* r_i	*Mass fraction civil transport* f_i	*Mass fraction combat aircraft* f_i	*Typical structural application*
Strength	$\frac{\rho_n}{\rho_o}\frac{\sigma_o}{\sigma_n}$	25%	15%	Tension: attachments Compression: wing top skin undercarriage Shear: wing spar webs Bearing: composite skin Notched: composite skin
Stiffness:		45%	45%	
Stiffness bending	$\frac{\rho_n}{\rho_o}\frac{E_o}{E_n}$			Aeroelastic: control surfaces Bending: Outer wing
Stiffness buckling	$\left(\frac{\rho_n}{\rho_o}\frac{E_o}{E_n}\right)^{1/3}$			Buckling: skin panels, stringers
Stiffness crippling	$\left(\frac{\rho_n}{\rho_o}\frac{E_o}{E_n}\frac{\sigma_o}{\sigma_n}\right)^{1/4}$			Crippling: wing top skin spars
Minimum Gauge	$\frac{\rho_n}{\rho_o}\frac{t_n}{t_o}$	15	20	Min. gauge: composite skin, helicopter fuselage outer wing, honeycomb
Fatigue	$\frac{\rho_n}{\rho_o}\frac{\sigma_o}{\sigma_n}$	15	20	Fatigue: lower wing, fuselage

reductions achieved by reductions in material density tend to reduce acceleration-induced load levels and unless the structure is aggressively re-optimized there will tend also to be an associated reduction in working stress levels leading to service life improvements plus mass reductions, hence the great attractiveness of lightweight materials.

In the following sections the relative performance of selected materials is discussed in terms of the properties outlined above. Consistently the effects of materials' density are accounted for by the use of specific property levels as opposed to absolute values. In general terms design allowable values for selected properties are used as opposed to typical levels. This is in an attempt to circumvent difficulties of comparison arising because of the marked differences found from material to material in scatter of property levels, in sensitivity to notches, or to the degrading effects of a natural environment that might include the effects of water vapour on an organic composite or corrosion on a metal. In making comparisons of the relative performance of fibre-reinforced composites allowance has to be made for the degree of alignment of the fibres in the loading direction. This is accomplished by the use of arbitrarily selected levels of alignment between 10% and 55% at a constant level of loading of 60% by volume of fibre in the polymeric matrix. The influence of cost of manufacture and material is addressed separately.

1.2.1 Density and Minimum Usable Gauge

Without doubt one of the most important physical properties affecting structural mass is that of density. The densities of current manufacturing materials employed in airframe and aeroengines are plotted (Figure 1.6). Notable lightweight materials are aluminium–lithium alloys, magnesium, polymeric matrix composites and honeycomb structures.

An important property related to material density is the minimum gauge of skinning material that can be produced and safely used. Thin skinning material has to be resistant to corrosion and inadvertent mechanical damage. Obviously, the safety of application will depend upon the nature of the structure and its working environment. Aluminium alloy sheet is used structurally in cladding aircraft fuselages in gauges as thin as 0.7 mm, perhaps a little thinner on helicopters and as thin as 0.1 mm in specialized honeycomb and space applications. Clearly, the ability to roll and chemically etch very thin sheet metals or to

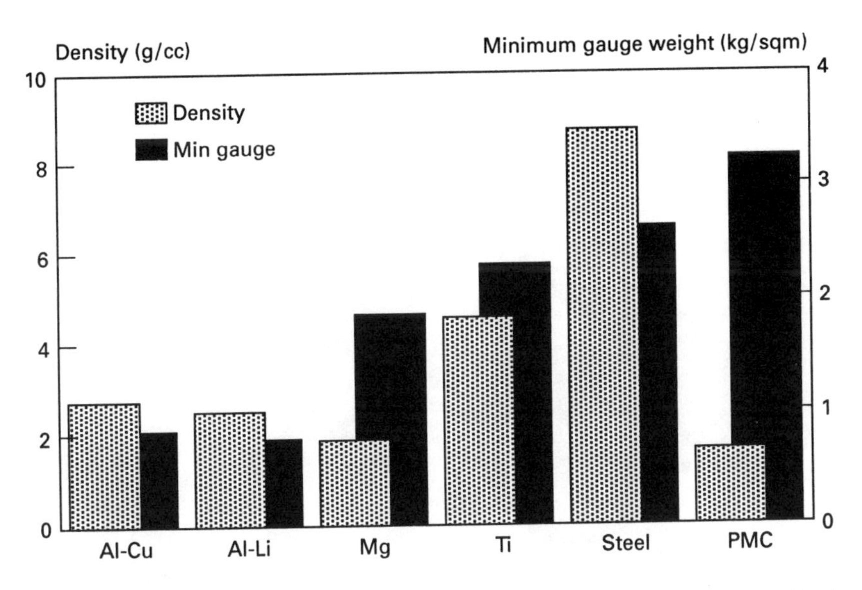

Figure 1.6 Densities and minimum useable gauges for a minimum 10-ply polymeric composite material.

produce very thin walled non-metallic honeycomb is important as are the properties including strength, fatigue strength and corrosion resistance of the resultant sheet.

Polymeric matrix composites or hybrid metal composite laminates present a special problem in the respect of minimum gauge requirements in that any laminate should contain a minimum number of plies to ensure consistency of property in any direction, the required degree of isotropy in property, and sufficient robustness of material. It is arguable as to the exact number of plies that might be deemed acceptable but it could be ten for a polymeric matrix composite and five for a hybrid laminate. Since the constituent plies of laminated materials themselves have a minimum producible thickness of perhaps 0.2 mm, a minimum composite gauge is achieved of 2 mm for polymeric matrix composites and of perhaps 1 mm for laminates. Clearly, the weight per unit area of sheet at minimum gauge differs in relative ranking from the simple consideration of density and aluminium alloys now seem very efficient (Figure. 1.6). This may explain, in part, the continued application of aluminium alloy

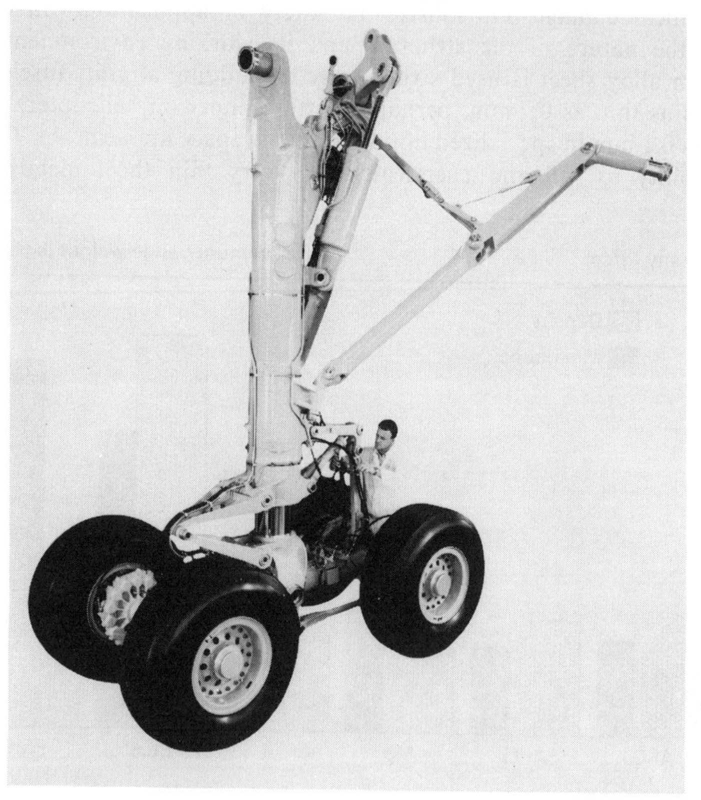

Figure 1.7 Large undercarriage – space minimized by high strength steel. (Courtesy Dowty Aerospace)

to the thin areas of aircraft and helicopter skinning in preference to composite application.

It should be noted, however, that there are applications, typified by undercarriage fittings and to a certain extent in an engine, where materials of maximum density such as ultra high strength steels might be chosen to minimize the volume of the component in question, assuming for the present that the levels of specific mechanical properties are adequate. The difficulty in accommodating the sheer size of very large undercarriages in fuselage or wing structures should not be underrated (Figure. 1.7).

1.2.2 Tensile Strength

The mechanical strength of a material whether under conditions of tensile, compressive, shear, bending or buckling loading can strongly influence its applicability to any particular design. In simple terms, engineering materials with a wide range of absolute tensile strengths are available. However, in terms of specific strength levels it becomes apparent that the strongest versions of the more advanced materials used in aerospace construction are generally mutually competitive in terms of their highest usable strength levels (Figure 1.8). In this example specific strengths are described as specific allowable strengths based on 75% of the minimum ultimate tensile strength of a metallic material or the

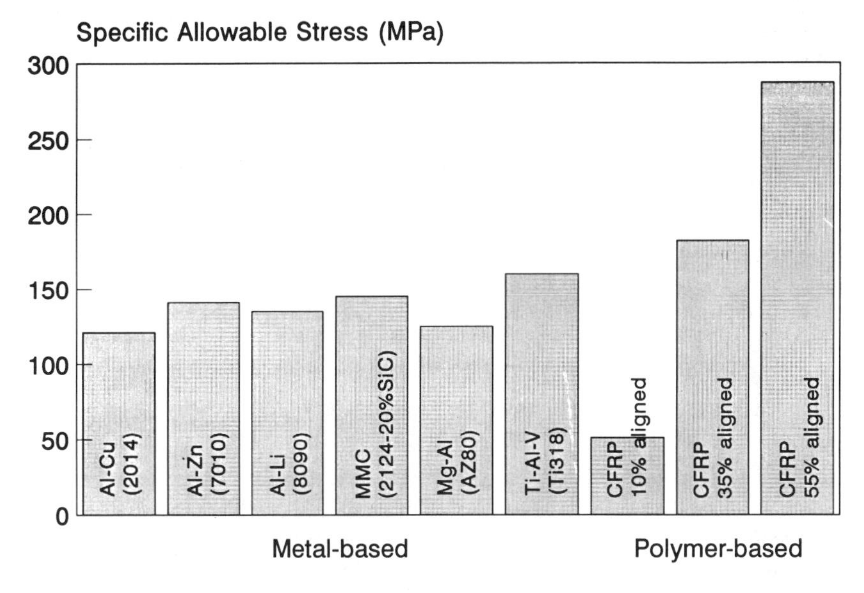

Figure 1.8 Specific maximum allowable strengths for 75% minimum UTS for metals and 0.5% strain for PMC.

specific stress level at 0.5% strain in a polymeric composite. That is, the strongest aluminium–zinc alloys are competitive with the ultra high strength steels and high strength titanium-based alloys. The extreme levels quoted for polymeric matrix composites reinforced by carbon fibres generally refer to anisotropic levels of property obtained in the direction of primary fibre alignment and quasi–isotropic laminates are more comparable. If high specific strength was the primary design requirement then any of these materials could be made competitive and a secondary requirement such as minimum gauge or maximum density might be important or, as shall be seen later, fatigue strength and corrosion resistance.

1.2.3 Limit loads, proof loads and ultimate design loads

In the design of strength critical aerospace structures account is taken of the variability in loads applied to any one structure throughout its life and of the variability in the strength of the structure stemming from natural variability in structural geometries, manufacturing practices and in degrading factors such as moisture retention or corrosion. Three critical structural load levels are defined, namely the limit load, the proof load and the ultimate design load. For any particular scenario or loading case selected, the limit load represents the highest load level experienced by the structure throughout its lifetime, encompassing the variability in service conditions. The proof and ultimate design loads for the same loading case are the limit load factored by a 'proof factor' or an 'ultimate factor' of typically 1.125 and 1.5 respectively and represent the minimum load levels below which the structure should not fail by unacceptable levels of deformation or ultimately by collapse and fracture. All the variabilities in materials performance under the detrimental effects of corrosion and manufacturing structural tolerances are accommodated in the proof and ultimate factors, while the natural scatter in material strength would be accommodated by matching the factored load levels against statistically derived minimum strength levels for the material. In certain designs typical of civil transport aircraft dominated by fatigue requirements, the proof factor may be 1.0, making the limit load and proof load levels identical.

Further reserve factors may be deliberately added to the proof and ultimate factors included in the design for both the onset of plasticity and for the ultimate failure case and indeed may occur naturally because of the inherently good performance of materials or structures. Typically, an ultimate factor of 1.5 might be specified but non-optimized metallic structures may well realize reserve factors of 2 when tested to destruction.

This combination of the two requirements to satisfy proof load and ultimate failure conditions, in effect, specifies a relationship between the yield and ultimate strengths of a material that must be obtained. Conventional metallic materials demonstrate a relationship between applied stress and resultant strain that is broadly characterized in the two regions of elastic and plastic strain response. In macroscopic terms a metallic structure would not be allowed to deform plastically nor to break such that there is a requirement both for a high plastic yield strength and a high ultimate failure strength. In general terms, an alloy used in civil transport aircraft will require a yield strength that is 67% of its ultimate tensile strength, while for a British military aircraft an alloy yield strength of 75% or more of the ultimate strength is required to accommodate the prescribed proof and ultimate factors. The natural relationships between minimum levels of yield strength, represented in the case of aluminium alloys by the minimum level of the 0.2% proof stress, and ultimate strength are plotted against the requirements for civil and military structures (Figure 1.9). It can be seen that the softer alloys used for damage tolerant structure, predominantly in civil aircraft, tend to be critical in yield strength while military high strength alloys tend to be critical in ultimate strength.

Fibre-reinforced polymeric (PMC) and metal matrix (MMC) composites again require special consideration at this stage. There is little argument that the longitudinal properties of the fibres used to reinforce modern aerospace structural composites are exceptionally high. However, most real structures require a combination of near isotropic in-plane performance and a balance of strengths in tension, compression

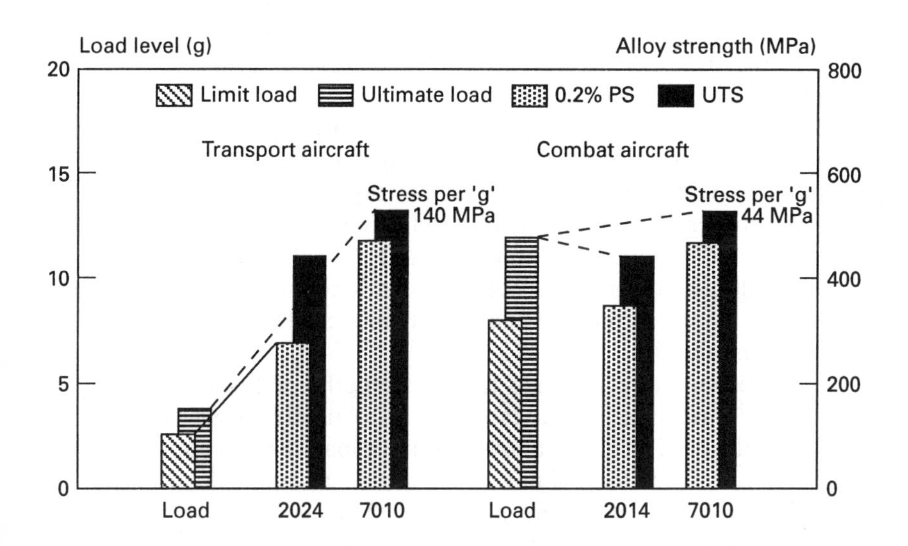

Figure 1.9 Loading levels and Al-alloy strength with limit loads of 2.5 g for civil and 8 g for combat aircraft. NB Ultimate loads are 1.5 × limit load.

and shear, for example. This ensures that a composite material with any of the available combinations of fibres and metallic or non-metallic matrices is likely to require alignment of the fibres in several selected orientations. To some extent these orientations and loadings can be tailored to suit the design in question assuming that all the appropriate loading conditions are known. Comparison of a material that has been semi-optimized for a specific design with more general mechanical properties of isotropic materials is therefore difficult. For example, in a conventional metallic design, it is likely that having met the strength requirements in the major axis of loading in either tension or compression, shear strength or any 'off-axis' strength requirements are easily met. In a fibre-reinforced composite structure this is not likely to be the case and attention has to be focused on the optimization of transverse and intermediate angles of fibre orientation. Since the performance of the weakest orientation is likely to affect the weight of the structure it is legitimate to make comparison of the properties of the fibre-reinforced materials in terms of their major and minor axes of loading simultaneously with those of the worst in-plane properties of the quasi-isotropic metals and metal matrix composites. This is achieved by the consideration of the performance of composites with selected loading levels in the direction of test or examinatiton, e.g. 10–55% (Figure 1.8). In this example the data for 10% loading might represent that for a transverse ply in comparison with the 55% used in the major axis of alignment and loading. There will, of course, be a contribution to the performance of the major axes by the fibres oriented at variable selected intermediate angles. Typically this will be significantly less than the vectored contributions of the fibre *per se*, making the prediction of ultimate strength a complex issue prone to scatter.

While prediction or measurement of 'first-ply' failure can be used to increase the accuracy of prediction of composite strengths for subsequent comparisons with metallic materials, the expedience is made of a comparison between the minimum allowed strengths specified for the isotropic metallic materials at limit load and the calculated stresses at the levels of elastic strain permitted at limit load in the fibre-reinforced composites, based on the more accurately modelled elastic moduli for the composite materials.

The large scatter in ultimate failure strengths of the fibre-reinforced composites can be accommodated by a relatively large safety factor applied to the level of strain allowed at limit load. For example, a typical value of 0.5% strain at limit load might be applied to a tension dominated failure mode in the major axis. This compares with an actual level of strain to failure that may well exceed 1%. This large safety factor also accommodates the damage likely to be suffered by the matrices of PMC under hot wet conditions and it incorporates the very damaging effects of the inevitable presence of stress concentration at notches. The comparative metallic performance would be taken to be the allowed specific stress level at limit load taken as 75% of the minimum ultimate tensile strength of the metal in the case of a military structure. The comparisons presented (Figure 1.8) of specific design allowable stresses and strains give a more realistic comparison of the relative performance of metallic materials and polymeric matrix composites.

The difficulty is compounded in the transverse orientation of uniaxial composites including fibre-reinforced metal matrix composite, where the failure strain may be as little as 0.1%! This reduces the safe working stresses normal to the fibre orientation to perhaps as little as 50 MPa. Two approaches to this problem are current, namely, the tailoring of the orientations of fibre reinforcement to try to meet the loading requirements in all orientations or the use of stronger matrices to raise the general level of isotropic performance. In the latter case interesting examples of hybrid composites are emerging in which the matrices, both metallic and polymeric, used for fibre-reinforced composites are themselves reinforced with particulate matter to improve their transverse properties.

1.2.4 Compression strength

Significant portions of an aircraft structure are dominated by compressive design requirements including the upper surface of the wing and associated internal spars, or internal fuselage structure such as floor beams and support struts. It will be seen that a combination of stiffness and strength is required to minimize weight in these types of structure, the exact balance depending upon the levels of load to be applied. With a metallic design it may be important that the structure does not buckle elastically or collapse plastically in a crippling mode. It is extremely unlikely that compressive strength will be critical in the metallic case, provided crippling is controlled, and the development of high compressive yield strength has been the major issue. In the case of aerospace aluminium alloys, high strength aluminium–zinc alloys have dominated such applications for the last five decades. A steady increase in yield strength has been achieved with these alloys over this period. In general terms the high strength metallic materials outperform the polymeric matrix composites in compression because the onset of localized buckling and subsequent delamination further reduces the allowable strains in the composites to perhaps 0.4% (Figure 1.10). However, a characteristic of the aluminium–zinc system, or for that matter high strength titanium alloys, is that they have relatively poor specific stiffness and are outperformed in their turn by polymeric composites, metal matrix composites or in this respect aluminium–lithium alloys. That is, while the compressive

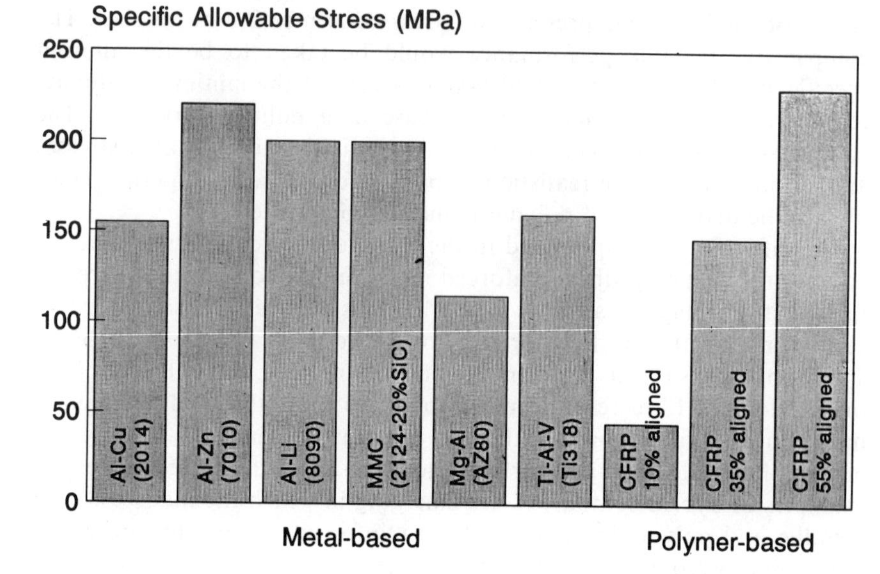

Figure 1.10 Specific compressive strengths with 0.1 % PS for metals and 0.4 % strain for PMC.

yield strength of a metal or the allowed compressive strain of a composite are important considerations, their resistance to elastic deformation and buckling must also be considered.

1.2.5 Specific stiffness

Approximately half the structure of an aircraft can be found to be critical in terms of stiffness, either in terms of its resistance to extension, bending, buckling or crippling. Here crippling is taken to imply permanent plastic deformation of an elastically buckled structure. Specific stiffness is therefore a materials property that has a strong influence on the choice of material during initial design studies. Specific stiffnesses are plotted for selected materials (Figure 1.11) and a specific comparison of resistance to elastic buckling is made based upon the parameter $E^{1/3}/\rho$ to represent the performance of long thin panels of material.

Comparisons are made in two parts, the first illustrating the relative performance of isótropic or quasi-isotropic materials (Figure 1.11), primarily engineering metals, and in the second plot (Figure 1.12) examples are given of the benefits of aligned fibre reinforcement against a standard value for isotropic aluminium–copper alloy. The improvements achieved with aluminium–lithium alloys and particulate reinforced materials are apparent, with gains in specific stiffness approaching 50%.

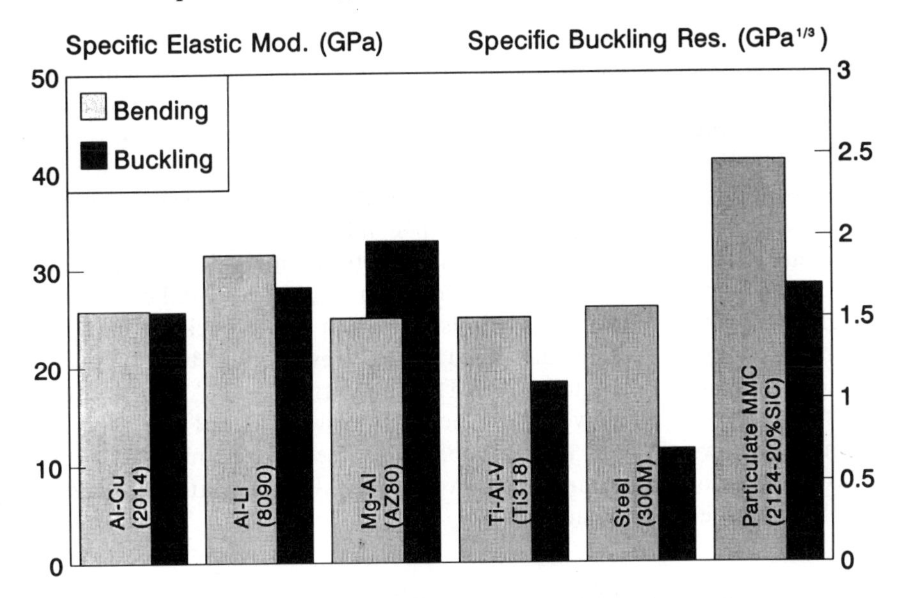

Figure 1.11 Specific elastic moduli for isotropic metals.

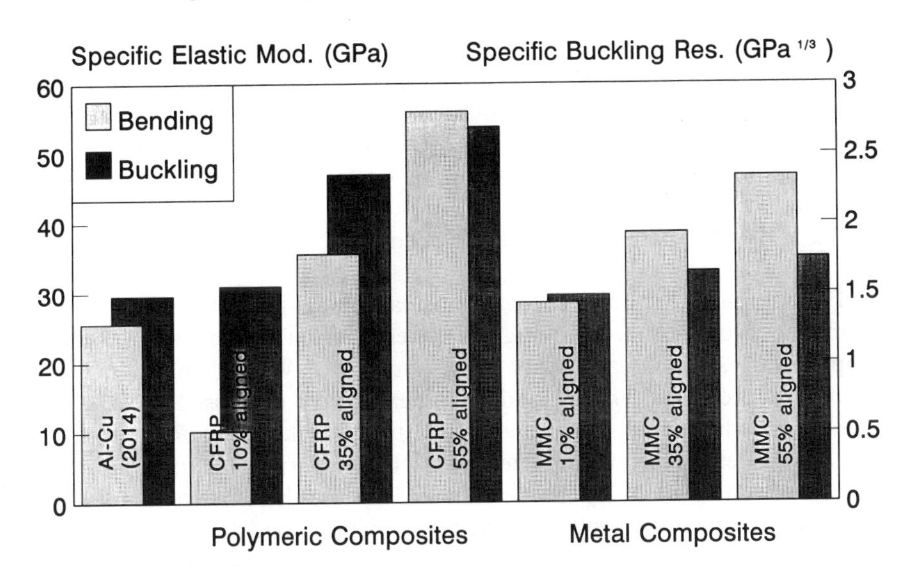

Figure 1.12 Specific elastic moduli for anisotropic fibre reinforced composites. (60 % fibre volume in PMC and 50 % fibre volume in MMC.)

In terms of the resistance to buckling, it is obvious that the materials with lowest density should perform well and the relative advantage of magnesium-based alloys shows clearly.

Fibre-reinforced composites, whether based on metallic or non-metallic matrices, are dominated by the degree of alignment of the fibre reinforcement. Clearly, as loading requirements become more isotropic the benefits of the fibre-reinforced composites diminish (Figure 1.12). However, it should also be noted that the fibre-reinforced composites based on non-metallic matrices perform relatively well in this idealistic comparison of resistance to buckling, because of their inherently low densities.

A further difficulty in the comparison between metallic structure and fibre-reinforced composites now occurs in that the metallic material might well be used at strain levels at which local buckling may be permitted while buckling may need to be prevented in the composite materials to prevent delamination. The comparison of $E^{1/3}$ ρ therefore is biased artificially in favour of the composite material in terms of comparative weight. For example, provided the metallic material is not plastically deformed, it may be allowed to buckle locally at perhaps 25% of the ultimate design load.

1.2.6 Strength and stiffness in compression structures

It is frequently the case that a combination of mechanical properties will be considered to be critical early in a design study and that a comparison of single property levels is inadequate. An example can be found in the requirement for high strength coupled with good stiffness in compressively loaded struts or wing panels. It is of little value to resist elastic buckling if plastic deformation sets in at a low value of strain. Similarly, in metallic tension-dominated structures, high strength may be of little value if the associated use of high working stresses produces premature fatigue failures.

To exemplify this point comparisons can be made of the behaviour of some modern materials that might be selected for a compressively loaded application such as an upper wing structure. Within the one wing structure the compressive loading index (often expressed as a compressive loading index P/L per unit width of plank, where P is end load and L the length of unsupported panel between ribs, for example) will fall from wing root to tip. Section thicknesses are therefore reduced in an efficient design from root to tip. Inboard, where loads are high, section strength is likely to dominate the situation but moving outboard the stiffness becomes progressively more important as the depth of the wing box and the section's thickness of the box skin materials are reduced. The balance between strength and stiffness requirements will vary significantly with aircraft weight such that, in general terms, a light aircraft wing will be dominated by stiffness and a large transport aircraft by strength.

It follows from these requirements that no one material will be completely optimized for a continuous upper wing skin. For example, by

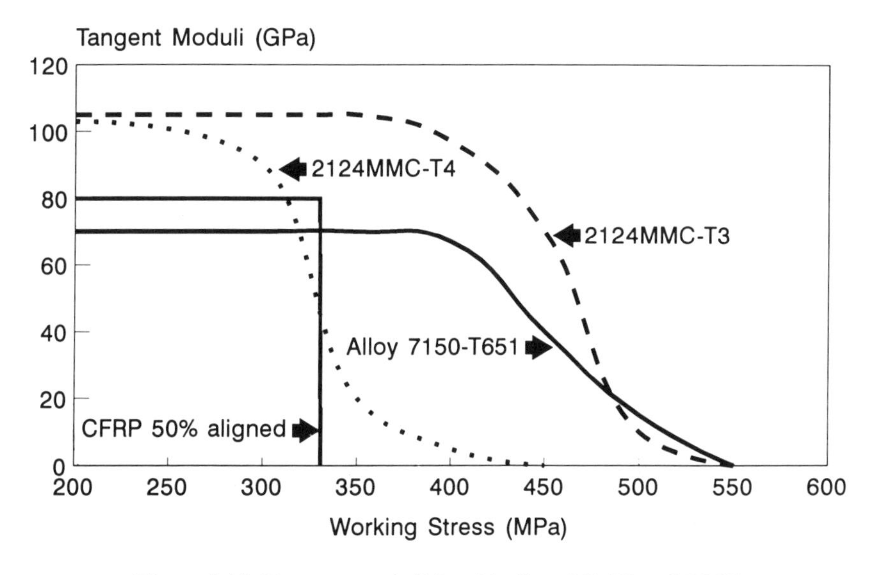

Figure 1.13 Tangent moduli for Al-alloy, MMC and PMC.

describing the elastic and plastic performance of selected metallic materials in terms of their specific tangent moduli at increasing levels of applied stress, one measure of their relative efficiencies in the two regimes can be obtained. Tangent moduli are readily obtained from compressive stress–strain curves (Figure 1.13) It can be seen that in terms of elastic performance, particulate reinforced composites perform efficiently, the latter being competitive with a notional polymeric matrix composite, but that the situation changes as the materials are more highly stressed with the strong aluminium–zinc alloys outperforming the polymeric composites. A tendency for the relatively low compressive yield strength of T4 particulate metal matrix composites to adversely affect their efficiency has been reported and can be seen here. A high strength aluminium–lithium alloy or an improvement in the allowable compressive strains limiting the performance of the polymeric composites would obviously be beneficial.

The comparison can be extended to compare the weights of the selected materials resistant to elastic buckling and plastic crippling under increasing levels of compressive loads applied to long, encastré rectangular panels of a constant length (e.g. 1 m) representing wing panel sections between ribs that are of sufficient 'minimum gauge' and are just thick enough not to buckle, yield or break. It can be seen that the carbon fibre reinforced PMC is most efficient at low levels of compressive loading index (Figure 1.14) typical of a light aircraft, but that because of the severe constriction of an allowable compressive strain of only 0.4% the

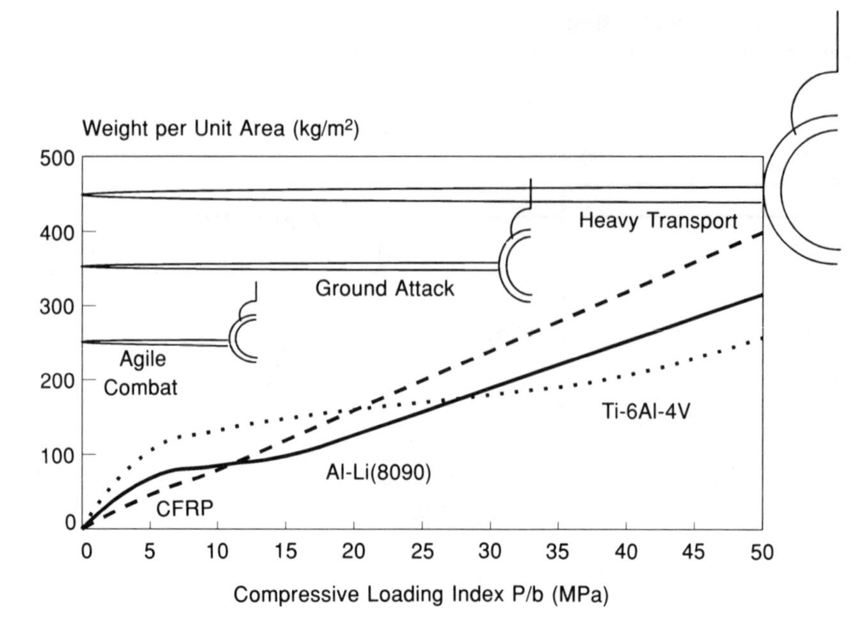

Figure 1.14 Resistance to buckling and crippling for the minimum weight of panel under compressive load.

metallic materials are more efficient at the higher loading levels. Conversely, the stiffness of titanium alloys is so low as to preclude its efficient use except in highly loaded parts.

1.2.7 Ductility

Ductility in a metal could be considered to be an important property in two respects. Firstly, most wrought metals are finally reduced to components by some form of forming operation and those, in particular, that involve stretch forming require adequately high ductility. Secondly, the onset of plasticity in a metal can assuage the effects of local stress concentrations, thereby strongly affecting the strength of a material in a built-up structure. In these respects ductility divides metals and composites, in particular, into two distinct fields, namely those with conventional levels of metallic ductility and those that are brittle.

Typical aerospace manufacturing methods for metallic material involve large operations such as the stretch forming of sheet to make curved fuselage panels where deformation strains may be up to approximately 6%. Aluminium alloys would normally be formed in a solution treated and quenched condition and then allowed to naturally age or be artificially aged up to the required strength level. That is, the structural component would be formed in metal in a soft condition and then be fully heat-treated to achieve service property levels. This type of approach will be precluded in the case of polymeric or metal matrix composites that remain inherently brittle or in the case of rapidly solidified metals and intermetallic alloys where the application of elevated temperature to achieve conformity of structure may destroy microstructural properties achieved by the very nature of the rapid solidification. In practice, aluminium alloys will be in a stretched and aged condition when found in large panels Since some of these operations involve large strains good ductility in the as quenched condition is required and normally achieved in conventional aluminium alloys. However, those varieties exhibiting strong natural ageing tendencies are frequently required to be stored under refrigerated conditions to preserve the good ductility of the quenched condition.

Small detailed parts containing large local strains may be formed by a process using multiple solution treatment and quenching operations between forming increments to restore the initial ductility to the heavily strained metal. This practice is usually concluded with a final resolution treatment and ageing but without the benefit of interspersed cold work producing material in a consistent T4 or T6 condition. Alloy compositions and conditions of heat treatment are detailed in Chapter 3.

Finally, metallic materials in both thin sheet and thicker plate forms can be hot formed by a range of techniques at different temperatures and strain rates. These vary from the slow hydrodynamic conditions of superplastic forming at temperatures near the metal solidus and strain rates of typically 10^{-3}/s, through slow strain rate creep forming, where the creep is conducted at the usual ageing temperature of the metal to result in a fully aged component after forming, to the rapid hydrostatic conditions of the hot fluid cell.

Two features need to be emphasized, firstly that temperature has a dramatic effect on the ductility of aerospace metallic materials and that the properties of any formed metal reflect not only its thermal heat-treatment history but also its mechanical manipulation.

Particulate reinforced metal matrix composites require further consideration at this point because they potentially introduce significant manufacturing difficulties. Early attempts to produce particulate MMC invariably produced materials with a tendency towards brittle behaviour. However, it is recognized that some of this difficulty was related to the presence of inadvertent defects in the material and in part to the addition of a hard reinforcement. For example, the manufacturing routes employed to produce isotropic MMC materials involve processes such as powder metallurgy, stir casting or liquid phase deposition and as such all would carry with them a significant potential for the inclusion of chance defects that will embrittle the transverse orientations. This feature was one principal deficiency in the repeated attempts to establish powder route aluminium alloys in the aerospace structural market. However, there seems little doubt that a fine reinforcement of silicon carbide particles in a conventional aluminium alloy can perform in a ductile manner. The ductilities, of approximately 15%, quoted for 2124 alloy reinforced by SiC in the as-quenched T4 condition, although not entirely comparable to those of the unreinforced alloy, are quite adequate for forming processes. Moreover, this ductility varies with temperature and heat treatment practice, much as it would in a monolithic unreinforced alloy, suggesting that hot forming is possible. Metal matrix composites and their properties are considered in more detail in Chapter 8.

It has to be recognized that polymeric matrix and metal matrix composites reinforced by long fibres and monofilaments exhibit little ductility, if any, as a composite. Manufacturing techniques for fibre and monofilament reinforced composites will therefore be limited to casting routes or those involving final assembly techniques such as diffusion bonding or curing of pre-impregnated laminates. Thermoplastic resin matrices for fibre reinforcement offer perhaps a compromise between the relative ease of application of a thermosetting matrix and the hot plasticity of a metal. However, hot deformation techniques in either a metallic or thermoplastic PMC tend to lead to difficulties in controlling fibre continuity and dispersion. While the intractable nature of the brittle material may seem to limit its application to those products where the laying-up of pre-impregnated laminates is possible, it should be noted that this manufacturing route may well prove cost effective in employing the minimal amount of material and minimizing the assembly of substructures.

1.2.8 Ductility and notch sensitivity

To develop full strength and toughness in any material used in a structural component requires some ductility or strain alleviation by microcracking local to stress concentrations that may be present as either macroscopic geometrical features or as microscopic features such as the edges and corners of internal defects or reinforcing phases.

The question of the level of ductility required to alleviate structural stress concentrations is a major issue. It is the case that metallic materials are frequently used with mechanical joints requiring fasteners or attachments and that, despite a stress concentration factor of perhaps 3 local to the fastening hole, the ductile metallic material will develop the full strength of an unnotched coupon of similar net section. This capability is not developed in brittle composites, whether polymeric or metallic in nature, where a reduction in strength at least of the same magnitude as the stress concentration might be anticipated (Figure 1.15). It would seem therefore that they will be treated as an elastic material without the ability to accommodate internal stress concentrations and defects by

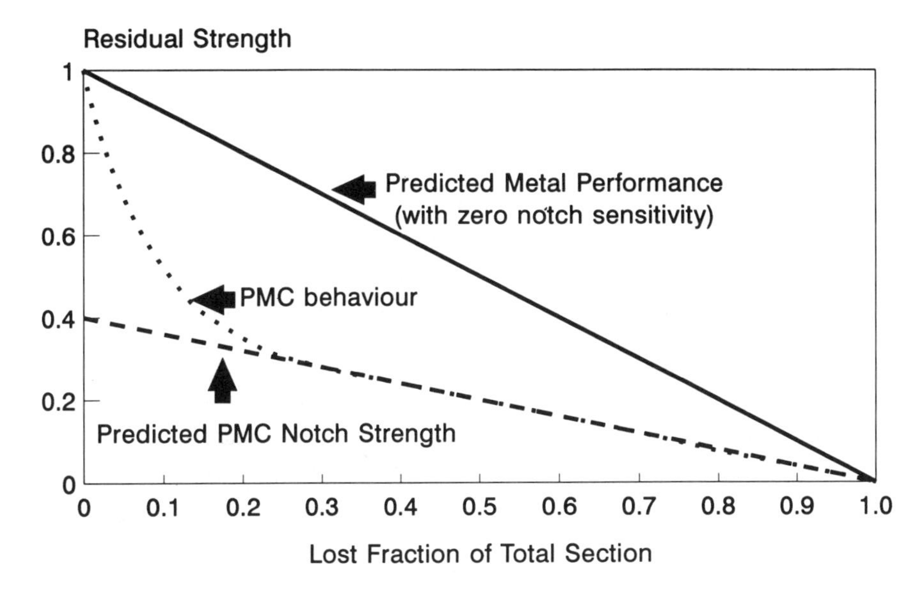

Figure 1.15 Notched strengths of metal and PMC with a stress concentration Kt = 2.5.

macroscopic plastic deformation and with limited accommodation by microcracking mechanisms. An approach of limiting allowed strain levels has to be adopted, as previously described for the polymeric composite materials. Consequently, due to inherent variations in composite quality in terms of fibre performance and distribution, reinforcement content and defect concentrations, large safety factors will necessarily be applied to engineering properties such as strength, notch strength and, possibly, stiffness. In the latter respect, it is noted that a statistical basis is required for the description of the elastic modulus of high performance fibres *per se*.

The metallurgist has become accustomed to ignoring the effects of stress concentrations on static (not fatigue) strength. This attitude will need to be changed as the application of brittle metal matrix composites becomes a reality.

Specifically arguing the value of plastic deformation in the alleviation of stress concentrations in ductile materials, care should be exercised in simple comparisons of materials based on ductility or total strain to failure derived from coupon tests. Consideration should be taken of the strain hardening performance of the material and its influence on the alleviation of local elastic stress concentrations. For example, although 2024–T3 has a lower strength than many of the aluminium–zinc alloys such as 7075–T651, the relatively low proof strength of the 2024 allows the dissipation of the stress concentration by plastic deformation such that the notched strength of the two alloys is similar.

1.2.9 Fracture toughness and notch sensitivity

An associated attribute to that of notch sensitivity is the fracture toughness of a material. That is, in addition to an ability to plastically deform when heavily loaded to assuage stress concentrations, a sound engineering material, when elastically loaded, must exhibit good residual strength in the presence of cracks, defects and notches, i.e. it should be tough. Individual components used in well-defined applications may require a specific level of fracture toughness to be achieved to enable cracks of a specified length to be contained. A prime example of this is the requirement for the skin of a pressure cabin to withstand the presence of a crack of approximately 0.5 m in half-length, the separation of two fuselage frames, at the limit load. In more general terms, a high level of fracture toughness is regarded as a good measure of a material's inherent quality. The well-established laws of linear elastic fracture mechanics dictate that as higher working stresses are employed in a material of given toughness, critical crack lengths will reduce according to the relationship:

$$\text{Critical crack length, } a_c = \text{constant. } K_{1c}^2/\pi\sigma^2$$

where K_{1c} and σ are the plane strain fracture toughness (K_c for plane stress states) and applied stress respectively and the constant term will reflect the crack geometry in a sample of nominally infinite dimensions. High strength materials used efficiently are likely to be highly stressed and consequently demonstrate short critical crack depths. This feature is therefore important in the critical examination of the performance of high strength metals including metal matrix composites offering significantly improved strengths.

To exemplify the effects of operating stress level, material strength and fracture toughness, critical crack lengths can be calculated for a notional large component containing a surface defect. This component is considered to have been manufactured in either a high strength, low toughness, material typifying a particulate reinforced metal matrix composite or in a weaker, tougher, conventional aluminium alloy. Each component when stressed to increasing fractions of its strength (represented by the minimum ultimate tensile strength) will fail at a reducing critical crack size and the combination of relatively low toughness and improved strength can be seen to produce relatively poor critical crack depths (Figure 1.16) but not completely beyond the range used in conventional alloys. However, it can be seen that the critical crack depths for the particulate reinforced composite are such that, even in the absence of an external notch, defects as small as 1.2 mm will start to reduce the effective strength of the material while defects of the order of 8 mm in size are required to have the same effect on unreinforced alloy.

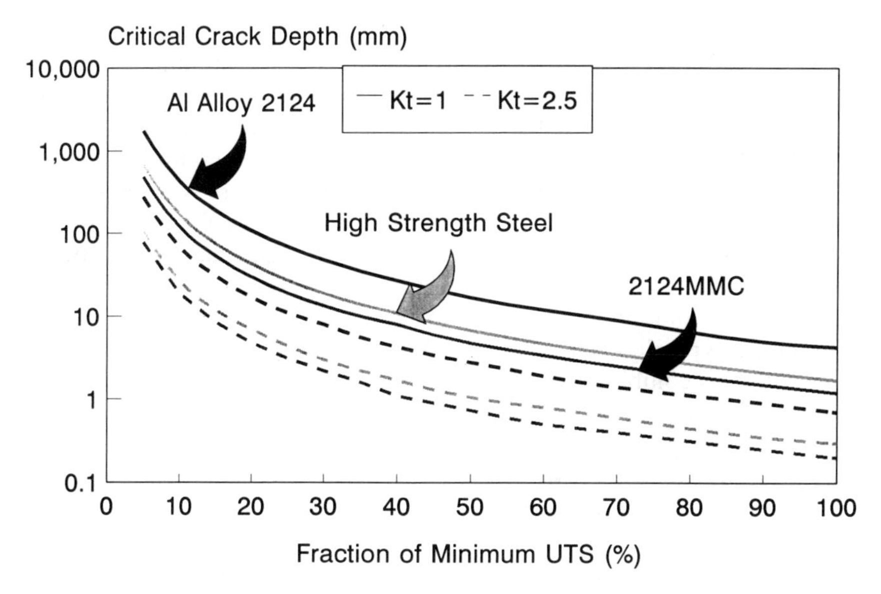

Figure 1.16 Critical crack sizes for plain (Kt = 1) and notched (Kt = 2.5).

The 1.2 mm defect size for reinforced material coincides well with the minimum registered equivalent size used in ultrasonic non-destructive evaluation of aerospace metals. That is, for conventional alloys, this prediction indicates that defects that are of sufficient size to cause a reduction in the strength of the material by failure before the onset of plasticity are well above the threshold size at which ultrasonic examination readily reveals their presence and above which specifications require their reporting, potentially for the rejection of the material. To achieve the same compatibility between non-destructive evaluation and the ability to fully realize the material's strength will require a significantly higher standard of cleanliness in the metal composites, where the situation is marginal, and a matching improvement in the resolution of non-destructive evaluation techniques.

When an external notch or defect is superimposed, it can be seen that in the presence of a stress concentration, typically of a factor of 2.5, critical crack sizes for the brittle metal composite at limit load reduce to less than 0.3 mm before plasticity can set in at approximately 75% of the tensile strength or the limit load. Defects commonly found in composites such as porosity or clusters of reinforcement are likely to be of this order. It is not surprising therefore to find that the metal composites are notch sensitive, unlike their conventional alloy counterparts, and that full material strength will not be realized in composites containing stress concentrations unless the fracture toughness is significantly increased.

Even the ability of the conventional alloy to contain defects up to the 1.2 mm size is marginal at stress levels of 75% of the minimum ultimate value, when a local geometrical stress concentration is superimposed. This is the level of the limit load and approaches the onset of plasticity with the exceedance of the metal's yield strength, whereupon ductility and notch sensitivity become more critical than fracture toughness.

A special mention should be made here of the performance of the ultra-high strength steels used in undercarriages and aerospace attachments and fittings. These alloy steels also possess very high strengths, *c.* 2000 MPa and, in proportion, relatively modest values of fracture toughness, typically 100 MPa $m^{1/2}$. As with all highly stressed strong materials, this combination of mechanical properties produces very small critical crack depths at limit load level of the order of 3 mm reducing to 0.5 mm in the presence of a stress concentration such as a bolt hole. However, unlike the MMC discussed, these high strength steels are ultra-clean and there is ample evidence that their strength is not compromised by the occurrence of chance defects at stress concentrations in components and that the fatigue strength is high. However, these small critical crack depths preclude the use of the damage tolerant crack containment approach, yet to be described, and fatigue lives are set on the basis of a safe life technique.

Values published for the fracture toughness of fibre-reinforced metals are not numerous but those that exist tend to show a trend in which poor interfacial bonding between matrix and fibre can produce good toughness by fibre pull-out, at least along the fibre axis, and composites with good interfacial bonding tending to show poor toughness. However, the combination of properties exhibited by fibre-reinforced metals, namely very high compression strength, good wear resistance and thermal stability, tend to make them attractive for applications where fracture toughness may not necessarily be an important asset.

1.2.10 Wear resistance and bearing strength

As the major structural properties such as strength and stiffness, fatigue strength are increasingly improved in new materials, so secondary requirements become all the more important. Minimum gauge aspects of polymeric composites have already been stressed but other properties such as wear resistance and bearing strength need to be considered when matrices of intrinsically low hardness are employed, such as the polymeric composite and the adhesively bonded laminate. It may prove to be the case that enhanced wear resistance can be obtained by the application of metal matrix composite materials to moving parts in the internal combustion engine, for example, or in any rotating or reciprocating machinery. The ceramic reinforcements added to metal matrix composites are usually very hard materials and in themselves very resistant to wear. However, it may also be noted that they are frequently employed as cutting and grinding media, i.e. they are likely to produce rapid wear in contacting materials. To achieve satisfactory wear resistance the performance of the system as a whole has to be considered. Nevertheless, the intrinsically high hardness and surface bearing strength may well prove attractive, particularly when additions are perhaps made to materials with relatively poor performance in this respect, such as the aluminium and titanium alloys used in aerospace.

1.3 PROPERTIES AFFECTING COST OF OWNERSHIP

It must be recognized that the cost of ownership of an aircraft, military or civil, throughout its service life may exceed the cost of purchase by at least a factor of 2. Much of this cost is incurred in the maintenance of an airworthy structure, i.e. in the inspection for and replacement of damaged and worn parts. The resistance of a material to degradation by fatigue and corrosion therefore has a strong influence on the in-service costs of an aircraft.

1.3.1 Fatigue strength and crack growth resistance

Most metallic structures are prone to fatigue damage and this is especially so in the lightweight structures employed throughout the aircraft industry. Traditionally, the fatigue process can be divided into two distinct phases, those of crack initiation and crack growth. In most practical engineering situations fatigue crack initiation occurs at the component surface in the vicinity of geometrical stress concentrations. Surface quality and avoidance of local stress concentrations are therefore of the utmost importance. The fatigue cracking process in metals requires applied stresses to produce local plastic strains and it is not surprising, therefore, to find that the stiffer, stronger materials are more resistant to fatigue crack initiation. Design against fatigue has developed over many decades. While attempts to produce **fail-safe** structures, in which the failure of any one part in a load path is accommodated by the loading of alternative, hitherto possibly redundant structure, have essentially been abandoned, two principal lifing techniques remain in vogue. These are the **safe-life** and **damage-tolerant** design approaches.

(a) Safe-life fatigue design

The basic principle of safe-life design against fatigue is that a prescribed level of residual strength, typically 80%, should remain in a component or structure at the satisfactory completion of a period of service, or safe life. As we shall see, the sensitivity of the fatigue process to stress concentrations and overall stress levels is such that significant scatter in performance exists in terms of the resistance of metallic structures to the initiation of fatigue damage. For this reason, the demonstrable fatigue life of the structure or component must exceed the safe life by a significant prescribed factor for safety. The magnitude of the factor will depend upon the rules of the regulating authority, and the nature of the demonstration of achievable fatigue life which may be by calculation or structural test. Typically, the factor is set statistically to allow a 97.5% probability of the occurrence of one failure, judged as the onset of detectable cracking, in 1000. The factor can be set in terms of numbers of cycles, in which case a value of 5 may often be found, or in terms of fatigue strength level, where a factor of 2 on the stress level will produce a similar rate of failure. It is also possible to accommodate a combination of safety factors on both life and fatigue strength by employing a safe fatigue *S–N* curve to allow for the natural shape of the S–N curve.

In general terms the fatigue life of a built-up structure containing stress concentrations typical of a fastened assembly is inversely proportional to the seventh power of the range in fatigue stress. (Some authorities quote the fourth power.) This high level of stress dependence of the fatigue process leads to large reductions in fatigue life with relatively small increases in working stress levels and in many cases limits the applicability of materials to stresses much below the natural strength of the material. For example, a structure capable of withstanding one applied stress cycle of maximum range 525 MPa, typical of the strength of an aerospace aluminium alloy, will be limited to a stress level of 110 MPa in a pressure cabin requiring a fatigue life of 60 000 cycles (Figure 1.17).

This approach, although simplistic, gives a reasonable conservative approximation to service fatigue performance. It does not, as described, explain the increasing rate of damage accumulation with higher mean stress for any fixed stress range, nor that many materials possess an infinite fatigue life once the maximum stresses drop below some critical level associated with microplasticity or micro-cracking.

This type of rule can be extended in concept to include the prediction of the life reduction produced by the application of known numbers of stress cycles of variable amplitudes. In terms of a simplified Miner's Rule approach, failure will be predicted to occur when the sum of the fractional consumptions of fatigue life, predicted for each of the applied stress amplitudes, reaches unity. For example, following the data presented (Figure 1.17), 90 fatigue cycles of 300 MPa is equivalent in fatigue damage contribution to approximately 200 000 cycles at a maximum stress level of 100 MPa, both producing failure in the shown aluminium alloy. A combination of 45 cycles at 300 MPa and 100 000 cycles at 100 MPa would similarly fail the sample.

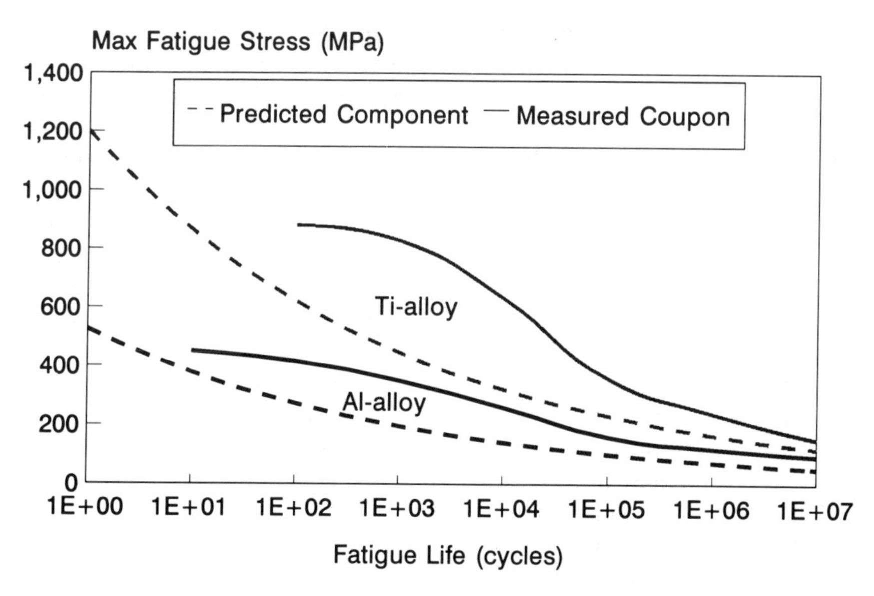

Figure 1.17 Fatigue strength of built-up components with predicted safe life and notched coupon results.

Moreover, the approach described here does not immediately incorporate load interaction effects on the initiation of fatigue cracking. The sequence of loadings suffered by an aerospace structure, although generally repetitive in nature, is likely to encompass a relative wide range in amplitude and frequency of occurrence of loadings. That is, while there will inevitably be a ground–air–ground cycle of a generally consistent magnitude in load, superimposed upon this will be a combination of manoeuvre and gust loadings occurring over a range of frequencies, mainly higher than the ground–air cycle. That is, each flight will contain a variable number of load excursions with variable amplitude with the occasional flight containing very severe loading extremes (Figure 1.1). These variable load excursions will combine to produce fatigue damage at rates that differ from the Miner's predictions. That is, the application of a load cycle of large magnitude will have a beneficial effect on the damage level produced by subsequent smaller cycles if the metal is, in effect, work hardened by the application of the high load. On the other hand, the application of many small cycles may exacerbate the damage caused by a large fatigue cycle. Empirical rules exist to predict these effects in engineering terms. They have little impact on the relative performance of the aerospace materials and are therefore considered to be beyond the scope of this book.

In terms of requirements for aerospace materials, two features need to be emphasized. Firstly, as can be seen (Figure 1.17), stronger alloys will tend to show higher fatigue strengths in approximate proportion to their strengths and it is common practice to normalize fatigue strengths in terms of ultimate tensile strength. Secondly, metallic materials are extremely notch sensitive under fatigue stressing conditions and the choice of alloy at any one strength level has a minor effect in comparison with the overriding effects of notches.

Polymeric fibre-reinforced composites, including fibre-reinforced aluminium alloy sheet laminates such as Arall (Aramid Aluminium Laminate), are generally claimed to exhibit excellent fatigue strengths. This is certainly the case for carbon fibre-reinforced thermosetting products stressed longitudinally under tension-dominated loading regimes. Because of the limitations in usable strain imposed by notched degraded strength requirements, infinite fatigue lives can be predicted in many polymeric composite structures at these allowed strain levels. However, as loading regimes are extended to those containing significant compressive contents and multiaxial loading requirements, including short transverse stressing, are imposed this good fatigue strength is thrown more into question (Figure 1.18). One characteristic of the fatigue failure of fibre-reinforced polymeric materials is that failure tends to be catastrophic by the linking of many sites of local microcracking, rather than by initiation and growth of a long crack.

The situation is as yet unclear with regard to the data available on the fatigue strengths of fibre-reinforced metals, seemingly because the difficulties experienced in transferring loads to the composite samples has

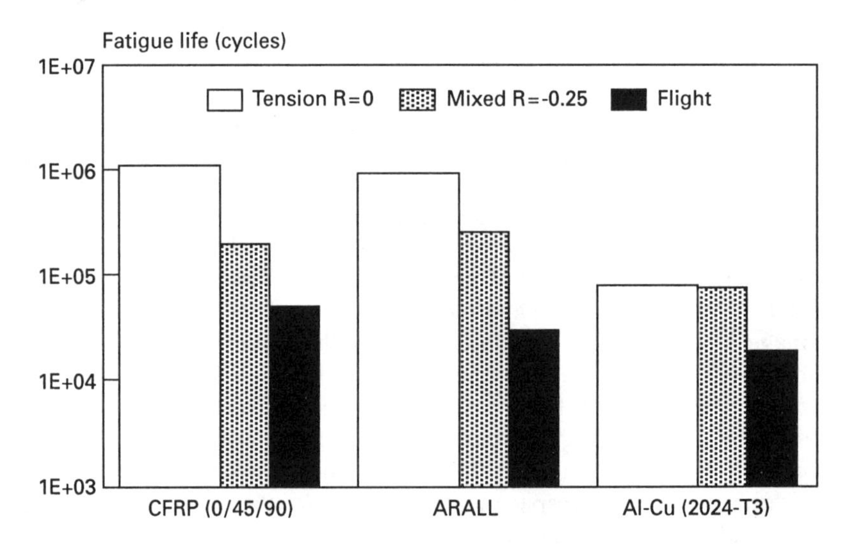

Figure 1.18 Relative fatigue life of CFRP, ARALL and alloy with pinned double lugs loaded in three conditions [3].

limited testing capabilities and the bulk of the data published pertains to flexural loadings. While the flexural fatigue strengths of fibre-reinforced composites stressed longitudinally can appear to be very good, the key issue of the performance of the component in the critical area of high load transfer, i.e. joints, has not been addressed satisfactorily.

(b) Damage-tolerant design approach

In this chapter the description of the damage-tolerant design approach is limited to the control of the relatively slow growth of fatigue cracks in metallic structures, but the principle could also be applied, for example, to the failure of polymeric matrix composites by progressive delaminations or the loss of residual strength in metal by stress corrosion cracking.

The basic principle behind the concept of damage-tolerant design is to recognize that damage will exist *ab initio* in manufactured items and that more will occur during subsequent service life. Where the inevitability of initially damaged structures has been accepted, the approach requires that the rate of accumulation of damage is predictable and controlled to prevent unacceptable loss in residual strength or, at worst, catastrophic failure. For a metallic structure suffering fatigue damage, a damage-tolerant design would incorporate both the resistance to fatigue crack growth and the resistance to the onset of rapid fracture represented by the material's fracture toughness. A specific example is provided by the case of fastened joints in an aircraft structure where the approach would require the assumption that pre-existing cracks potentially exist at each fastener hole and that the most highly stressed hole contains a pre-existing defect of a prescribed size dependent upon the nature of manufacture of the component and its inspectability. It would then be required to demonstrate that the growth of the cracks to an extent that unacceptable loss in strength occurs at limit load requires more than the declared service life of the component by a prescribed factor. This safety factor will be selected according to the inspectability of the structure in question but, typically, may be a factor of 2.

This approach poses severe problems when the realistic scenario of the contemporary development of multiple cracks from neighbouring fastener holes is considered with the consequent rapid consumption of remaining sections far exceeding the rate produced by the growth of a single crack from one defect. In this respect the resistance of the metal to the growth of short cracks at relatively low values of stress intensity factor is all important. This is one of the attractive features of aluminium–lithium alloys typified by 8090-T81 where good resistance to crack growth is found at low levels of stress intensity factor (Figure 1.19). Comparison is made with the best of the aerospace aluminium sheet

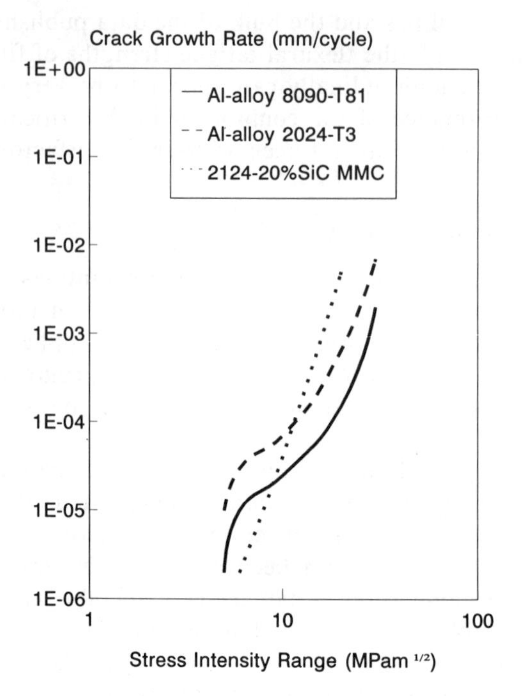

Figure 1.19 Fatigue of Al-alloys and MMC for 2 mm sheet, constant amplitude and R = 0.1.

alloys 2024–T3, traditionally used in pressure cabin skinning, and with a modern metal matrix composite, particulate reinforced 2024 (2124) alloy which shows a high fatigue strength but rapidly accelerating crack growth. That is, in the case of particulate and whisker-reinforced aluminium alloys, two characteristics have emerged, at least for the composites investigated to date. These are an improved resistance to crack growth at low levels of stress intensity factor in the threshold region and a relatively poor performance at high levels of stress intensity factor, where the influence of the reduced fracture toughness is seen. It would seem likely that the improved fatigue strength of the stiffer composite materials is reflected in this improved threshold performance, although the influence of residual stresses, induced in the material during the heat treatment, on the threshold behaviour, needs to be investigated.

While it has just been argued that the fatigue period required to grow cracks may be dominated by the early growth of short cracks, damage-tolerant design methodologies also require the demonstration of the residual strength of structures containing long cracks or large readily detected defects and hence place some reliance on the toughness of the material. Perhaps the most striking example of this requirement is the ability of large pressure cabin skins to withstand the presence of very long cracks without failing catastrophically. A typical requirement might be for the pressure cabin to withstand the limit load level with two complete bays of skin cracked between the supporting frames. Linear elastic fracture mechanics tends to be applied to such design and qualification exercises despite their dubious validity in the near-plastic conditions in such tough materials. In general terms, plane stress fracture toughness of sheet material is expressed as a K_c value derived by tangency from a crack resistance curve, while valid plane strain K_c values are used for thick section material. Many aircraft materials are used at thicknesses where transition levels of toughness between the plane strain and plane stress are obtained to some extent compounding the analytical problems.

(c) Resistance to fatigue cracking and fracture toughness

Once again the combination of two properties in balance is required, this time to achieve damage tolerance in a metallic structure. Both resistance to fatigue cracking and to rapid tearing are simultaneously needed.

The combination of good fracture toughness and resistance to fatigue crack growth to produce damage tolerance can be illustrated by calculating the fatigue crack growth life left in a structure assuming the presence of initial defects of increasing size (Figure 1.20). Lives predicted for the growth of cracks under constant amplitude stressing for damage-tolerant

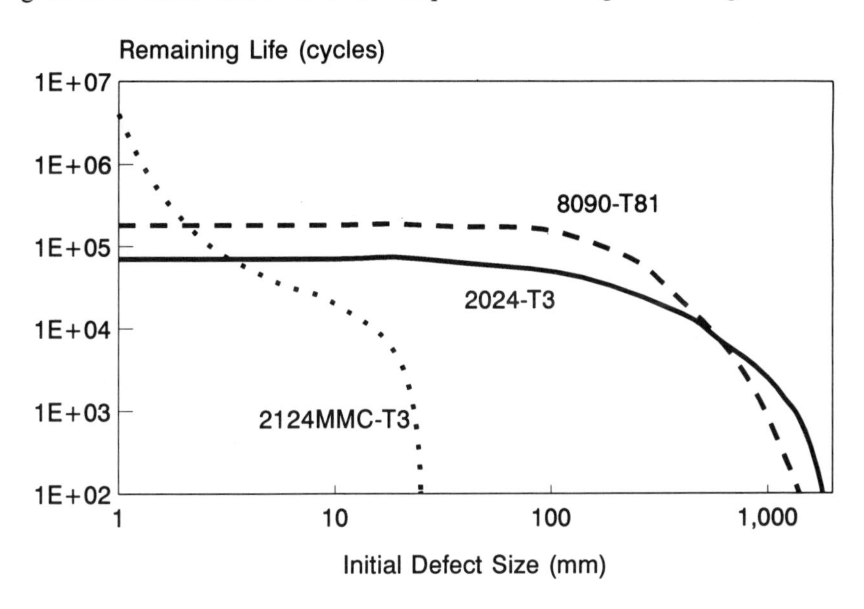

Figure 1.20 Residual fatigue life in Al-based sheet for constant amplitude, peak stress of 100 MPa and R = 0.1.

aluminium–copper alloy 2024, aluminium–lithium alloy 8090 and for metal matrix composite reveal consistent trends for a stress range of 100 MPa typical of a pressure cabin application. Although the MMC has much higher fatigue strength than the damage-tolerant aluminium–lithium, its poor fracture toughness and rapidly accelerating crack growth leaves no damage tolerance once the initial flaw size exceeds 2 mm. A similar but less marked trend is seen in a comparison of the 8090 with the 2024 where the higher toughness but poorer crack growth resistance of the 2024 again leads to a crossing of the curves at long crack sizes.

As with fatigue crack initiation, account must be taken of load interaction effects in the crack growth regime. Although fatigue crack growth data generated under cycles of constant amplitude will tend to predict the crack growth contributions of individual loading cycles within a programme of loads of varying magnitude reasonably well, two forms of 'load interaction' must be accounted for. These are the beneficial effects of an occasional high peak load, the plastic damage at the crack tip retarding subsequent crack extension, and the crack sharpening or accelerating effects of compressive loading cycles or many small amplitude loads delivered at high frequency. Numerical means are available for the analysis of damage rates under such complex loadings but improvements in predictive capability are continuously sought.

It is not necessary in this chapter to develop the techniques for crack growth analysis under variable amplitude loading but it is important to indicate that the response of differing materials to complex loadings can vary significantly. For example, polymeric-reinforced composites and laminates tend to perform very well under tension dominated constant amplitude loadings but are found much less resistant to compression dominated or high frequency loadings, the almost exact reverse of the situation for conventional metallic alloys. With conventional aluminium alloys, differing responses of growing fatigue cracks to variable amplitude loading must also be considered in the available alloys. Notably, the damage-tolerant aluminium–copper and aluminium–lithium alloys show very marked crack growth retardation effects with the occasional peak load. This effect is not so strongly apparent with aluminium–zinc alloys and consequently, even though they have excellent fracture toughness, the aluminium–copper and lithium alloys are more damage tolerant.

The description of the growth of fatigue cracks in aligned fibre and monofilament-reinforced composites has not yet apparently reached a satisfactory state in which comparison can be made with monolithic materials. The anisotropic nature of aligned composites, the variable nature of the bonding between fibre and matrix, the difficulties inherent in transferring loads to the fatigue samples and the residual stresses in the samples combine to produce micro-cracks that are discontinuous, that do not grow or that grow in planes and directions that do not

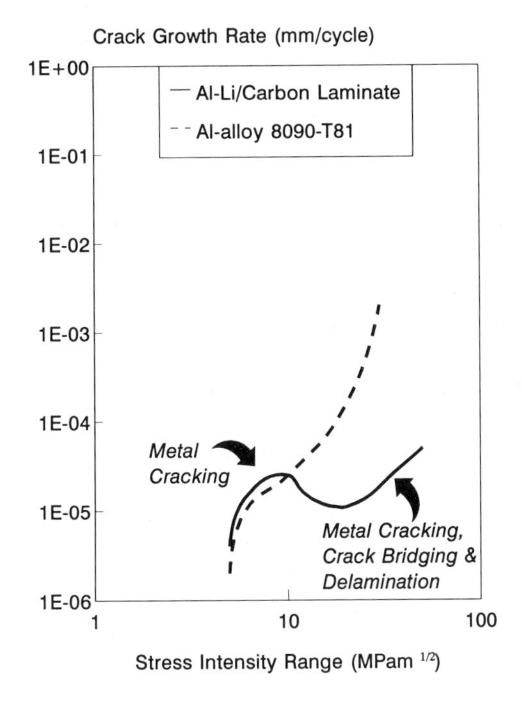

Figure 1.21 Fatigue of Al-alloy laminate for 2 mm sheet, constant amplitude and R = 0.1.

correspond to fatigue cracking in conventional alloys. Nevertheless, if an analogy is drawn between the fibre and monofilament-reinforced composites and adhesively bonded laminates of the ARALL type, then it can be anticipated that under certain circumstances, such as a weak fibre matrix interface, fibre-reinforced composites will produce extremely good resistance to crack growth by a damage bridging mechanism. Mention should be made here of the extremely good resistance of adhesively bonded laminates of aluminium alloy sheet reinforced with layers of fibre-reinforced polymeric composite. This type of hybrid material achieves the advantages of the crack bridging capability of fibre-reinforced materials (Figure 1.21), but retains the basic toughness and damage tolerance of the metallic sheet. Difficulties are encountered, however, as with the polymeric materials, in the description of damage rates in terms of the parameters of linear elastic fracture mechanics.

1.3.2 Strength and damage tolerance in tension structures

The combination of good tensile strength and high damage tolerance, or fatigue resistance, is a further example of the need to balance properties of materials in an optimized manner, and the importance of this balance between strength and resistance to fatigue cracking in metallic materials needs to be emphasized. Safe fatigue life and damage tolerance have been considered in detail in previous sections, but it should be indicated here that the efficient use of metallic materials in structures loaded predominantly in tension may well be limited by these fatigue considerations and not by static strength. For example, in safe-life terms, the minimum allowable value for the ultimate tensile strength of a conventional aerospace aluminium alloy might be in the region of 525 MPa and this may be equated to the ultimate failure for the most demanding loading case. Following the simple rules previously expounded the onset of significant plastic yielding will typically occur at 450 MPa, easily exceeding the proof load requirement of no significant plastic deformation at 0.75 of the ultimate stress equivalent to 394 MPa. Yet to achieve a satisfactory fatigue life the maximum operating stress level may be restricted to perhaps 350 MPa in a lower wing surface of a combat aircraft or 110 MPa in a pressure cabin skin.

Analogous arguments can be raised for a damage-tolerant approach when considering the requirement for acceptable critical crack sizes which, as previously indicated, will be controlled by the magnitude of the operating stresses and the fracture toughness and by the rates of growth of fatigue cracks dependent upon the fourth power of the range in fatigue stress. The limitations to design efficiency that can be imposed by fatigue are very apparent.

1.3.3 Fatigue strength enhancements

To circumvent these severe limitations imposed by fatigue cracking, much emphasis is now placed on the use of prior conditioning of attachment fastener holes to improve fatigue performance in metallic structures. Three features should be noted. Firstly, holes and surfaces can be cold worked to produce beneficial residual stress states enhancing fatigue lives by an order of magnitude (Figure 1.22). Secondly, interference fit fasteners can be inserted to achieve a similar effect, and finally the combinations of adhesive bonding with mechanical fastening can be used, especially with thin section sheet, to reduce stress concentrations at the fastener holes.

1.3.4 Corrosion performance

Metallic and non-metallic structural materials must resist the ravages of an aggressive environment for the entire lifetime of the aircraft, which may be as much as 30 years. Elevated and sub-zero temperatures, high humidity or rain, salt-laden coastal atmospheres, mechanical damage by

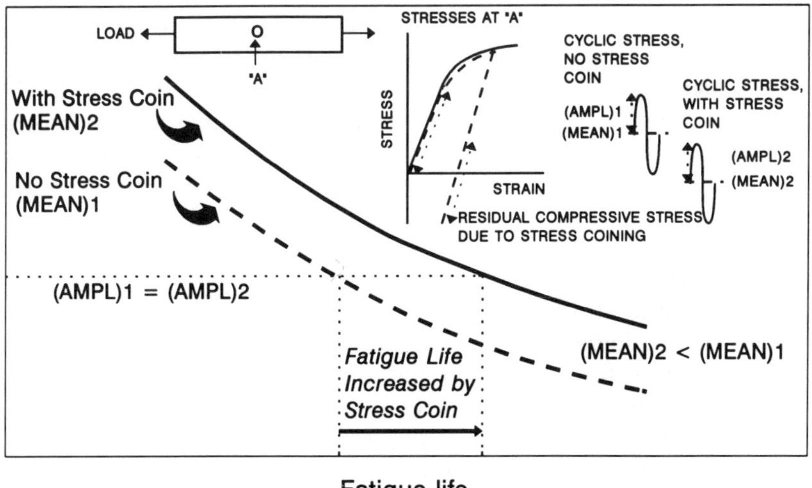

Figure 1.22 Effect of stress coining on fatigue life for residual compressive stress at fastener hole.

ice, ultraviolet radiation and atomic oxygen combine to produce deterioration in surface protection schemes and in the underlying materials of aerospace structures.

Corrosion in metallic structures can take the form of pitting of surfaces, the penetration of grain boundaries and the genuine cracking of the material, usually along the grain boundaries. Typically, general corrosion has to be extremely severe to reduce metallic sectional thicknesses to levels at which the strength of the structure is compromised but, more perniciously, the presence of even small corrosion pits can induce fatigue cracking.

In real terms, all forms of corrosion or environmental damage must be prevented from the start of aircraft service because the variability in rates of environmental degradation makes a damage-tolerant approach difficult, if not impossible. Hence, the demonstration of satisfactory residual strength in any one structure tends to be based on the assumption that data and tests on materials in dry laboratory conditions are generally applicable to real structures. This is non-conservative, even when the effects of a corrosive environment are accounted for in the crack growth period, because of the damaging effect of corrosion on the initiation of fatigue cracking. That is, while the effects of aggressive environments on the rate of growth of fatigue cracks are readily quantified, the influence of corrosion on the rate of nucleation of fatigue cracks at neighbouring fastener holes is difficult to predict. If a totally conservative damage-tolerant approach were adopted, in which every fastener hole was assumed to be corroded to the extent that fatigue cracking starts at the first day of service, then very severe loss in service life would be predicted and the aircraft structure would probably not have a viable life. This is not the real case, since most aircraft demonstrate very long service performances in very demanding environments relying on good protection schemes, including the wet assembly of joints, to prevent the initiation of corrosion and subsequent fatigue damage and on inspection of very tough structures to detect cracks, once initiated.

The more pernicious forms of corrosion attack found in metallic structures include exfoliation, stress corrosion and hydrogen embrittlement with accelerating mechanisms of crevice corrosion and filiform attack to be considered.

Cracking by exfoliation corrosion and stress corrosion, although potentially dangerous, tends to be confined to short transverse orientations in the metal, a direction that normally is lightly stressed. In both cases susceptibility of the grain boundaries to electrochemical reactions coupled with crack opening stresses generate a tendency to crack. Exfoliation attack of surfaces and end grains is caused by corrosion products wedging open the attacked grain boundaries, while, in the related stress corrosion, externally applied stresses or residual heat treatment stresses in the metal provide the crack opening forces. The improved understanding of the mechanisms of these metallic problems has resulted in efficacious solutions dependent in nature upon product form and component. For example, susceptible aluminium alloys can be heat-treated to non-susceptible conditions (Figure 1.23) or can be quenched slowly or stretched after quenching to reduce residual stress levels to minimal values at which stress corrosion is no longer a problem.

While the control of corrosion and corrosion cracking in metallic aerospace materials has become a major success, it must be remembered that new alloys and materials can regenerate problems as they take their place in the selection of proven applications. Aluminium–lithium alloys are prone to stress corrosion attack when in inappropriate heat-treated conditions, although possessing a high natural resistance to corrosion, and the lessons learned with aluminium–zinc alloys may have to be reapplied to this new type of alloy.

The requirements for aeroengine materials in resistance to corrosion attack differ little from those of the airframe. It might be noted that much use is made of titanium and nickel-based alloys with inherently high corrosion resistances, but that problems may be exacerbated by elevated temperatures in wet environments during cooling phases of the engine duty cycle and by the ingestion of salt-rich atmospheres in combustion stages of the engines.

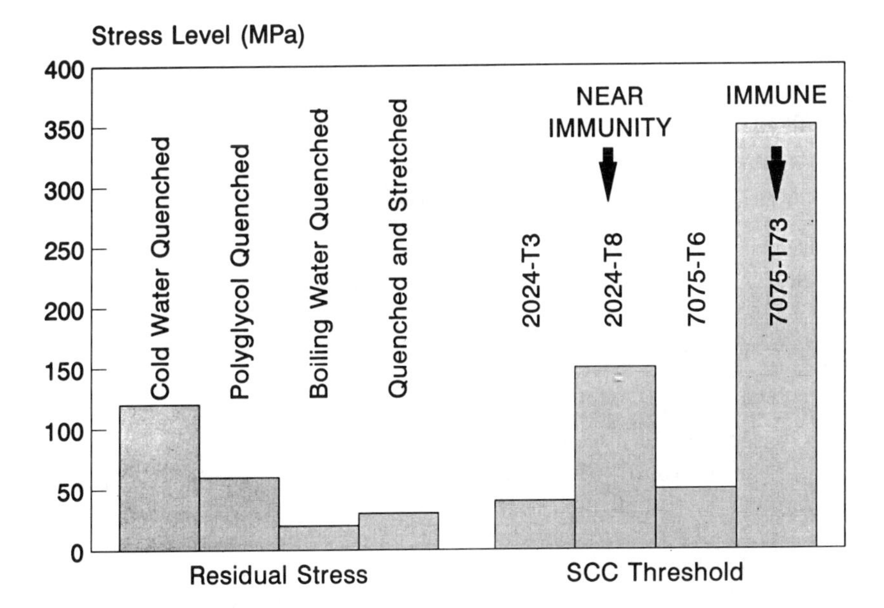

Figure 1.23 Residual stress and SCC in Al-alloys.

Experiments have continued to assess the corrosion behaviour of metal matrix composites following the standard practices of laboratory testing in accelerated environments typified by the use of alternate immersion in 3.5% NaCl or exposure to acidified salt fog atmospheres. These have been coupled with outdoor exposure trials in industrial and marine environments. Studies of the micro-mechanisms of corrosion in fibre and particulate reinforced aluminium alloys have been conducted by potentiostatic and polarisation techniques. It would seem to be the case that, in general terms, the electrochemical characteristics of the matrix alloys used in MMC construction will dominate the corrosion behaviour. However, some evidence has been found for the preferential attack of the matrix in the vicinity of SiC particulate reinforcement (rather than at the interfaces) in aluminium–copper alloys, suggestive of an influence of the particles on the precipitation characteristics of the matrix and hence on the local corrosion performance. Studies on alumina reinforced aluminium–silicon alloys indicated that traces of impurities, such as iron in the form of intermetallic particles, resulting from the production of the composite had a stronger influence on the corrosion performance of the composite than the alumina reinforcement *per se*.

It may be claimed that composites based on polymeric matrices are, in effect, immune to corrosion attack presenting real opportunities for the manufacture of long life structures. In practice allied problems such as the pick-up and retention of moisture and the degradation of mechanical performance under hot-wet conditions present similar constraints to the application of the material as does corrosion to metallic systems although the micro-mechanisms are different in nature.

1.3.5 Fatigue strength and corrosion resistance

No single material is ideally optimized for aircraft applications whether metallic or non-metallic in origin. The balance required between resistance to corrosion and fatigue in metallic materials provides a good example of the compromise required. For example, considering the major airframe aluminium alloys, the provision of very long service lifetimes virtually free from corrosion requires the use of protective schemes either based on the use of surface cladding of near pure aluminium or the use of anodised films loaded with corrosion inhibiting primers and paints. Although extremely efficacious in resisting corrosion, both these techniques reduce the fatigue strength of the underlying material (Figure 1.24) but not as much as would corrosion attack. In practice, however, allowances made for the presence of geometrical stress concentrations accommodate the detrimental effects of potential corrosion attack and all aluminium alloy structures are protected in the appropriate way.

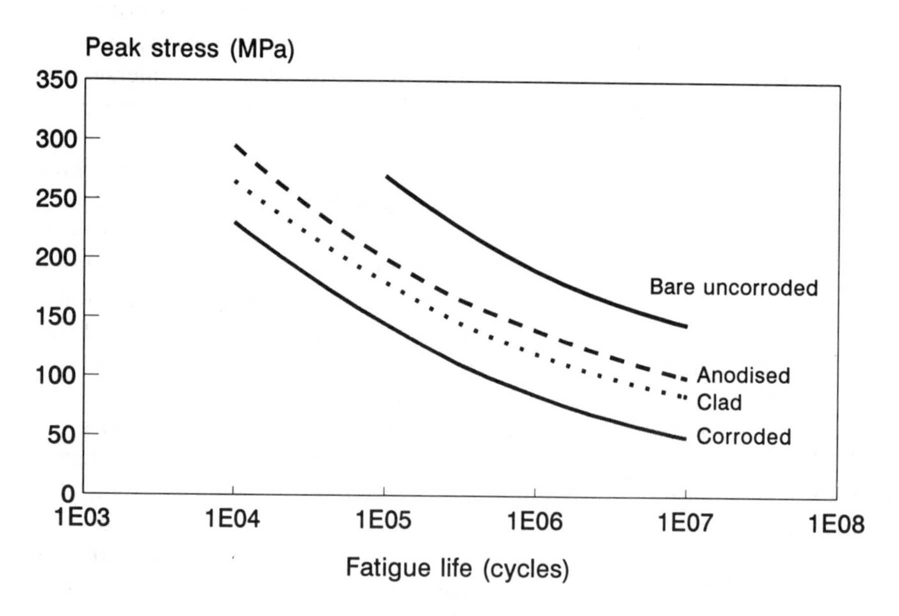

Figure 1.24 Effect of surface finish on fatigue life for Al-alloy 2014–T6 sheet at R = − 1 and Kt = 1.

1.3.6 High temperature performance

The majority of aerospace structures operate between temperatures of approximately – 50 °C to + 90 °C with limited specialized requirements extending this envelope for high-speed flight and for space applications. The lengths of exposure to elevated temperatures vary considerably, however, according to aircraft type and application (Figure 1.25). It is therefore often the case that it is the length of exposure at elevated temperatures rather than the temperature *per se* that is most demanding of material performance. This kinetic rather than thermodynamic requirement is well illustrated with early design requirements of the alloy used in Concorde construction where creep strains had to be limited to less than 0.1% in the required 10 000 hours of high-temperature exposure at modest stress levels of perhaps 150 MPa and temperatures of not more than 120 °C. Thus, strength at temperature was never a problem but creep resistance was.

The future of hypersonic airframes such as the National Aero Space Plane (NASP) places additional demands on structural materials. Structural designs for all areas of such a vehicle require high stiffness, thin gauge, product forms which can still be fabricated into loadbearing components and retain their performance at high temperatures and during thermal cycling. The high-temperature capability of superalloys must be combined with the density of titanium alloys, and potential materials include rapidly solidified titanium aluminides (TiAl or $TiAl_3$)

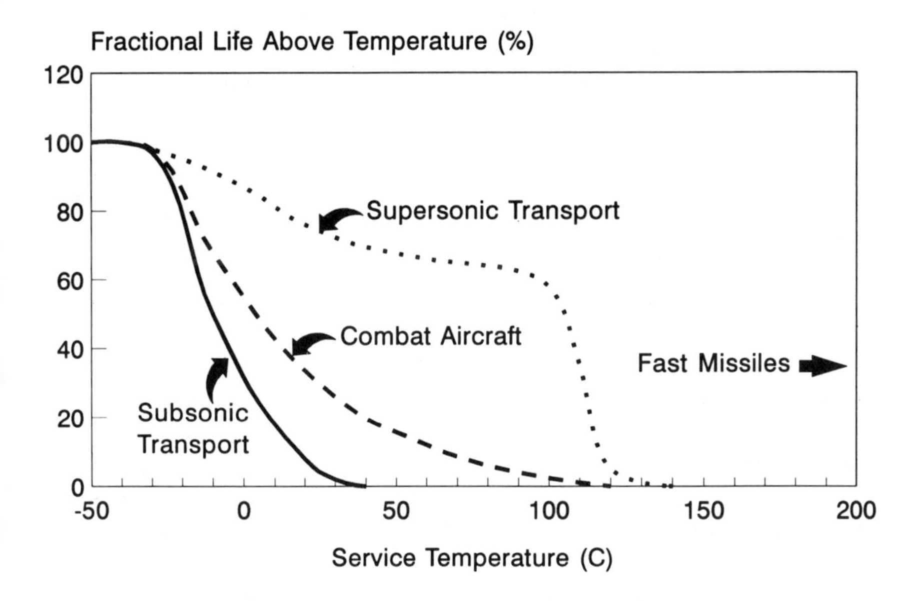

Figure 1.25 Thermal envelopes for aircraft with proportions of flying hours.

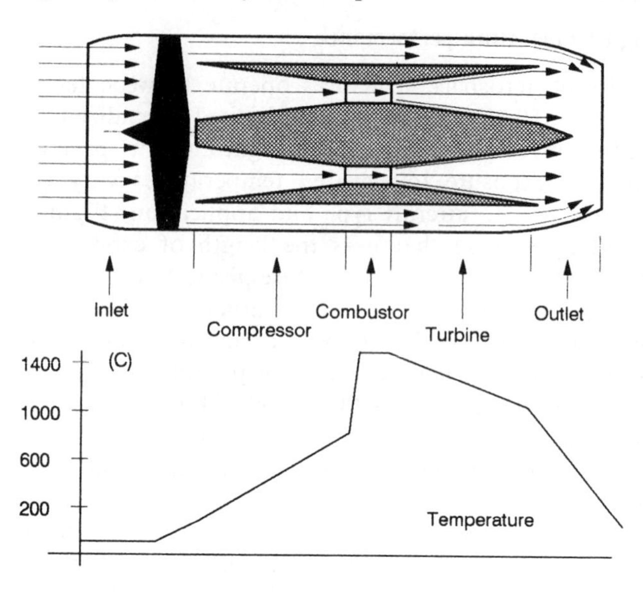

Figure 1.26 Typical engine temperature profile for transport aircraft. (Courtesy Rolls Royce plc)

reinforced titanium aluminides, carbon-carbon composites and ceramic matrix composites such as SiC-reinforced SiC. Much property evaluation is still required and many material inadequacies will have to be overcome prior to the commercial development of structures using such materials.

Requirements for applications in aeroengines are more demanding in terms of the temperature envelopes (Figure 1.26) but in engineering properties the similarity with airframe requirements is still found. That is, strength, creep resistance, fatigue strength, fracture toughness, corrosion resistance, etc. are still important in much the same way.

Improved performance of existing aeroengines is being achieved via extensive use of coatings (e.g. thermal barriers) to conventional metallic substrates. Further advances, however, in design and materials fabrication may lead to large rotating assemblies being much reduced in mass by the development of new components such as the integrally reinforced bladed ring or BLING fabricated from SiC fibre-reinforced titanium alloy. Nevertheless, future propulsion systems will require increased thrust-to-weight ratios, demanding materials having higher operating temperatures and higher specific strength and stiffness. It may be the case that ceramic-based materials will in part meet this challenge. Toughened monolithic ceramics may be considered if the component volume is small (e.g. helicopter engine turbines) and the design can be tailored to accommodate a brittle material behaviour. Ceramic fibre-reinforced composites (SiC in SiC) having increased design-allowable strengths at

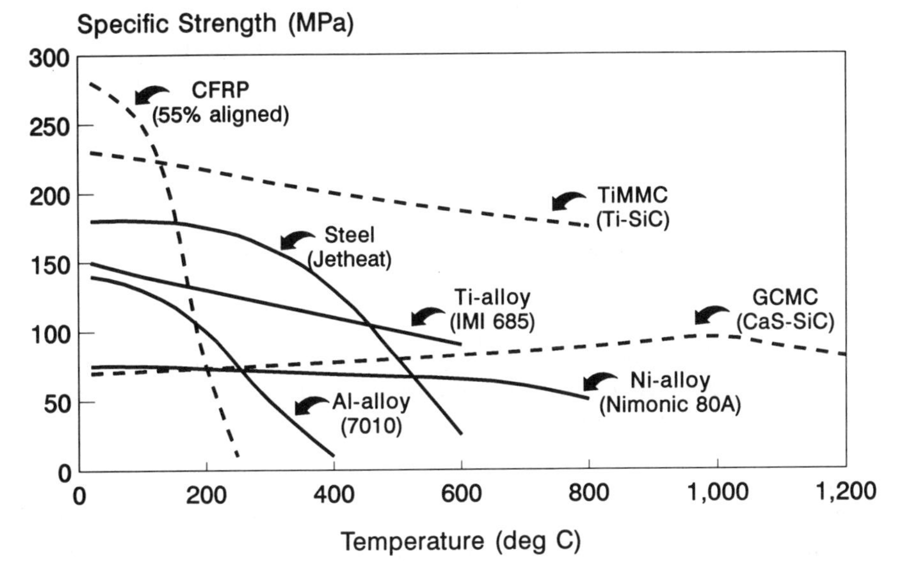

Figure 1.27 Variation of design stress with temperature for typical specific tensile stresses.

high service temperatures have been demonstrated but their commercial development remains for the future.

Strength requirements for elevated temperature applications are conventionally characterized in two ways, namely, recovered, cold, strength after prolonged periods of exposure at elevated temperature and the instantaneous strength at temperature after prescribed periods of prior soaking at the operating temperature. Both characterize materials properties as a function of temperature of application reasonably. Comparison of the retention in specific strength of selected lightweight aerospace alloys can be made (Figure 1.27). It can be seen that degradation occurs for all the metallic materials with elevated temperatures but, as might be anticipated, the materials with higher melting points tend to maintain their strengths to higher temperatures.

However, while strength at temperature provides an indication of the relative performance of materials, it does not provide an absolute guide to the resistance of the material to very slow strain rate creep damage under sustained load. The Concorde requirement to limit creep strains to less than 0.1% deformation in the prescribed period was very demanding, but necessary, to limit permanent extension of the aircraft over its lifetime. At the time of the design the aluminium–copper alloy 2618 was chosen because of its high resistance to creep deformation. Little exists 30 years later to better it, other than refined variants of the same aluminium–copper system.

In terms of the well-established Graham and Walles methodology, the description of the creep curve where the creep strain, ε, accumulated in time, t, is described in the following formula where C, β, and κ are constant:

$$\text{Creep strain } \varepsilon = \sum C\sigma^{\beta} t^{\kappa} (T^{1} - T)^{-20\kappa}$$

When σ and T, denoting stress and temperature respectively, are constant, the creep equation can be reduced to an Andrade form where:

$$\varepsilon = at^{1/3} + bt + ct^{3}$$

The very high temperature and stress dependence of the creep strain process, *c.* power 20, ensures that the weight and operating envelopes of supersonic transport airframes and portions of aero engines will be limited by creep performance.

In addition to long-term resistance to creep deformation, selected high-temperature alloys may be required to demonstrate high creep strength and ductility at stress concentrations such as fastener holes. That is, while general field strains may be limited to 0.1%, for example, deformation at stress concentration must not locally exceed the creep strength or ductility of the materials. The onset of breakaway or tertiary creep must be resisted in these practical circumstances. This feature has proven much harder to predict and, to a certain extent, alloys with good resistance to primary and secondary creep may show poor tertiary characteristics. The relative behaviour of materials in terms of fracture, or creep rupture, may be predicted from time-temperature-stress extrapolation procedure of which the Larson–Miller parameter (ϕ) is given by

$$\phi = T(20 + \log t_{\mathrm{r}})$$

where t_{r} is the creep rupture time. The stress dependence of the Larsen–Miller parameter for titanium- and nickel-based alloys is shown in Figure 1.28; the creep performance of titanium is strongly influenced by alloy composition and thermomechanical processing (Chapter 3).

Perhaps one most important consideration in the justification of the high cost of the development of metal matrix composites or intermetallic alloys is that the new materials may produce a much enhanced performance at elevated temperatures, thereby extending the operation envelopes of the lighter materials. For example, in this respect aluminium alloys, suitably reinforced, should provide a natural competition for the heavier titanium alloys, and titanium MMCs, in their turn, provide a substitute for iron- and nickel-based high-temperature materials.

In assessing the high-temperature performance of metal matrix composites important distinctions have to be drawn, once again, between the reinforcement and matrix types, and account has to be taken of both

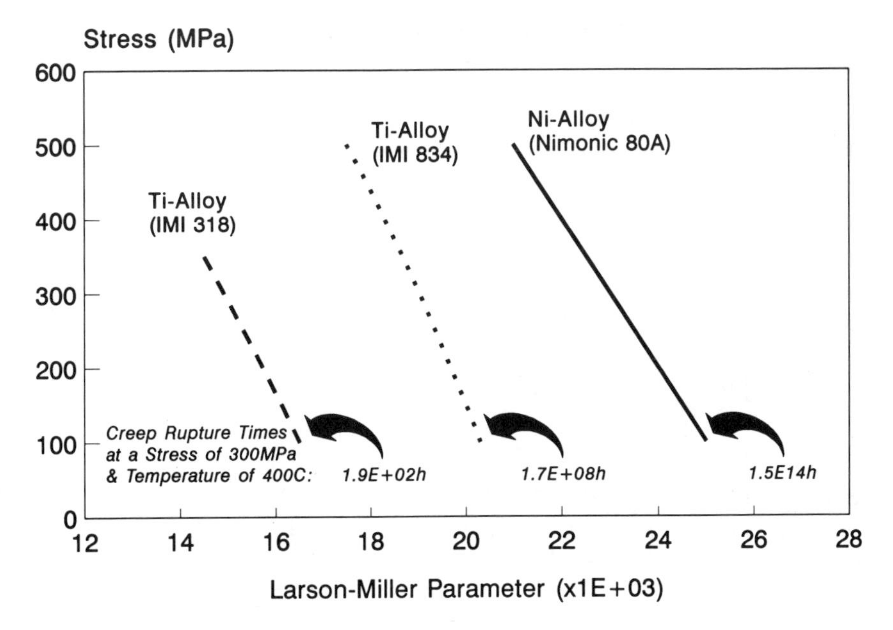

Figure 1.28 Creep rupture performance of Ti and Ni alloys.

instantaneous strength improvements at temperature and on recovery and of aspects of the long-term stability, since once again improvements in high-temperature strength may not be matched by those in creep strength.

It would seem that the elevated temperature strengths of whisker-reinforced materials are marginally superior to those of the particulate varieties, perhaps the consequence of a better combination of improved load transfer on a micro-scale and reduced interaction with matrix precipitation for the long, thin shape. However, the elevated temperature performance of long fibre and monofilament-reinforced matrices seems significantly improved, at least in terms of strength and stiffness. Once again it should be noted that the ductilities and transverse properties of the materials in these categories are so low as to be limiting features for many structural applications. However, appropriate redesign of a specific component to take advantage of the directional properties may still lead to significant weight savings, especially when thermal performance is required, for example, in the development of a continuously reinforced engine disc or bladed ring and potentially in structural applications to space planes.

1.3.7 Dimensional stability and thermal properties

It may also be the case that, in addition to high-temperature strength and resistance to creep, dimensional stability *per se* is important; that is, that

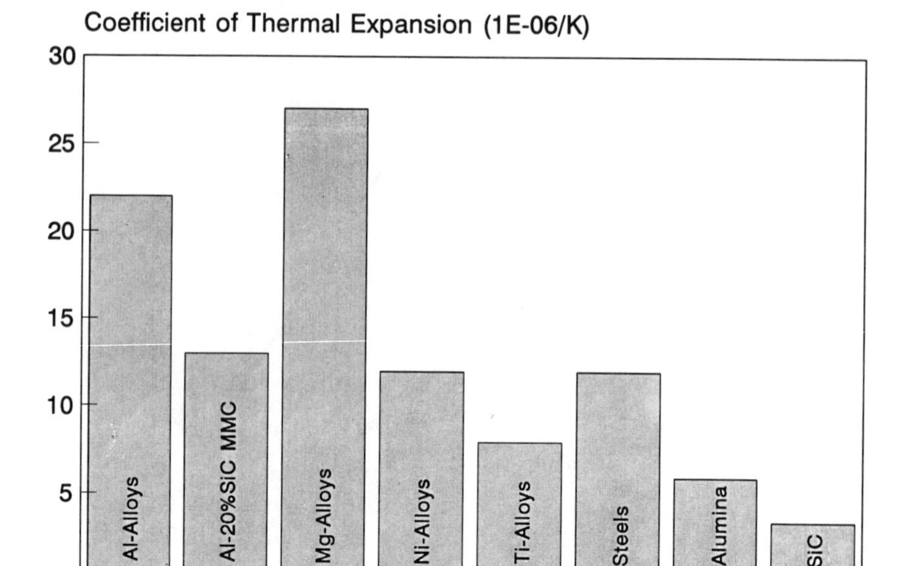

Figure 1.29 Thermal expansion.

the coefficients of thermal expansion are required to match a particular target. The performance of metal matrix composites and the ability to tune the thermal expansion behaviour of these materials to some extent should be mentioned at this point.

It has been clearly demonstrated that low coefficients of thermal expansion are invaluable in matching thermal strains in electronic components and in their mounts. These can be obtained by the incorporation of relatively large fractions of ceramic reinforcements in, for example, aluminium alloys. It would seem quite feasible to produce lightweight aluminium alloy based components with the coefficients reduced below those heavier materials conventionally used for applications where low expansion is required, i.e. beryllium, invar, steel, etc. (Figure 1.29). These aluminium-based materials would still exhibit the required relatively high thermal and electrical conductivities, making them most attractive for electronic base and packaging applications.

Other associated thermal requirements should perhaps be mentioned. For example, good thermal conductivity may be required in aerospace structures for dissipation of heat through the structure that is heated aerodynamically or by engine exhaust. In general terms, high conductivity and heat content are an advantage for steady state conditions, while for specialized requirements such as resisting extreme temperatures for short periods of time, in the event of a fire or accident, low conductivity may be preferred. Missile structures provide special requirements in this aspect.

In the sense that most airframe structures are aluminium–copper or aluminium–zinc alloy, most of the materials employed are similar in their thermal properties. Aluminium–lithium alloys demonstrate a drop in thermal conductivity by a factor of approximately 2, sufficient to require reconsideration of heat-treatment times, and these alloys are found to provide better resistance to burn through under extreme conditions. The reduced thermal conductivity of metal matrix composites may make them attractive for applications such as fire walls, traditionally made in titanium, but with the potential to much reduced weight if an aluminium-based metal matrix composite could be substituted.

1.4 COST-EFFECTIVE DESIGN

Critical aspects that have been considered so far are the technical properties, such as the levels of specific strength and specific stiffness that can be achieved, and the service performance of new materials under fatigue and corrosion conditions. To these important considerations must be added the aspect of cost. This can now be extended to include the unit cost of material to introduce the concept of cost-effective manufacture. That is, in aerospace terms, if two materials exhibit the same strength, the lighter one will be chosen and, if two materials exhibit the same specific strength, the cheaper one will be chosen, all other aspects being considered equal. These two concepts can be considered in combination by plotting potential mass savings in selected real applications against specific cost for a range of high-performance aerospace materials taking conventional aluminium alloy structure as the base line (Figure 1.30).

Cost of ownership is probably more important than initial cost of build, but much harder to assess accurately. An aircraft lasting 30 years in service may well accrue twice its initial cost in service and maintenance at the original financial rate. A real aspect of this consideration is the cost and efficacy of repairs that might be applied during service lifetimes. Following the practical experiences of operators, it can be seen (Figure 1.31) that a major problem with polymeric composite structure is the high cost of repairs and that the difficulty in repairing what is essentially a monocoque structure leads to quantized jumps in repair cost as increasingly larger pieces of structure have to be replaced. Conversely, the cost of repairs to metallic structure is only weakly dependent upon the size of the damage zone.

The repair, renovation and replacement aspects of new materials such as the metal matrix composites, intermetallic alloys and adhesively bonded laminates have yet to be addressed in a determined way.

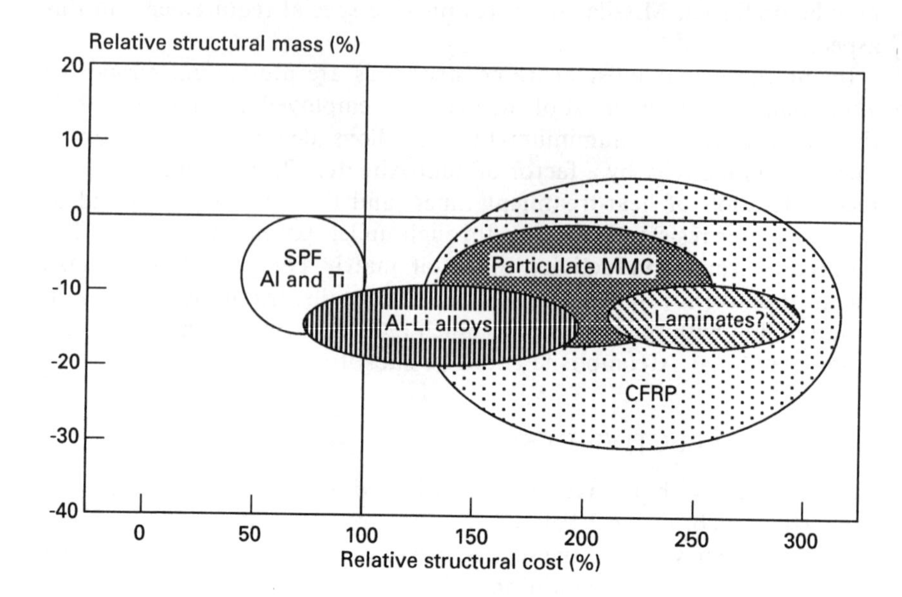

Figure 1.30 Cost effectiveness of new materials.

Ultimately, however, it may prove to be the key property and conventional metallic alloys will again emerge as consistently the most cost effective.

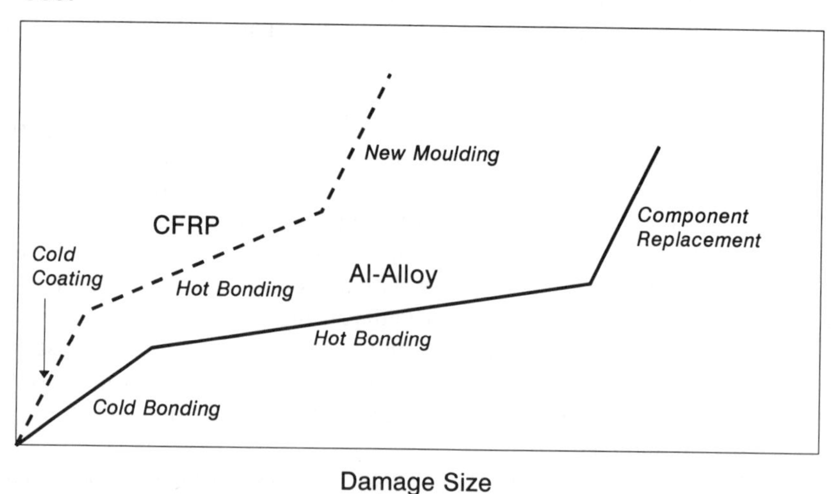

Figure 1.31 Comparison of repair cost – a schematic illustration for wing repair [4].

1.5 CONCLUDING REMARKS

The requirements developed for aerospace materials are effectively the result of 50 or more years of practical experience. Thus the prime importance of fatigue and corrosion resistance is a result of the natural susceptibility of the lightweight aluminium alloys that are in predominant use. If the application of laminated materials such as plywood or modern conventional composites had persisted or grown, then this chapter would have been dominated by requirements such as resistance to delamination, moisture absorption, or inadvertent damage, for example. The history of aerospace materials developments during these last five decades reveals that minor changes remaining within type have been achieved relatively easily, taking perhaps ten years to introduce a new alloy composition or heat treatment practice, for example. Changes out of type, for example from metallic design to polymeric composite, have proved almost impossibly difficult, taking as much as 30 years and requiring the development of an entirely new engineering environment.

Against this perception, some casual judgements can be made on the promise and problems offered by new materials such as metal matrix composites, ceramic matrix composites or intermetallic alloys. It has been clearly shown that all of the examples chosen will generate improved specific properties in some form, and provided that attention is paid to the choice of matrix etc. high specific stiffness can be matched by improved strengths. Therefore structure that is critical in the sense of mechanical stability, i.e. the resistance to bending, buckling and crippling, may significantly benefit from the application of these new materials, especially at elevated temperatures. However, many of those now offered seem to develop high levels of performance by the suppression of plastic deformation, producing a compromise between truly ductile metallic performance and the elastic polymeric composite. It remains to be seen whether the new requirements and design tools developed for polymeric adequate ductility and toughness will be found to enable the application of these near brittle materials in an engineering environment.

In the same light, the development of an environment in which the performance of anisotropic fibre-reinforced materials can be successfully accommodated, at least in thin-walled structures, is perhaps the major achievement of the polymeric composites. The property improvements offered by fibre-reinforced metallic and ceramic composites seem even more alluring, especially when high-temperature requirements are present. Again, it remains to be seen whether the design tools developed for fibre-reinforced polymerics can be applied to these new brittle anisotropic materials.

It is clearly the case, for either the major metallic systems or the developed conventional polymeric composites, that small improvements made against perceived inadequacies will generate significant increases in applicability while radical new structural requirements such as space aircraft are needed to enable the development of radically new materials.

REFERENCES

1. Dorey, G. (1987) *J. Phys. D (App. Phys.)*, **20**, 245.
2. Eckvall, J. C., Rhodes, J. E. and Wald, G. G. (1982) ASTM STP 761, 328.
3. Nowack, H. *et al.* (1991) *Mat. Sci. Monograph* 72, 495.
4. Thorbeck, J. (1991) *Mat. Sci. Monograph* 72, 323.

2

Aluminium alloys: physical metallurgy, processing and properties

P.J. Gregson

2.1 INTRODUCTION

The development of the aerospace industry during the twentieth century has relied heavily on the availability of aluminium and its alloys. Since their introduction into Zeppelins during World War I and both civil and military aircraft since World War II, aluminium alloys have remained the dominant materials for the construction of subsonic airframe structures. Currently, wrought aluminium alloys account for greater than 70% of the materials usage for a modern transport airframe on account of their low density, good mechanical properties and corrosion resistance at ambient temperatures. Their availability in many product forms (Figure 2.1) together with ease of fabrication ensures that aluminium alloys continue to be selected for structurally efficient airframes [1]. Similar requirements have led to extensive use of aluminium alloys for structural applications in satellites and space launch vehicles since the late 1950s.

In all of these situations the requirements of the aluminium alloys have become increasingly exacting. Improved performance is required in military airframe structures and economic efficiency in civil applications. These demands may be addressed by developing new alloys having enhanced specific strength and specific stiffness, and by use of advanced fabrication techniques to reduce the weight of built-up structures. Some extended temperature performance required in high altitude, high performance airframes and in engine components may be achieved through application of advanced processing of specially designed alloys. This chapter will describe the processing, structure and properties of conventional aerospace aluminium alloys and proceed to discuss advances in

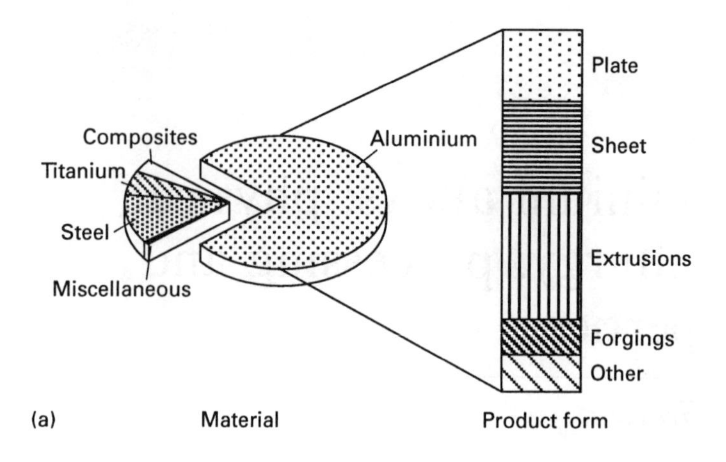

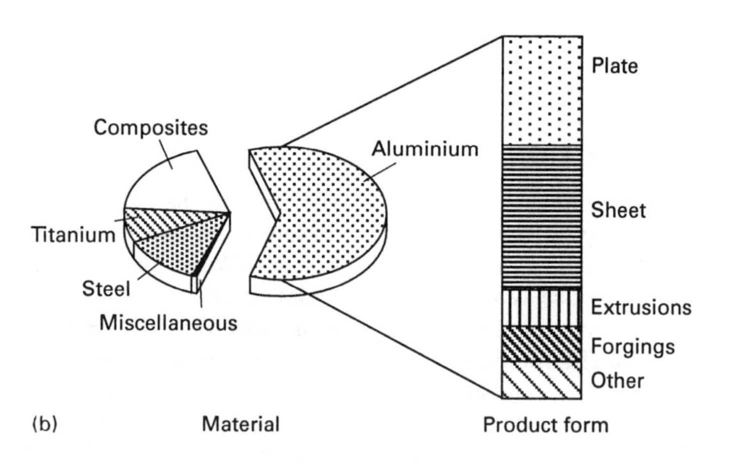

Figure 2.1 Approximate distribution of Al-alloy used in large transport aircraft: (a) USA and (b) European manufacturers.

alloy development, processing and fabrication technology which have been exploited to ensure that aluminium alloys remain a dominant low-temperature structural material in aerospace applications.

2.2 ALUMINIUM ALLOYS: PROCESSING AND PROPERTIES

Historically the wide range of aluminium alloys has been separated according to three categories, as follows:

- **Casting alloys** which are based principally on the Al–Si and Al–Mg alloy systems. These may be used in some critical fittings (e.g. engine pylons), but are more generally used in small, non-loadbearing applications (e.g. components of control systems). In view of their relative low tonnage use in structural aerospace applications, these alloys will not be considered further in the present chapter.

- **Wrought non-heat-treatable** alloys such as Al–Mn and Al–Mg, the properties of which are controlled by a combination of solid solution strengthening and dislocation structures introduced by cold work. The performance of these alloys is inadequate for loadbearing aerospace applications.
- **Wrought heat-treatable alloys** such as Al–Cu, Al–Zn–Mg and Al–Li–Cu–Mg which require heat treatment to develop their high strength potential via precipitation hardening. These alloys are the primary metallic materials of choice in airframe and other structural aerospace applications, and will be addressed in this chapter.

To assist classification, the international alloy designation system (IADS) has been adopted to identify alloy composition and temper (Figure 2.2). Most of the aluminium alloys in airframe applications are of the 2xxx and 7xxx series with specific tempers selected for design property requirements. For example, fuselage structures of civil aircraft are generally constructed from 2024–T3 alloy skins having excellent damage tolerance attached to 7075–T6 alloy stringers and longerons in which the high strength of this alloy can be exploited. Further discussion

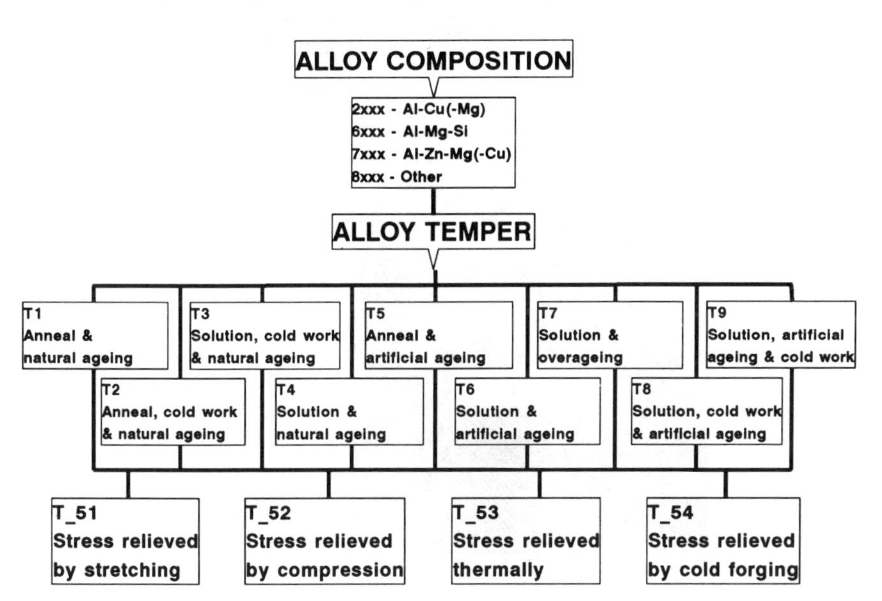

Figure 2.2 Al-alloy and temper designation (IADS) for heat-treatable aluminium alloys.

of heat treatment will follow a description of the fabrication of conventional aerospace alloys.

2.2.1 Alloy production

Over 90% of the aluminium alloys used in aerospace applications are fabricated via the ingot-casting route, and the semi-continuous direct chill (DC) casting process (Figure 2.3) is the dominant method for the efficient manufacture of rectangular or round section billets. Low melting point major alloying additions (e.g. Mg, Cu, Zn) are added to the molten charge as pure elements, while high melting solutes (e.g. Ti, Cr, Zr, Mn) are added in the form of master alloys. During casting, a solid shell of metal is formed in the water-cooled mould, and further solidification occurs as the ingot is withdrawn from the mould and is impinged by a water spray. The ingot is subsequently scalped to remove inverse segregation in the surface regions, and to improve surface quality. To limit hot-cracking the cast ingot grain size is controlled by small additions of Ti–B; there remains debate as to the precise role of heterogeneous nuclei of $TiAl_3$ and TiB_2 particles in this inoculation process. Besides the

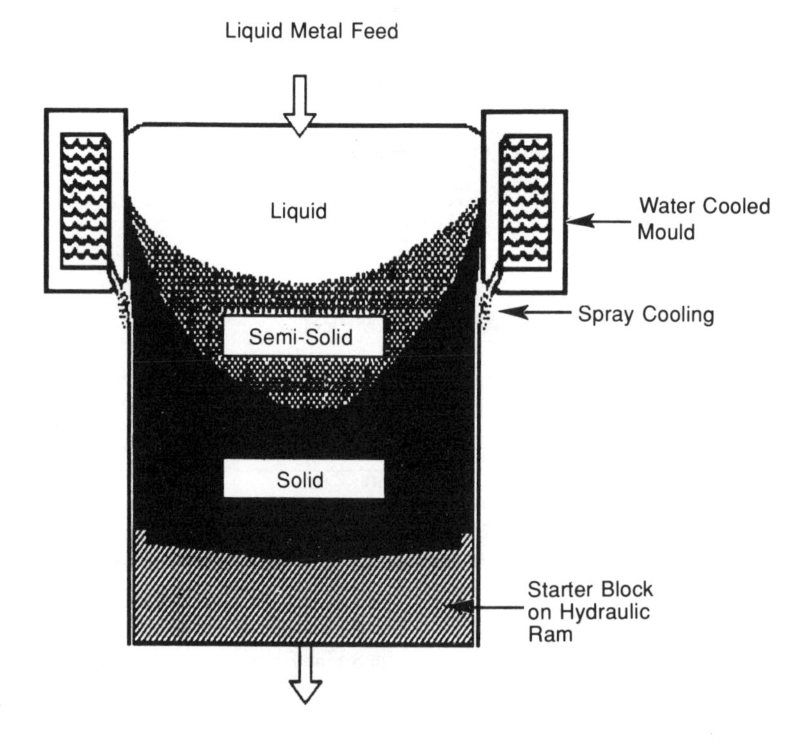

Figure 2.3 Vertical direct chill (DC) casting process.

Figure 2.4 Microsegregation within as-cast grain structure of 7xxx alloy: anodized and viewed under polarized light (courtesy P. J. E. Forsyth).

macrosegregation inherent in large cast ingots, constitutional microsegregation occurs within the dendrite cells to give the characteristic cored appearance (Figure 2.4). The solute redistribution can be predicted by the Scheil equation

$$C_s = C_0(1 - f_s)^{k-1}$$

where C_s = solute concentration in the solid, C_0 = average alloy composition, f_s = fraction solid and k = partition coefficient. A finer cell size, promoted by a high rate of solidification, results in less microsegregation.

The ingot is subsequently homogenized at a high temperature to reduce the compositional variation and remove the more soluble phases (e.g. $CuAl_2$) which form during eutectic solidification. The insoluble compounds (e.g. $FeAl_6$) remain, and are deleterious to properties (section 2.3.1). Control of heating rate and homogenizing time and temperature allows precipitation of submicron intermetallic particles (e.g. $MnAl_6$, $ZrAl_3$) which control the grain structure during subsequent fabrication. The billet may then be hot rolled to plate and subsequently cold rolled to sheet product, forged close to shape, or extruded to the required section. During hot rolling, the surfaces of the billet may be roll-clad with high-purity aluminium for corrosion protection.

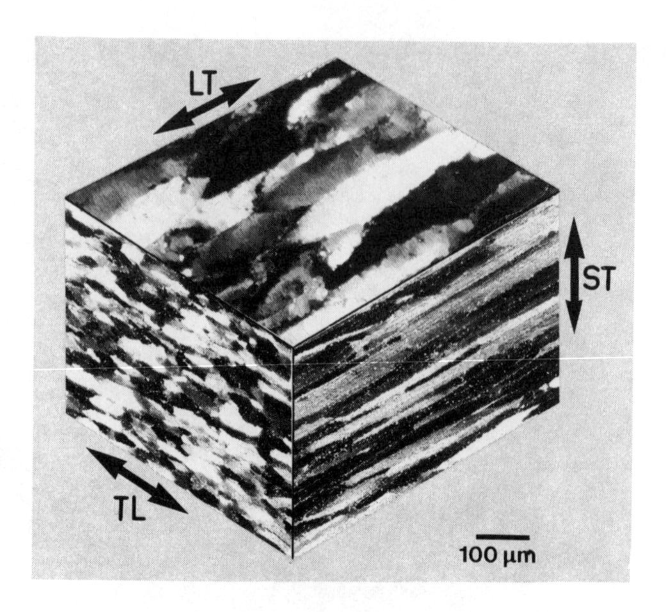

Figure 2.5 Grain structure of a wrought aluminium alloy (8090) including definition of the three principal directions: anodized and viewed under polarized light (courtesy V. J. Bolam).

In addition to effecting the necessary shape change, these fabrication processes incorporating a combination of thermal and mechanical treatments have a profound influence on the structure and properties of the alloy. During hot deformation an elongated grain structure is established (Figure 2.5) and this is accompanied by the development of preferred orientations or textures as crystallographic planes orientate themselves in response to deformation. Hot-rolled aluminium alloys may exhibit a range of texture components from $\{110\}\langle 112\rangle$ to $\{112\}\langle 111\rangle$ while a strong $\langle 111\rangle$ fibre texture is usually evident in extrusions. The strength of the individual texture components is controlled by the processing parameters and by the presence of constituent particles and solute additions. The texture is further influenced by the formation of sub-grains as a result of dislocation rearrangement within the deformed grains. This directionality of grain structure and underlying crystallographic texture leads to anisotropic mechanical and corrosion properties, especially where grain boundaries have a dominant effect.

During hot deformation, equiaxed sub-grains are formed, their size being influenced by the effect of alloying additions on dislocation mobility, and by the deformation variables. The sub-grain diameter, d_s, varies with strain rate ($\dot{\varepsilon}$) and the temperature T (K) according to

$$d_s^{-m} = a + b \ln [\dot{\varepsilon} \exp (Q/RT)]$$

where a and b are empirical constants, R is the gas constant and Q is the activation energy (similar to that for self-diffusion) [2]. The strengthening imparted by the sub-grain structure is usually described in terms of a modified Hall–Petch relationship

$$\sigma_y = \sigma_0 + k' d_s^{-p}$$

where σ_0 is the strength of the substructure-free material, k' is the strength coefficient of the sub-boundary (~ 0.05 MN m$^{-1.5}$ in comparison with ~ 0.15 MN m$^{-1.5}$ for grain boundaries) and p is constant in the range 0.25–0.5 [3]. The establishment of a well-developed sub-grain structure, characteristic of alloys having a high stacking fault energy, reduces the driving force for recrystallization, and dynamic recrystallization is likely to occur only during those deformation processes imparting a very high strain rate, e.g. extrusion. For other product forms recrystallization may occur during subsequent annealing treatments; this process is stimulated by high dislocation densities in the vicinity of coarse constituent particles [4] but retarded by small particles which impede the nucleation and growth of recrystallizing grains, and prevent grain growth after recrystallization [5].

In addition to defining grain size and shape, substructure and texture, the thermomechanical processing can be designed to control the dislocation density within the matrix of the alloy, and at the sub-grain boundaries. Besides their direct effect on alloy performance, the dislocations will influence the nature and distribution of the fine strengthening precipitates which form during subsequent heat treatment. These structural phenomena, together with their role in developing the properties required in the specific alloys, will be discussed in detail later.

DC cast wrought aluminium alloys will remain an important structural material, particularly in large transport aircraft where large section sizes are required (e.g. the wing cover of a transport aircraft may be machined from a single alloy plate). However, there are applications for which other manufacturing routes may be more cost effective or result in a material with improved performance. Where appropriate, the production of near net shape components may be of substantial benefit from an economic and environmental viewpoint.

Squeeze casting, employing a pressure of typically 200 MPa to the liquid metal during solidification, may provide a cost-effective route for small components traditionally manufactured from forgings. In this process the grain structure can be controlled, and casting defects eliminated, to give properties comparable to their wrought counterparts [6]. Spray deposition processes incorporate the benefits imparted by rapid solidification processing (RSP): Liquid metal is atomized, and the metal droplets are quenched (> 10^5 C/s) to below the solidus as they build up on the casting. Spray-formed alloys experience a rapid transition through the liquidus to the solidus, giving a fine stable grain size and a refined constituent microstructure on account of the small dendrite cell size (~ 0.1 μm compared with ~ 50 μm for DC casting). However, further cooling below the solidus is relatively slow, thus limiting the extended solubility possible in the production of metal ribbons or powders.

The powder metallurgy (PM) route provides an alternative approach for the production of near-net shape or semi-finished products (Chapter 12). Powder production may involve mechanical alloying to introduce hard sub-micron particles of oxides or carbides, or gas atomization to take advantage of RSP (including extended solid solubility as a result of the rapid quench to room temperature). This route offers a further range of alloys of unique compositions, but high cost, size limitation and quality control issues will restrict applications to those for which the improved performance is essential.

2.2.2 Alloy fabrication

Although different fabrication philosophies are adopted by the various airframe manufacturers (for example, the Europeans traditionally favoured machining of intregral parts from plate since the material cost of conventional aluminium alloys is low and they are readily machined using numerically controlled equipment, while the Americans use proportionately more sheet and extruded products – Figure 2.1), the aluminium alloys are available in a wide range of relatively cheap and readily available product forms. The heat-treatable aerospace alloys are in general not weldable as a consequence of susceptibility to hot-tearing in alloys with high Cu + Mg content, and airframe structures have traditionally been fabricated via extensive use of mechanical fasteners. There are, however, notable exceptions, such as the Al–Cu–Mg alloy 2219 which is widely used in welded space structures, and Al–Li–Mg alloys used in welded airframes and rocket fuel tanks. Nevertheless, substantial structural and economic efficiencies could be effected by exploiting superplastic forming and diffusion bonding of alloy sheet and by adhesive bonding.

In the superplastic condition, alloy sheet can be formed under the application of an inert gas pressure (Chapter 9). Accurate control of forming pressure and back pressure is required to limit cavitation at the grain boundaries and triple points in aluminium alloys. The mechanisms proposed to explain superplastic deformation involve diffusion, dislocation motion, grain boundary sliding and grain rearrangement [7]. These require the alloy to possess a fine structure (grain-size) which is stable at high temperatures. In aluminium alloys this is normally achieved by using a dispersion of small particles of the metastable, cubic phase Al_3Zr which prevents grain growth during superplastic deformation at temperatures up to ~ 500 °C. Alternatively, the alloy may be thermochemically processed to achieve a very fine grain size; this may involve an overageing process to develop coarse particles which promote intense deformation zones during warm rolling, leading to recrystallization during subsequent solution treatment. Superplastic forming may be used in combination with diffusion bonding to produce complex monolithic structures with a consequent reduction in structural mass and fabrication costs; this process is discussed in Chapter 10. Similar benefits can be achieved through the use of advanced structural adhesives in the fabrication of airframe structures; this is the subject of Chapter 11.

2.2.3 Alloy heat treatment

The properties of the high-strength aluminium alloys used in aerospace applications are dictated principally by the presence of second phases, and predominant amongst these are the fine particles which form during precipitation-hardening (or age-hardening) after fabrication. In these alloys Mg, Zn, Li and Cu exhibit high solid solubility in aluminium at elevated temperatures, but this decreases with decreasing temperatures such that the equilibrium solid solubility at room temperature is less than the solute content of the alloy (Figure 2.6). Heat treatment usually involves solution treatment at a high temperature within the single phase region, followed by rapid cooling (or quenching) to a low temperature to obtain a solid solution supersaturated with these elements and with vacancies. Ageing at either room temperature (natural ageing) or some intermediate temperature (artificial ageing) results in the decomposition of the supersaturated solid solution to produce precipitate particles which may form heterogeneously at preferential sites (e.g. sub-grain boundaries and dislocations) or homogeneously throughout the matrix. In the latter case the decomposition of the supersaturated solid solution (α_{ss}) usually occurs by the following sequence:

$$\alpha_{ss} \rightarrow \text{solute atom clusters} \rightarrow \text{intermediate precipitate} \rightarrow \text{equilibrium precipitate}$$

The solute atom clusters (usually a few atom planes in thickness) are fully coherent with the matrix and are often referred to as Guinier–Preston (GP) zones after the researchers who confirmed their existence. The metastable intermediate precipitate, which may form at the sites of stable GP zones or nucleate heterogeneously, is considerably larger than the GP zone and is partially coherent with the lattice planes of the matrix. The final equilibrium precipitate is larger still and is completely incoherent with the parent lattice.

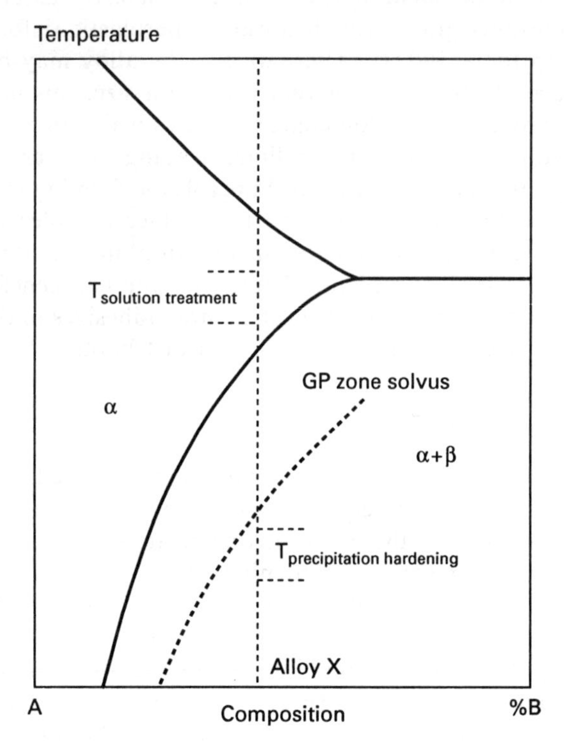

Figure 2.6 Schematic phase diagram including GP zone solvus indicating heat-treatment temperatures for alloy X.

Current understanding of the precipitation in high-strength aluminium alloys is based on theories for the thermodynamic stability of clusters and intermediate precipitates [8] and for the kinetics of nucleation and growth of solute atom clusters [9]. Homogeneous precipitation involving GP zones depends upon the ageing temperature and the concentration of vacancies, which, in turn, will be effected by the solution treatment temperature and the cooling rate during the subsequent quench. The hypothetical phase diagram in Figure 2.6 contains a metastable solvus line which, for a given alloy, dictates a critical temperature (T_c) below which the nucleation of GP zones may occur. This critical temperature may be displaced to higher temperatures when the vacancy concentration is increased. A low vacancy concentration such as that in the vicinity of grain boundary will restrict homogeneous precipitation and lead to a vacancy depleted precipitate free zone (PFZ). Alternatively, heterogeneous nucleation and growth of equilibrium phases can occur at grain boundaries, thus producing a solute depleted PFZ. Both types of PFZ are restricted by lowering the ageing temperature, which increases the driving force for homogeneous precipitation and decreases the nucleation and growth of equilibrium precipitates. In order to retain minimum heat-treatment times a double ageing process may be employed: a low temperature treatment to promote GP zones is followed by a higher temperature to give a fine dispersion of precipitates from the pre-existing zones. Additionally, the formation of a particular precipitate may be influenced by other solute elements, or by trace element additions. This is usually associated with vacancy interactions and a consequent change to the kinetics of the process or with modification to the interfacial energy of the particles and associated movement of T_c.

2.2.4 Alloy properties

(a) Strength

The mechanical properties of the high-strength alloys are controlled by a complex interaction of microstructural features: coarse intermetallic constituent particles, sub-micron dispersoid particles, fine precipitates, and grain size, shape and associated substructure together with the nature and strength of any crystallographic texture.

The increase in strength during age-hardening of heat-treatable alloys is associated principally with the interaction of mobile dislocations with the fine precipitates. Zones and metastable precipitates which are coherent with the matrix are cut by dislocations during plastic flow. In the case of aluminium alloys these particles strengthen the material by a combination of three important mechanisms [10, 11]:

- *Stacking fault hardening* for which the strength increment from the particles is

$$\Delta\sigma_{sf} \propto \frac{f^{0.3}\, r^{0.5}}{G^{0.5}\, b^2}\, \gamma_{sf}^{1.5}$$

  where f and r are the particle fraction and radius respectively, G is the shear modulus, b the Burgers vector and γ_{sf} is the stacking fault energy. This may be particularly important for ordered precipitates with low coherency strains and low modulus mismatch.
- *Coherency hardening* where the interaction of the dislocations with the stress field of the coherent precipitates (ε_c) contributes an increment of strength

$$\Delta\sigma_c \propto \frac{G r^{0.5} f^{0.5}}{b}\, \varepsilon_c^{1.5}$$

- *Modulus hardening* associated with the difference in moduli between particle and matrix. This generates an increment of strength

$$\Delta\sigma_E \propto \frac{Gb}{\lambda}\left(1 - \frac{E_1^2}{E_2^2}\right)^{0.5}$$

where λ is the interparticle spacing and E_1 and E_2 are the modulus of the soft and hard phases respectively.

The combination of these cutting mechanisms leads to increased strength with increasing particle size and volume fraction. With time, particle coarsening occurs leading to increased particle size and interparticle spacing; the particles now resist cutting and the dislocations are forced to bypass the particles. The strength provided by the Orowan hardening is given by

$$\sigma_0 \propto \frac{Gb}{\lambda} \ln \frac{r}{b}$$

This depends strongly on λ^{-1} and decreases with increasing r. A schematic illustration of the alloy strength as a function of precipitate radius (which increases with ageing time) is given in Figure 2.7. When the particles are sheared by moving dislocations, the size of the particle on the slip plane is reduced, resulting in a local decrease in resistance to further dislocation glide; deformation becomes concentrated into intense slip bands and there is little work-hardening. When Orowan looping occurs, the slip distribution becomes more homogeneous, and work-

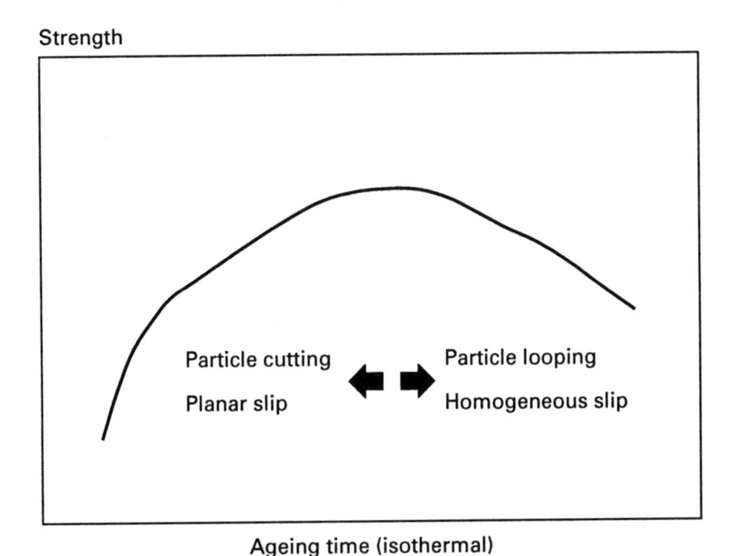

Figure 2.7 Effect of precipitation hardening (ageing) on mechanical strength, indicating the deformation mode.

hardening is increased. By modifying the deformation mode, the presence of these fine precipitates has an important influence on properties other than strength.

While precipitation-hardening is the dominant strengthening mechanism in these heat-treatable alloys, the contribution from solution strengthening may be significant, particularly in underaged tempers. This contribution is derived from the combined effect of a change in modulus and in the misfit strain in the vicinity of the solute [12]. The increment of strength provided by solid solution hardening is given by

$$\Delta\sigma_{ss} \propto G\,\varepsilon_c^{1.5}\,c^{0.5}$$

where ε_c is the combined modulus and misfit contribution, and c is the atomic concentration.

Precipitation-hardening of aerospace aluminium alloys is usually undertaken in the range 100–200 °C. Continued exposure to these temperatures during ageing or, more important, during service life, will lead to overageing and a consequent reduction in strength. Thus, alloy choice and temper selection must be appropriate for the operating conditions which the material experiences during its service life.

(b) Toughness

Since the introduction of damage-tolerant design to take account of the presence of cracks and other flaws in fabricated airframe structures, it is the fracture toughness of the alloy which has become the most important material parameter for many applications. The presence of the coarse brittle constituent particles which are not dissolved during homogenization (section 2.2.1) is deleterious to toughness; control of impurity elements such as Fe and Si is important for the aerospace alloys so that adequate toughness can be achieved. The influence of the finer dispersoid particles on toughness is more complex. Their presence *per se* is considered to be deleterious since decohesion of the matrix/particle interface leads to premature microvoid nucleation. Since the stress required for microvoid nucleation is proportional to $r^{-0.5}$, it might be expected that finer dispersoids (e.g. Al_3Zr) would be beneficial, and there is evidence that alloys containing Zr are more resistant to fracture than those containing Cr or Mn for grain-refining purposes. However, the transition metal additions also influence grain size, shape and crystallographic texture. By suppressing recrystallization and limiting grain growth, and thus promoting transgranular fracture, the dispersoids are beneficial to toughness [13], but if the grain structure of the wrought alloy is such that fracture involves the failure of planar grain boundaries (e.g. in the ST orientation), the fracture toughness may be reduced.

In general, the toughness of aluminium alloys decreases as the strength is increased by heat treatment. The fine coherent particles promote planar slip which may result in early crack nucleation within the slip band or at the grain boundaries where large stress concentrations will be established by the strain localization. This effect will, of course, vary with the nature and size of the precipitate; thus, the optimum balance of toughness and strength may be associated with different temper treatments for different alloys, as discussed later.

(c) Stress corrosion

Another important design criterion for structural aerospace applications is the resistance of aluminium alloys to environment-induced fracture or stress corrosion cracking (SCC) [14]. This pernicious failure mechanism occurs at grain boundary sites, and is usually attributed to either highly localized anodic dissolution of grain boundary regions or the embrittlement associated with ingress of atomic hydrogen at grain boundary sites where diffusion is rapid. The precise mechanism will depend on the particular combination of alloy and the environment, but the general dependence of SCC susceptibility upon alloy heat treatment is illustrated in Figure 2.8. During ageing, grain boundary precipitation and/or the development of PFZs adjacent to the grain boundaries will result in either the zones or the boundaries being anodic with respect to the grain interiors, thereby promoting localized grain boundary attack. The micro-

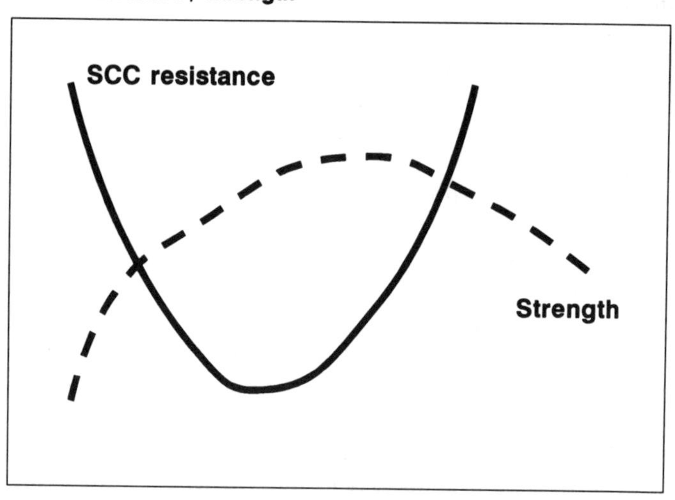

Figure 2.8 Dependence of SCC resistance upon ageing treatment.

structure will also influence the role of atomic hydrogen on the deformation and fracture process. Additionally, the stress concentrations generated where intense slip bands impinge upon grain boundaries will contribute to the intergranular failure; this will be reduced as the precipitates coarsen and the deformation becomes homogeneous.

Surface treatment is usually required to reduce corrosion-related degradation of the alloys. Sheet clad with nearly pure aluminium may negate the need for further protection, as in polished fuselage skins, but heat-treatment practices must be carefully controlled to prevent diffusion of solute elements from the alloy into the clad skin. Furthermore, the presence of the soft skin may promote the initiation of fatigue cracks, thereby reducing (up to 50%) the fatigue strength of the alloy. An alternative approach employs anodizing to provide an adherent oxide protective coating and the subsequent use of chromate-containing primers. However, the anodizing treatment will itself reduce the fatigue strength by an amount depending upon alloy composition, grain structure and heat treatment, and on the nature of the film formed during chromic acid, phosphoric acid or sulphuric acid anodizing.

(d) Fatigue

By its effect on the deformation mode, the fine microstructure would also be expected to influence significantly the fatigue properties of the heat-treated alloys. Planar slip will lead to the early initiation of fatigue cracks. Furthermore, continuous dislocation cutting of fine precipitates during cyclic loading may result in localized precipitate dissolution as the particles become smaller than the critical size for thermodynamic stability; the establishment of soft, persistent slip bands will be detrimental to fatigue performance [15]. In this case a more uniform slip distribution promoted by incorporation of dispersoid particles (e.g. Al_6Mn) or by the use of more stable precipitates would be expected to be beneficial. Alternatively, promotion of slip band cracking may increase crack branching and crack path tortuosity, leading to increased fatigue crack closure and a reduction in the effective stress intensity at the crack tip (Figure 2.9). In these circumstances overageing to provide incoherent strengthening precipitates, thereby homogenizing deformation, may be advantageous to SCC resistance but lead to a decrease in the fatigue crack propagation resistance.

(e) Creep

At elevated temperatures the creep performance of the alloy will assume significance, and for the aerospace aluminium alloys this will depend on the nature and concentration of solute atoms, age-hardening precipitates

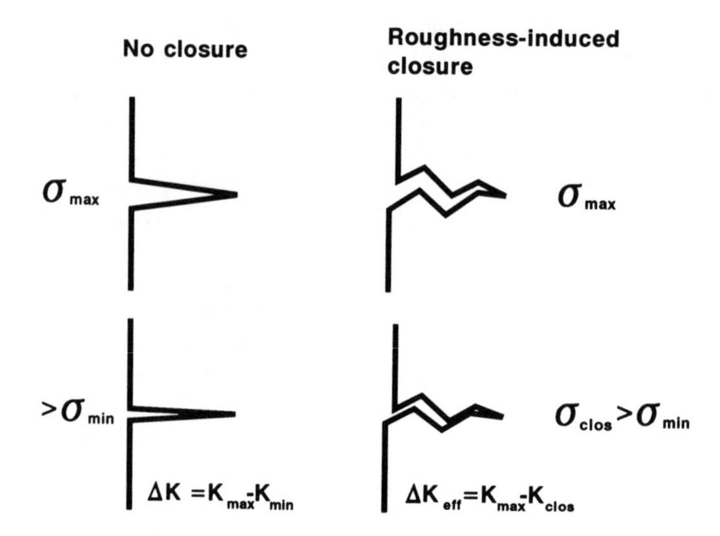

Figure 2.9 Schematic illustration of roughness-induced fatigue crack closure.

and dispersoid particles. Solute atoms restrict dislocation climb to varying degrees, thereby restricting the creep processes. The age-hardening precipitates may be beneficial but they are unstable at elevated temperatures, and so it is the presence of a dispersion of fine intermetallic particles (e.g. $FeNiAl_9$), stable at the required service temperature, which is used to control the creep resistance of aluminium alloys. These principles are exploited in the dispersion strengthened alloys (section 2.4.2).

(f) Summary

Structural aerospace designs require optimization of properties for specific applications. For this reason a host of different tempers employing single age-hardening treatments, duplex ageing and thermomechanical processing have evolved for conventional aluminium alloys (2xxx and 7xxx), and the more recently developed lithium-containing alloys. The rationale behind these treatments will be discussed in the next section.

2.3 CONVENTIONAL AEROSPACE ALUMINIUM ALLOYS

2.3.1 2xxx Alloys

Duralumin (Al–3.5Cu–0.5Mg–0.5Mn), used in structural members of Zeppelin airships, was the first heat treatable aluminium alloy based on the Al–Cu system – the 2xxx alloys. Since then many Al–Cu(–Mg) based

Table 2.1 Approximate composition of some 2xxx aerospace alloys

Alloy	*Cu*	*Mg*	*Zn*	*Mn*	*Cr*	*Si*	*Fe*
2014	3.9–5.0	0.2–0.8	0.3	0.4–1.2	0.1	1.0	0.7
2024	3.8–4.9	1.2–1.8	0.3	0.3–0.9	0.1	0.5	0.5
2124	3.8–4.0	1.2–1.8	0.3	0.3–0.9	0.1	0.2	0.3
2025	3.9–5.0	0.05	0.3	0.4–1.2	0.1	1.0	1.0
2219	5.8–6.8	0.02	0.1	0.2–0.4	–	0.2	0.3
2618	1.9–2.7	1.3–1.8	0.1	(0.9–1.2Ni)		0.2	1.1

alloys have been developed for aircraft construction, the composition of some of these being included in Table 2.1. In all of these alloys the presence of Si improves liquid metal fluidity and hence castability, and Fe reduces hot cracking in the ingot. However both form coarse intermetallic compounds (e.g. Al_7Cu_2Fe and Mg_2Si) which are detrimental to fracture toughness (e.g. the toughness of 2124–T8 is ~ 20% higher than that of 2024–T8); Fe and Si concentrations are therefore kept to a minimum. Ti serves to refine the cast grain structure and the Mn and Cr additions form dispersoid particles (e.g. $Al_{20}Cu_2Mn_3$ and $Al_{18}Mg_3Cr_2$) which control the wrought grain structure. Alternatively, the more efficient grain refiner Zr may be added to develop a very fine superplastic microstructure (e.g. Supral Al–6Cu–0.5Zr).

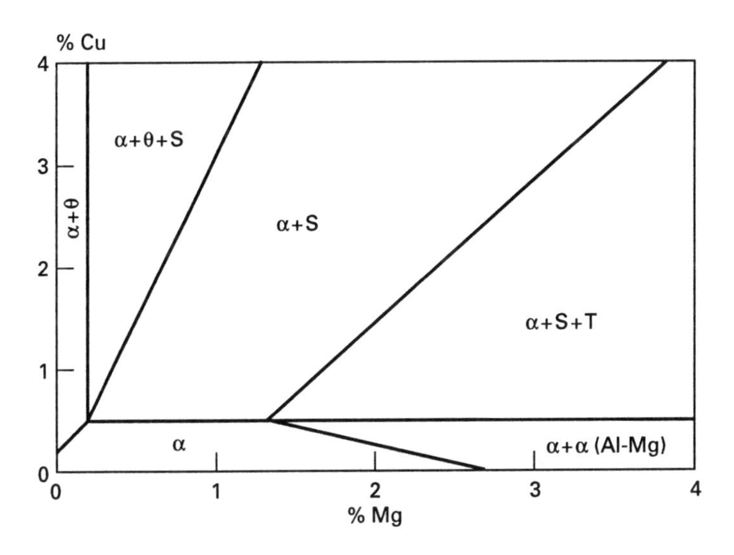

Figure 2.10 Isothermal section of ternary Al–Cu–Mg phase diagram at 190 °C: $\theta = Al_2Cu$, $S = Al_2CuMg$ and $T = Al_6CuMg_4$.

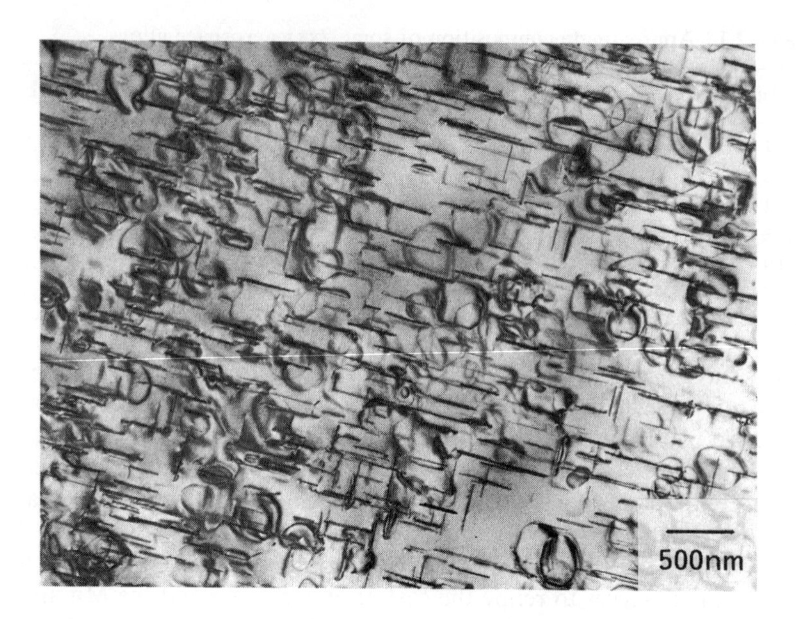

Figure 2.11 θ′ precipitates ($CuAl_2$) in Al–Cu alloy aged at 240 °C (courtesy G. W. Lorimer).

Cu and Mg provide the major alloying elements in the 2xxx alloys, and, through solution strengthening and precipitation – hardening, have a dominant effect on alloy properties. The isothermal section at 190 °C (~ age-hardening) of the Al–Cu–Mg phase diagram (Figure 2.10) indicates that in the alloys of commercial importance, precipitates of θ ($CuAl_2$) and S (Al_2CuMg) may form during heat treatment. θ is promoted by a high Cu:Mg ratio, and the ageing sequence (Figure 2.11) is summarized as

$$\alpha_{ss} \rightarrow \text{GB zones} \rightarrow \theta'' \rightarrow \theta' \rightarrow \theta$$

For alloys having a high magnesium concentration the precipitation process can be described as

$$\alpha_{ss} \rightarrow \text{GP(B) zones} \rightarrow S' \rightarrow S$$

Crystallographic details of these precipitates are summarized in Table 2.2, and the orientation relationships between matrix and precipitates are reviewed elsewhere [16].

Many 2xxx alloys demonstrate a significant strengthening response at room temperature; the natural ageing has been attributed to complex interactions involving GP zones, solute atoms and vacancies. Improved

Table 2.2 Crystallography of precipitates in 2xxx alloys

Precipitates	*Crystallography*
Al–Cu(–Mg) (high Cu:Mg)	
GP zones	Cu-rich zones (thin plates)
θ″	Fully coherent intermediate precipitates probably nucleated at GP zones
θ′ – $CuAl_2$	Semi-coherent plates on $(100)_\alpha$ generally nucleated at dislocations Tetragonal: $a = 0.404$ nm, $c = 0.580$ nm
θ – $CuAl_2$	Incoherent equilibrium phase which may nucleate at the surface of θ′ Body centred tetragonal: $a = 0.607$ nm, $c = 0.487$ nm
Al–Cu–Mg (low Cu:Mg)	
GP zones	Cu- and Mg-rich zones as thin rods along $\langle 100\rangle_\alpha$
S′–Al_2CuMg	Semi-coherent laths on $\langle 210\rangle_\alpha$ along $\langle 100\rangle_\alpha$ nucleated at dislocations. Orthorhombic: $a = 0.404$ nm, $b = 0.925$ nm, $c = 0.718$ nm
S–Al_2CuMg	Incoherent equilibrium phase of similar morphology to S′ Orthohombic: $a = 0.400$ nm, $b = 0.923$ nm, $c = 0.714$ nm

strength is obtained by elevated temperature ageing, and maximum strength is usually associated with precipitation of semi-coherent particles of θ′ and/or S′. The formation of these precipitates may be influenced by trace element additions: for example, Si additions restrict the condensation of vacancies and promote more homogeneous distribution of S′ precipitates (Figure 2.12). However, in 2xxx alloys the increased strength associated with artificially aged tempers is accompanied by a substantial degradation in fracture toughness, with the result that in many airframe applications the alloys are used in the naturally aged (T3 and T4) tempers. For example, on the Airbus A320 fuselage skins and stringers, bottom wing skin and stringers and wing spars and slats are fabricated from 2024-T3; these are but a few of the applications for these damage-tolerant alloys in primary aerospace structures.

Nevertheless, in these underaged tempers the alloys are susceptible to SCC due to the strain localization associated with dislocation cutting of the very fine strengthening zones and precipitates. Protection from the environment via cladding, extensive use of chromic acid anodizing, sealing and painting treatments, and wet installation of fasteners of the fuselage skins and internal structure is therefore essential, particularly in corrosion critical areas such as the bilge and door areas.

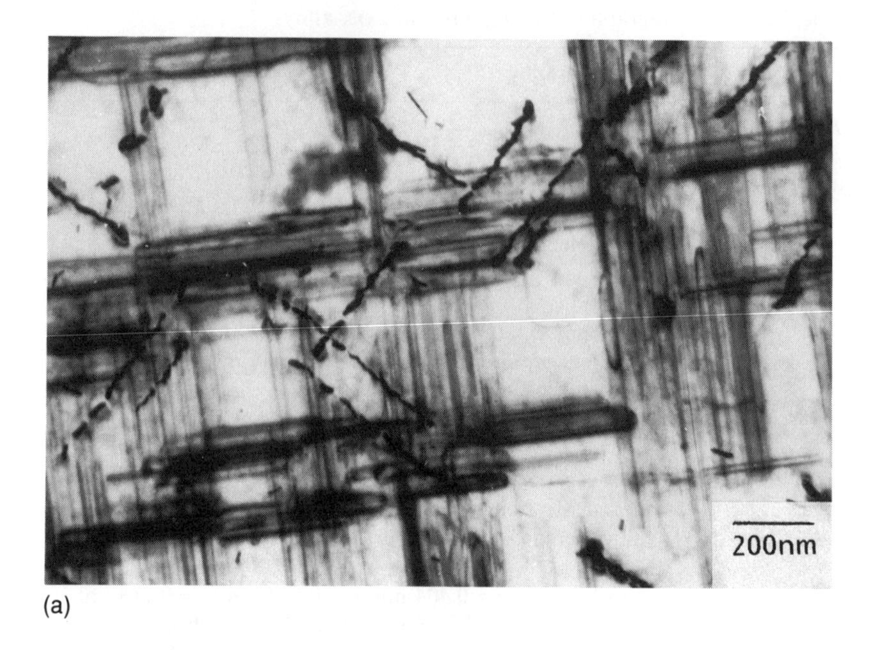

(a)

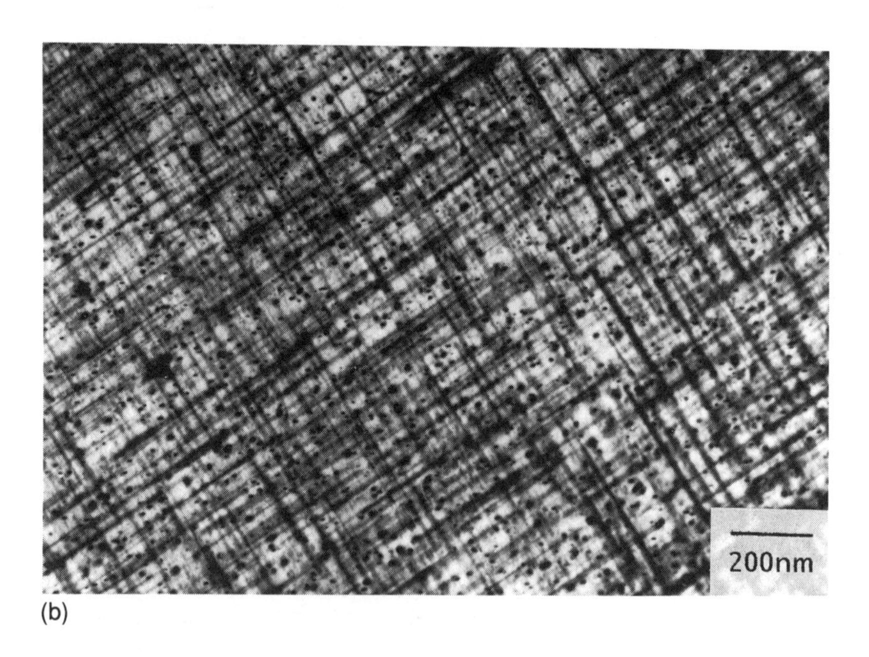

(b)

Figure 2.12 S' precipitates (Al_2CuMg) in (a) Al–2.7Cu–1.4Mg and (b) Al–2.7Cu–1.4Mg–0.2Si aged at 190 °C: ⟨100⟩Al orientation (courtesy G. C. Weatherly).

Table 2.3 Approximate composition of some 7xxx aerospace alloys

Alloy	*Cu*	*Mg*	*Zn*	*Mn*	*Cr*	*Zr*	*Si*	*Fe*
7001	1.6–2.6	2.6–3.4	6.8–8.0	0.2	0.2–0.3	–	0.4	⟨ 0.1
7010	1.5–2.0	2.2–2.7	5.7–6.7	0.3	⟨ 0.1	0.1–0.2	0.1	0.2
7050	2.0–2.6	1.9–2.6	5.7–6.7	0.1	⟨ 0.1	0.1–0.2	0.1	0.2
7075	1.2–2.0	2.1–2.9	5.1–6.1	0.3	0.2–0.3	–	0.4	0.5
7475	1.2–1.9	1.9–2.6	5.2–6.2	0.1	0.2–0.3	–	0.1	0.1

2.3.2 7xxx Alloys

7xxx alloys based on the Al–Zn–Mg(–Cu) system offer a greater response to precipitation-hardening treatments, with the potential for substantially enhanced strength. The compositions of some appropriate aerospace alloys are included in Table 2.3. Besides the major alloying elements, and the impurity and dispersoid additions discussed previously, some alloys (e.g. 7010 and 7050) contain Zr for more efficient grain refinement and reduced quench sensitivity, and improved strength and toughness is observed. The isothermal section at 200 °C of the Al–Zn–Mg phase diagram (Figure 2.13) reflects the thermodynamic driving force for the precipitation of $\eta(MgZn_2)$ and/or $T(Al_{32}(Mg, Zn)_{49})$ during heat treatment. Their precipitation sequences are summarized respectively as

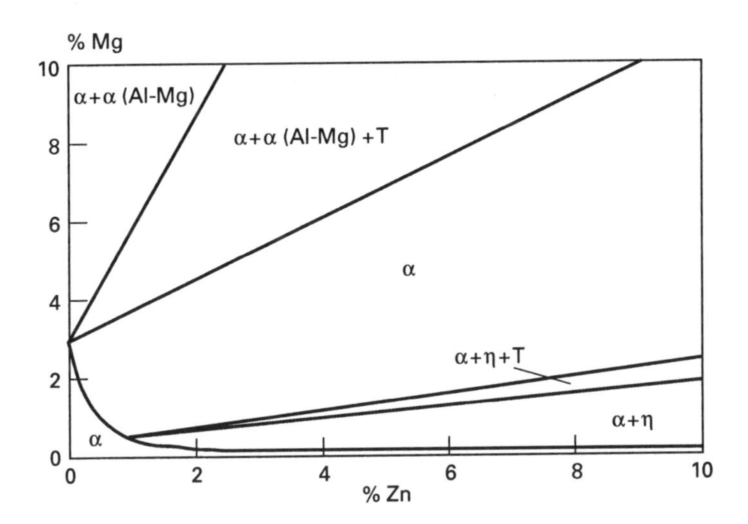

Figure 2.13 Isothermal section of ternary Al–Zn–Mg phase diagram at 200 °C: $\eta = MgZn_2$ and $T = Al_{32}(Mg, Zn)_{49}$.

Table 2.4 Crystallography of precipitates in 7xxx alloys

Precipitates	*Crystallography*
GP zones	Zn + Mg rich zones (spheres)
η'–$MgZn_2$	Semi-coherent platelets which may form from GP zones Hexogonal: a = 0.496 nm, c = 0.868 nm
η–$MgZn_2$	Incoherent equilibrium precipitates (plates or rods) exhibiting many orientation relationships with the matrix Hexagonal: a = 0.521 nm, c = 0.860 nm
T′–$Mg_{32}(Al, Zn)_{49}$	Semi-coherent precipitates of irregular morphology which may form instead of η' in alloys with high Mg:Zn ratios Hexagonal: a = 1.388 nm, c = 2.752 nm
T–$Mg_{32}(Al, Zn)_{49}$	Equilibrium precipitates which may form from η if ageing temperature > 190 °C, or from T′ in alloys with high Mg:Zn ratios Cubic: a = 1.416 nm

$$\alpha_{ss} \rightarrow \text{GP zones} \rightarrow \eta' \rightarrow \eta$$

and

$$\alpha_{ss} \rightarrow \text{GP zones} \rightarrow T' \rightarrow T$$

with the details of the different precipitates being included in Table 2.4. Cu additions provide further solution strengthening, additional precipitation by substituting for zinc in $MgZn_2$, and reduce the coherency of $MgZn_2$.

For those alloys with a high Zn:Mg ratio, the age-hardened microstructure (Figure 2.14) typically comprises:

- A homogeneous distribution of metastable η' in the alloy matrix; this may nucleate from the spherical GP zones which form during the earliest stages of ageing below the GP zone solvus (~ 150 °C for most alloys of commercial importance).
- Preferential precipitation of equilibrium η particles at the grain boundaries.
- Precipitate-free zones adjacent to the grain boundaries. These have been attributed to solute depletion associated with the early formation of the grain boundary precipitates, and to the vacancy concentration profile formed in these regions during the quench. Research on experimental Al–Zn–Mg alloys [16] has indicated that after a standard quench and 180 °C ageing the extent of the PFZ exceeds that of the true solute depletion, and that the introduction of an initial short low temperature ageing treatment (~ 100 °C) may lead to much of the PFZ being filled in with precipitates during subsequent ageing. Such

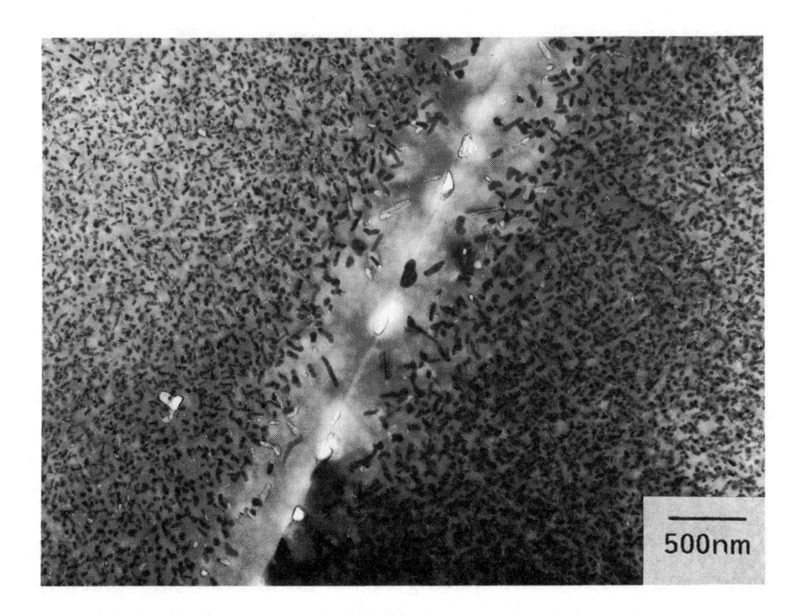

Figure 2.14 Matrix η′ and grain-boundary η precipitates ($MgZn_2$) in Al–6Zn–3Mg alloy aged at 180 °C (courtesy G. W. Lorimer).

changes are exploited in commercial practice as subtle temper treatments designed to produce an optimum balance of properties.

The 7xxx alloys respond to natural ageing, but unlike the 2xxx alloys they do not develop a stable naturally aged condition and so generally they are not used in the T3 or T4 tempers. Furthermore, pre-ageing deformation has little effect on the precipitation behaviour during heat treatment and so generally there is no advantage to be gained from a T8 temper. Consequently, the alloys have been used conventionally in the T6 or T7 temper employing ageing temperatures in the range 120–135 °C. However, in this condition the alloys exhibit a pronounced susceptibility to SCC, with threshold stress intensities (K_{Issc}) lower than those for 2xxx alloys [17]. There remains no consensus view as to the microstructural feature (e.g. grain boundary precipitates, matrix precipitation, PFZ) having most influence on the SCC behaviour. Nevertheless, Cu additions in excess of 1% are known to offer a substantial improvement in stress corrosion resistance, as does overageing. The latter has led to the development of two complex heat treatment practices to provide an improved balance of strength and SCC resistance:

- Since the alloys develop good strength at low ageing temperatures (120 °C) and improved SCC resistance by ageing at 170 °C, a duplex ageing treatment (T73) employing 4 hours at 120 °C and subsequent ageing at 170 °C (utilizing the nucleation of GP zones at the first ageing temperature to develop finer precipitates) is employed. The compromise of properties achieved in 7075 T73 gives an increase in the critical stress for SCC from ~ 50 MPa to ~ 300 MPa and a 15% reduction in strength compared with the T6 tempers. This heat treatment is now employed for a range of critical aerospace components.

- An alternative approach involves retrogression (or reversion) and reageing (RRA). From the T6 condition (24 hours at 120 °C) the alloy is heat treated (reverted) for a short time at elevated temperature (e.g. 5 min at 250–280 °C), cold water quenched and reaged at 120 °C. This process may be applied easily to thin sections, but for thicker sections extended reversion at a lower temperature may be necessary (e.g. 1 hour at 220 °C). The RRA treatment influences the nucleation of η' within the grains, but it is microstructural and microchemical changes in the vicinity of the grain boundary which are believed to impart the major beneficial effect to SCC resistance in the RRA condition. Some workers suggest it is the spacing of the grain boundary precipitation and their role in providing sites for the trapping of hydrogen which is the critical factor [18], while other researchers indicate the solute concentrations in the vicinity of the boundary have a more important influence [19].

These alloys find extensive application where their high strength can be fully exploited, especially in compression dominated structures. For example on the Airbus A320, keel elements and stabiliser fittings of the fuselage are of 7075–T7351 while the skins, stringers and spars of the upper sections of the wing are largely manufactured from 7010–T6.

2.4 ADVANCED AEROSPACE ALUMINIUM ALLOYS

2.4.1 Al–Li alloys

Lithium is unique amongst the more soluble elements in aluminium in that it provides the basis for solution and precipitation hardening while simultaneously reducing density (3% per wt% Li) and increasing elastic modulus (non-linearly from 66 to 80 GPa for Al–2.5 wt% Li). Structural weight savings achieved by density reduction far outweigh the savings gained from similar proportional improvements in other mechanical properties (e.g. strength, stiffness, fracture toughness and fatigue performance), and weight savings of approximately 15% have been achieved in demonstrator and prototype parts. These weight savings are attained currently at a cost premium of × 2 to × 3 for the raw alloy produced via the DC cast route. Efficiencies in production and fabrication may reduce this penalty in built-up structure; for example, superplastic formability

Table 2.5 Approximate composition of some lithium-containing alloys

Alloy	*Li*	*Cu*	*Mg*	*Zr*	*Si*		*Fe*	*Specific gravity*
1420	1.5–2.6	–	4.0–7.0	< 0.30		(Mn + Cr)		2.50
2020	1.1	4.5	–	–		(Mn + Cd)		2.71
2090	1.9–2.6	2.4–3.0	0.3	< 0.15	0.1		0.1	2.60
2091	1.7–2.3	1.8–2.5	1.1–1.9	< 0.10	0.2		0.3	2.58
8090	2.1–2.7	1.0–1.6	0.6–1.3	< 0.16	0.2		0.3	2.53
8091	2.4–2.8	1.8–2.2	0.5–1.2	< 0.16	0.3		0.5	2.54
Weld-alite	1.3	4.5–6.3	0.4	0.14		(0.4Ag)		2.73

of the Zr refined Al–Li based alloys is well established and development of a commercially viable diffusion bonding process may enable components to be produced by combined SPF/DB. Alternative production routes incorporating rapid solidification technology and mechanical alloying are being considered for specialist applications, notably those where elevated temperature performance is required. Nevertheless, for alloys with an intrinsically high solid solubility, the benefits of RS processing are likely to be limited, and the wide range of alloys of current commercial interest are manufactured by conventional routes. The compositions of selected alloys, together with their specific gravities, are listed in Table 2.5.

Two categories are evident: those employing small lithium additions to Al–Cu alloys to enhance properties such as strength (e.g. 2020 or Weldalite), and those with higher lithium contents (e.g. 8090 or 2091) which offer the maximum density reduction. The alloys are based largely on the recently developed quaternary Al–Li–Cu–Mg system, although the Russian alloy 01420, which has seen widespread use in military aircraft for 15 years, is based on the ternary Al–Li–Mg system, and alloy 2020, used for wing covers in the Vigilante aircraft in the 1960s, is of the Al–Li–Cu type. As with conventional 2xxx and 7xxx alloys, the lithium-containing alloys are heat-treatable and the resulting properties are controlled by a complex combination of precipitation reactions. These may occur in combination and be interdependant, but each reaction may be summarized independently as follows, with crystallographic information on the various phases being included in Table 2.6.

All alloys : $\alpha_{ss} \rightarrow \delta'(Al_3Li) \rightarrow \delta(AlLi)$

Al–Li–Mg : $\alpha_{ss} \rightarrow \delta'(Al_3Li) \rightarrow Al_2MgLi$

Al–Li–Cu (low Li:Cu) : $\alpha_{ss} \rightarrow \text{GP zones} \begin{array}{l} \rightarrow \theta'' \rightarrow \theta' \rightarrow \theta(CuAl_2) \\ \rightarrow T_1(Al_2CuLi) \end{array}$

Table 2.6 Crystallography of precipitates in lithium-containing alloys

Precipitates	*Crystallography*
δ' (Al_3Li)	Coherent spherical precipitates with ordered Li_2 structure Cubic: $a = 0.404$ nm
δ(AlLi)	Equilibrium precipitates which nucleate heterogeneously at grain boundaries Cubic: $a = 0.637$ nm
Al_2MgLi	Equilibrium rod-like precipitates with $\langle 110 \rangle_\alpha$ growth direction Cubic: $a = 1.99$ nm
T_1(Al_2CuLi)	Equilibrium thin platelets with $\{111\}_\alpha$ habit plane which nucleate at GP zones in Cu-rich alloys and at dislocations in Li-rich alloys. Hexagonal: $a = 0.497$ nm, $c = 0.934$ nm
T(Al_6CuLi_3)	Equilibrium particles which form along $\langle 110 \rangle_\alpha$ in the matrix but predominately at high angle boundaries. Body centred cubic: $a = 1.39$ nm
θ'', θ', θ ($CuAl_2$)	As in 2xxx alloys
S′, S(Al_2CuMg)	As in 2xxx alloys but precipitation kinetics influenced by Li additions

Al–Li–Cu (high Li:Cu) : $\alpha_{ss} \rightarrow T_1(Al_2CuLi)$

Al–Li–Cu–Mg : $\alpha_{ss} \rightarrow$ GP zones $\rightarrow S' \rightarrow S(Al_2CuMg)$

The details of these phase transformations have been reviewed elsewhere [20], but a summary of the precipitate phases observed in the age-hardened microstructures of alloys containing 2–3 wt% Li is presented in Figure 2.15. All alloys contain a high volume fraction of coherent δ' particles; these form homogeneously throughout the matrix, although preferential nucleation occurs at Al_3Zr sites – Figure 2.16. Precipitation of δ' provides a strengthening increment; this is associated chiefly with order-hardening and results in extensive planar slip during plastic deformation. Overageing to promote Orowan looping of the coarser δ' particles results in deleterious precipitation of equilibrium δ phase on the grain boundaries and the simultaneous formation of δ' PFZs adjacent to the boundaries. However, additional precipitation-hardening phases, notably $T_1(Al_2CuLi)$, provide increased strength and may contribute to slip homogenization. Precipitation of non-shearable $S'(Al_2CuMg)$ in alloys having significant additions of copper and magnesium (8090, 8091 and 2091) also contributes to strength and promotes more homogeneous deformation. However, the formation of this precipitate is influenced by the binding of vacancies to the lithium atoms in solution which, in turn, is controlled by the volume fraction of the

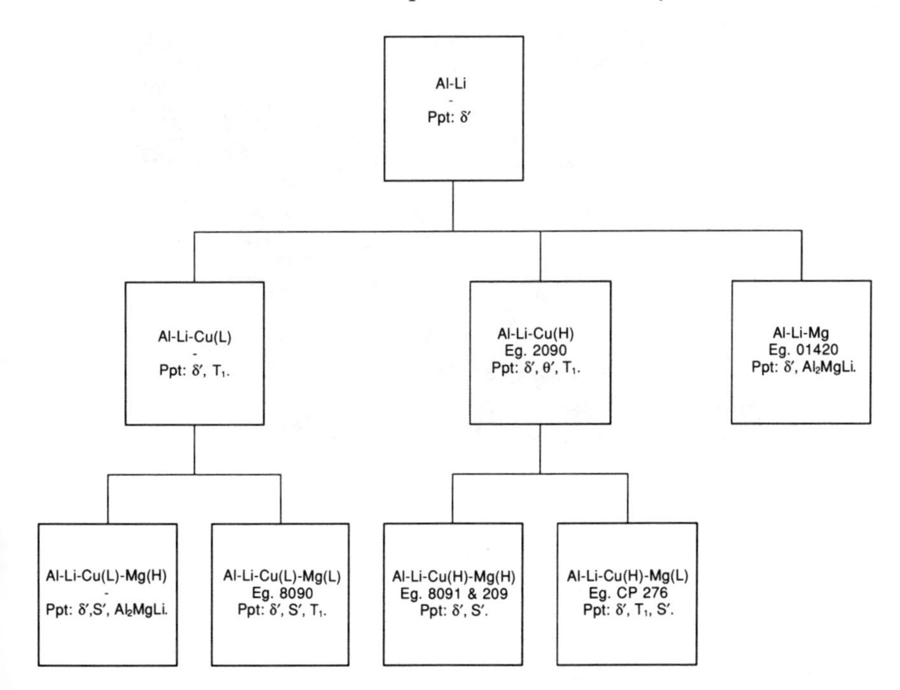

Figure 2.15 Strengthening precipitates in age-hardening Al–Li alloys: $\delta' = Al_3Li$, $T_2 = Al_2CuLi$, $\theta' = Al_2Cu$ and $S' = Al_2CuMg$. Equilibrium particles of $\delta(AlLi)$, $T_2(Al_6CuLi_3)$ and Al_2MgLi may also be present, as may a fine dispersion of $\beta'(Al_3Zr)$ particles (alloy content low (L) or high (H)).

lithium-containing precipitates (principally δ′) formed during natural or artificial ageing.

The overall strength of the alloys is controlled by these matrix precipitates, but the alloys exhibit substantial anisotropy in mechanical properties compared with earlier alloys, with a 20% reduction in tensile and shear strength at 40–60 degrees from the major working direction. The Zr in combination with Li renders the alloys extremely resistant to recrystallization, giving an unrecrystallized grain structure with a pronounced crystallographic texture. This unrecrystallized and textured grain structure, together with their propensity for planar slip, is responsible for the anisotropy and contributes to the poor fracture toughness of these lightweight alloys; the latter is exacerbated in the ST direction for which planar and weak grain boundaries lie perpendicular to the loading direction. Other effects which limit toughness are the formation of embrittling grain boundary phases, notably $T_2(Al_6CuLi_3)$ and $\delta(AlLi)$ during processing and/or heat treatment, strain localisation in PFZs at grain boundaries and the presence of impurities, particularly of the alkali–metal elements (eg Na, K), within the alloy. As a consequence

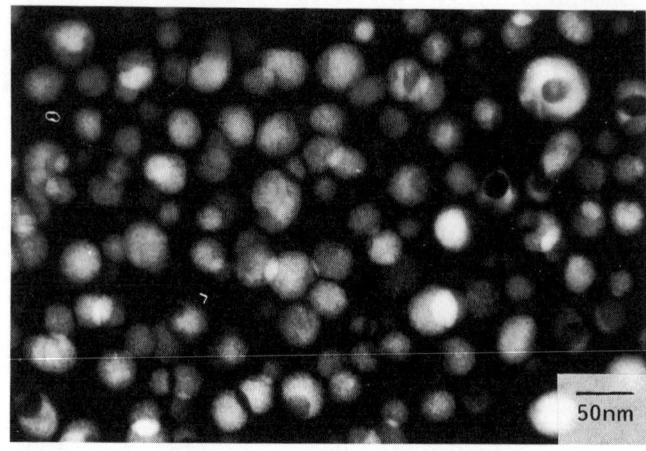

(a)

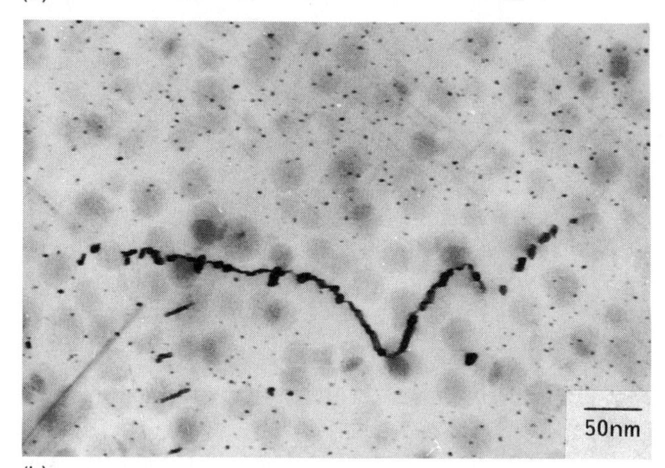

(b)

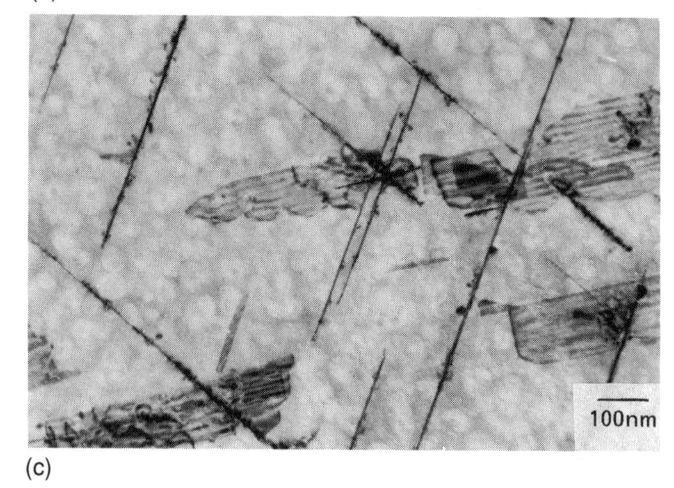

(c)

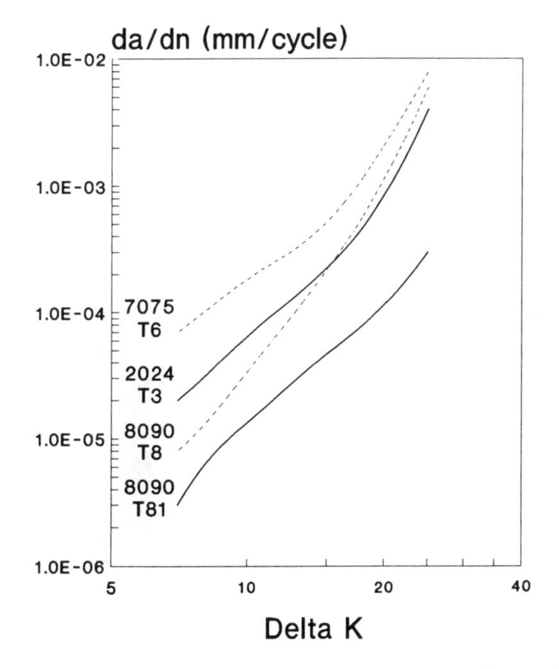

Figure 2.17 Fatigue crack growth rates of 8090–T81 (underaged) 2024–T351, 8090–T851 and 7075–T651 tested in laboratory air.

alloy composition has evolved to promote coprecipitation of δ′, T_1 and S′ while removing all unwanted impurities. TMP may be used to develop a fine recrystallized grain structure, and a work-hardened microstructure exhibiting improved strength and toughness, and heat-treatment procedures usually employing natural and low temperature ageing have been selected to give an optimum balance of strength and toughness.

While fatigue crack initiation may be promoted in the lithium-containing alloys due to the combination of planar slip and texture, the resistance to fatigue crack growth is substantially better than that of conventional aluminium alloys (Figure 2.17). This is attributed to increased elastic modulus and to the crystallographic nature of the crack path, resulting in crack deflection and roughness induced crack closure. However, for short cracks typical of those growing from neighbouring

---

Figure 2.16 Age-hardening precipitates in Al–Li based alloys: (a) δ′ precipitates in 8090 aged at 190 °C including precipitation at β′(Al_3Zr) particles: ⟨100⟩ Al orientation. (b) S′ precipitates in 8090 aged at 210 °C including fine homogeneously nucleated laths and coarse precipitates on a dislocation helix: ⟨100⟩ Al orientation (courtesy S. A. Court). (c) T_1 precipitates in Al–3.0Li–1.5Cu–0.1Zr aged at 170 °C: ⟨110⟩ Al orientation.

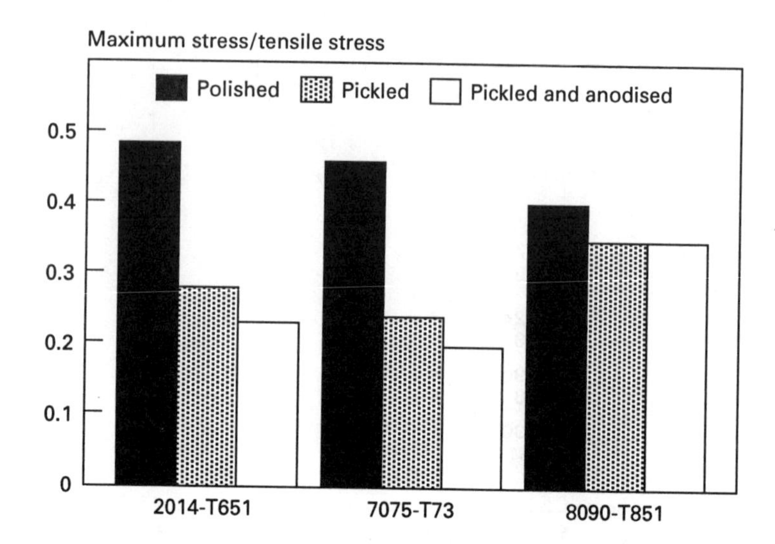

Figure 2.18 ST fatigue strength of alloys 2014–T651, 7075–T73 and 8090–T851 after sulphuric acid pickling and chromic acid anodizing surface treatments: ($R = -1$).

fastener holes, there is little crack tip shielding, and crack growth rates approach those of conventional alloys under similar conditions. Nevertheless, in general the fatigue behaviour of the lithium-based alloys is considered to be superior to that of alternative alloys, particularly since they are less susceptible to attack during typical surface treatments (e.g. anodizing) (Figure 2.18).

From the current data it would appear that corrosion and stress corrosion performance can at least match that expected of comparable 2xxx and 7xxx alloys. The general dependence of SCC resistance upon isothermal ageing is followed with the development of Cu-rich grain boundary precipitates being detrimental. Deformation prior to ageing, overageing or duplex ageing (e.g. 48 h at 150 °C followed by 4 h at 230 °C) promotes intragranular precipitation of the copper-containing phases; this reduces the local potential difference between matrix and grain boundary and increases the corrosion resistance. These effects may be overshadowed by the influence of grain structure: in unrecrystallized sheet, the alloys have a very high resistance to SCC in the plane of the sheet, even when the material is aged to a very high strength while short-transverse properties are inferior but match those of other alloys.

Table 2.7 Approximate monotonic mechanical properties of selected aluminium alloys

Alloy	*Temper*	*Solution treatment (°C)*	*Ageing treatment (h/ °C)*	*Yield strength (MPa)*	*Tensile strength (MPa)*	*Elongation (%)*	*Toughness* K_{IC} (LT) ($MPa\,m^{1/2}$)	*SCC resistance (A (Good)–D)*
2xxx								
2024	T351	495	–	325	470	20	40	C
2024	T851	495	12/190	415	450	6	25	B
2014	T451	505	–	290	425	18	–	C
2014	T651	505	18/160	415	485	12	22	C
7xxx								
7075	T651	460	12/135	505	570	11	28	C
7075	T73	460	20/120+8/175	435	495	11	32	B
7010	T7651	475	12/170	450	525	10	24	B
Li-containing								
2091	T8X(UA)			370	460	15	40	–
8090	T81(UA)	530	12/150	360	445	11	45	C
8090	T6	530	32/170	400	470	6	35	C
8090	T851	530	32/170	455	510	7	30	C
2091	T851			475	525	9	25	C
2090	T83			510	565	5	–	–
8091	T851	530	32/170	515	555	6	22	C

Note: These 'typical' LT properties are averages for a wide range of product forms, section sizes and processing routes and may not be exactly representative of any particular produce or size.

To replace conventional aluminium in aerospace applications, the lithium-containing alloys must offer the correct balance of properties together with ~ 10% weight saving. To meet this challenge three alloy types can be defined, and approximate values for the monotonic properties of these and conventional alloys are given in Table 2.7:

- Damage-tolerant alloys such as 2091–T8X and 8090–T81 (underaged) which might replace 2024–T3 in damage-tolerant structure such as fuselage skin and wing skin for transport aircraft and rotorcraft.
- Medium-strength high stiffness alloys such as 2091–T851 and 8090–T6/T851 which should be superior to 2014–T6 or 2024–T8 in medium strength and stiffness critical structure such as control surfaces, stringers and fuselage frames.
- High-strength alloys such as 2090–T83 and 8091–T85 offering improved resistance to buckling compared with 7075–T6/T73 for compression-loaded structures such as upper wing skins and undercarriage parts.

To date, the quaternary alloys have been used in a wide range of demonstrator components on civil and military aircraft. These have been selected to assess the performance of the alloy and to demonstrate potential fabrication advances. For example, the suitability of 8090 for SPF to net shape will reduce the cost of the material in built-up structure (particularly if post-form resolution treatment can be avoided), and the alloy shows excellent potential for DB (Chapter 10). Their widespread incorporation into fuselage skins and stringers and wing sections for the full-scale fatigue testing programme for Airbus A330/340, and the extensive use of forgings, plate and sheet throughout the cabin structure of the Westland Augusta EH101 should lead to substantial quantities of these alloys being included in future generations of civil, and, to a lesser extent, military airframe structures.

There is also a future for lithium-containing alloys developed for specialist application; the development of the weldable high-strength alloy Weldalite has led to early incorporation of the alloys into the large welded cryogenic fuel tanks for the NASA space programme. It may also be used in the welded aluminium structure (currently 2219–T851) for the ESA spacelab project.

2.4.2 Alloys incorporating rapid solidification processing (RSP) and vapour quenching (VQ)

Solution-strengthening and precipitation-hardening aluminium alloys, including 2xxx, 7xxx and lithium-containing alloys, generally show an improved combination of strength, toughness and stress corrosion resistance as a result of the refined grain structure and extended solid solubility associated with RSP (Chapter 12). However, apart from specially selected applications, the benefits exhibited by these materials are generally offset by excessive production costs.

Other alloys produced by this route include the dispersion-strengthened alloys developed for improved elevated temperature performance. The introduction of fine oxides and carbides (typically < 100 nm) into the matrix of a solution-strengthened or precipitation-hardened matrix can be successfully achieved via the compaction of mechanically alloyed powders, but it is the introduction of RSP which has enabled the development of alloys strengthened by a high volume fraction of fine, uniformly distributed intermetallic particles. The alloys are generally based on additions of transition metals, principally Fe, which exhibit high liquid solubility and low solid solubility, and form thermally stable intermetallic particles which are resistant to coarsening at temperatures of up to 500 °C (e.g. alloy 8009 (Al–8.5Fe–1.7Si–1.3V) is strengthened by ~ 30 vol% of 30 nm $Al_{13}(Fe, V)_3Si$ silicides) [21]. The alloys exhibit superior strength retention at elevated temperatures to wrought alloys (Figure 2.19) and approach that of Ti–6Al–4V when compared in specific

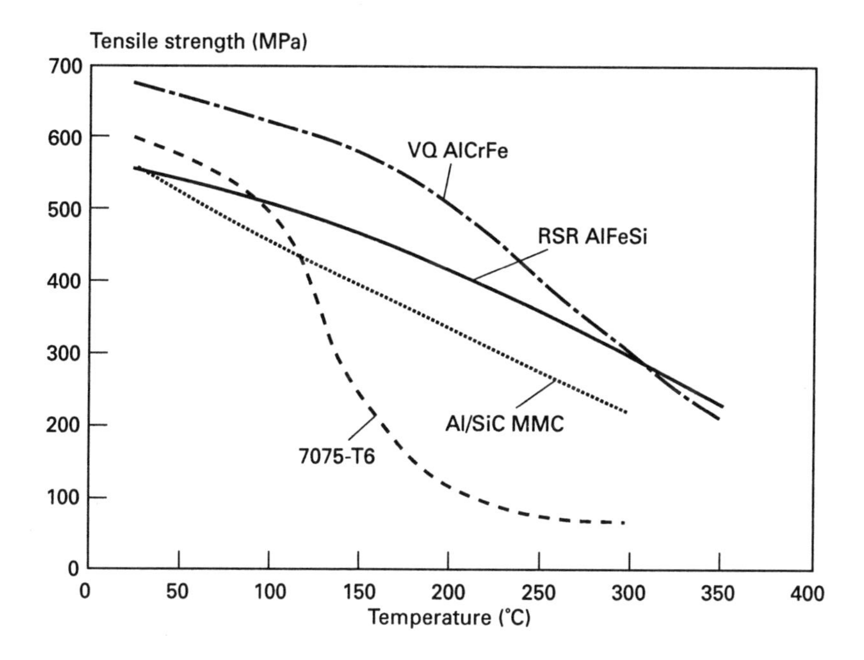

Figure 2.19 Elevated temperature properties for conventional alloys, RS and VQ alloys, and particle-reinforced MMCs.

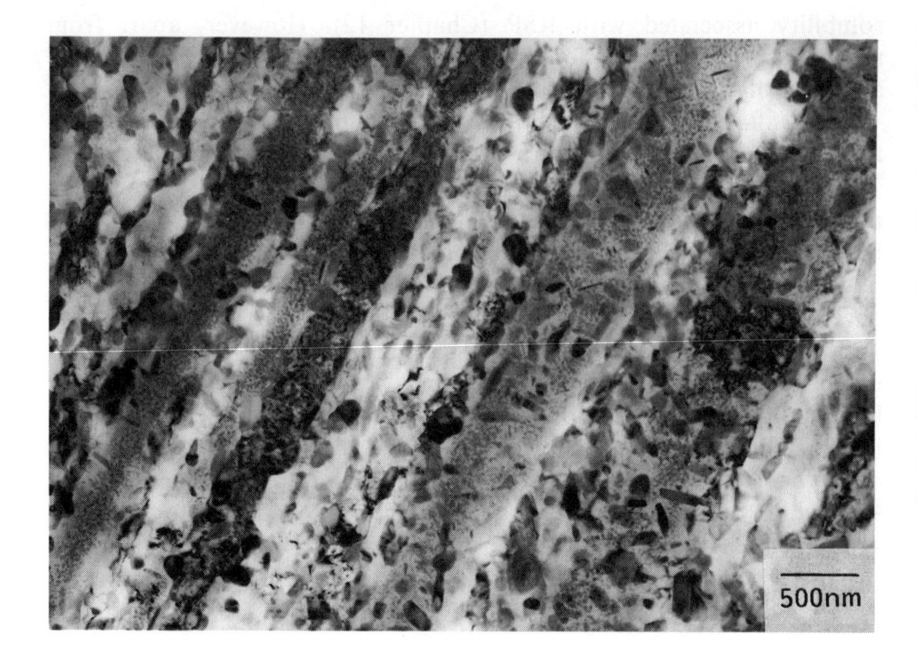

Figure 2.20 Microstructure of VQ Al–7.5Cr–1.2Fe alloy after 480 h at 350 °C showing stable fine columnar grains, Cr_2Al_{13} particles and fine Fe_3Al precipitates (courtesy C. J. Gilmore).

terms (i.e. σ/ρ). Also of interest is the comparison with SiC particle-reinforced aluminium metal matrix composites showing improved elevated temperature stability of the dispersion-strengthened alloys. The combination of elevated temperature, strength and stability with good fatigue and stress corrosion resistance makes these alloys attractive for airframe components which are strongly aerodynamically heated or in the vicinity of the engine. Their incorporation into 'cool' engine components offers further potential weight savings.

An alternative approach for the development of alloys incorporating high concentrations of transition elements involves an atom-by-atom quench from the vapour phase. This vapour-quenching route (VQ) has been used to produce Al–Cr–Fe alloys having excellent room temperature strength and stiffness [22] on account of the high Cr supersaturation and precipitation hardening from Al_7(Cr, Fe) particles (Figure 2.20). At temperatures < 300 °C the VQ alloy substantially outperforms the RS alloys and this alternative fabrication route for the development of non-equilibrium alloys deserves further consideration, although production rates are likely to remain low.

2.5 CONCLUSIONS

Aluminium alloys continue to be used in large quantities for the construction of airframe structures. Exploitation of the lightweight Al–Li based alloys together with further development of advanced fabrication technologies (e.g. SPF/DB) will lead to improved structural efficiency and the continuing use of precipitation-hardened alloys in many airframe applications. The excellent thermal stability of the dispersion-strengthened alloys produced by rapid solidification processing may be exploited as the temperature envelope experienced by new aerospace vehicles expands. Further development in the form of particle-reinforced alloy composites and fibre-reinforced alloy laminates may result in materials exhibiting enhanced specific stiffness and damage tolerance as discussed in Chapter 8.

REFERENCES

1. Peel, C. J. (1986) *Mat. Sci. Tech.*, **2**, 1169.
2. Jonas, J. J., Sellars, C. M. and McG. Tegart, W. G. (1969) *Met. Rev.*, **15**, 1.
3. McElroy, R. J. and Skopiak, Z. C. (1972) *Int. Met. Rev.*, **17**, 175.
4. Humphreys, F. J. (1977) *Acta Metall.*, **25**, 1323.
5. Stewart, A. T. and Martin, J. W. (1975) *Acta Metall.*, **23**, 1.
6. Chadwick, G. A. (1986) *Met. and Mater.*, **2**, 693.
7. Edington, J. W., Melton, K. N. and Cutler, C. P. (1976) *Prog. Mat. Sci.*, **21**, 61.
8. Lorimer, G. W. and Nicholson, R. B. (1966) *Acta Metall.*, **14**, 1009.
9. Pashley, D. W., Jacobs, M. H. and Veitz, J. T. (1967) *Phil. Mag.*, **16**, 51.
10. Gleiter, H. and Hornboger, E. (1967) *Mater. Sci. Eng.*, **2**, 285.
11. Decker, R. F. (1973) *Met. Trans*, **4**, 2495.
12. Fleischer, R. L. (1963) *Acta Metall.*, **11**, 203.
13. Thompson, D. S. (1975) *Met. Trans.*, **6A**, 671.
14. Holroyd, N. J. H. (1989) *Proc. Environment-Induced Cracking of Metals*, 311, National Association of Corrosion Engineers, Houston, Texas.
15. Vogel, W., Wilhelm, M. and Gerold, V. (1982) *Acta Metall.*, **30**, 21.
16. Lorimer, G. W. (1978) in *Precipitation processes in solids*, 87, Met. Soc. American Institute of Metallurgical Engineers, New York.
17. Speidel, M. O. (1975) *Met. Trans.*, **6A**, 631.
18. Scamens, G. M. (1978) *J. Mater. Sci.*, **13**, 27.
19. Doig, P., Flewitt, P. E. J. and Edington, J. W. (1975) *Corrosion*, **10**, 347.
20. Flower, H. M. and Gregson, P. J. (1987) *Mat. Sci. Tech.*, **3**, 81.
21. Gilman, P. (1990) *Met. Mater.*, 505.
22. Gardiner, R. W. and McConnell, M. C. (1987) *Met. Mater.*, 255.

FURTHER READING

Polmear, I. J. (1989) *Light Alloys*, Edward Arnold, London.

Mondolfo, L. F. (1970) *Aluminium Alloys: Structure and Properties*, Butterworths, London.

Vasudevan, A. K. and Doherty, R. D. (eds) (1989) *Aluminium Alloys: Contemporary Research and Applications*, Academic Press, London.

Starke, E. A. and Sanders, T. H. (eds) (1986) *Aluminium Alloys: Their Physical and Mechanical Properties*, EMAS, Warley, UK.

Aluminium Lithium Alloys: Proceedings of International Conferences – 1st: AIME, 1981; 2nd: AIME, 1983; 3rd: Inst. Metals, UK, 1985; 4th: *J. Physique*, France, 1987.

Fine, M. E. and Starke, E. A. (eds) (1986) *Rapidly Solidified Powder Aluminium Alloys*, ASTM.

New Light Alloys, Lecture Series No. 174, Advisory Group for Aerospace Research and Development, NATO (1991).

3

Titanium alloys: production, behavior and application

J. C. Williams

3.1 INTRODUCTION

Lightweight materials constitute the backbone of the competitive advantage of modern aerospace structures. These materials are important because they enable achievement of high structural efficiency which is essential for high performance under a variety of operating conditions. The selection of materials and of material conditions for lightweight structures is typically a problem bounded by multiple constraints, requiring a detailed understanding of all the materials characteristics and all aspects of mechanical behavior. The current material choices available to the designers of high performance structures include high strength steel or aluminum alloys, polymer matrix composites and titanium alloys. The factors that determine which of these represent the best choice include: the operating temperature, the design limiting property (strength, stiffness, fatigue, etc.), the volume or space available, cost considerations, the operating environment, fabrication and other manufacturing requirements, the intended service lifetime of the component and the total number of components required (lot size). Obviously, this is a formidable list of constraints, but for a given application, any of these can affect the ultimate choice of material. One objective of this chapter is an attempt at discussing how Ti alloys compare to other materials under these constraints and how such comparisons affect the usage of Ti alloys for high performance aerospace structures, especially aircraft and aircraft engines. Other objectives include a general description of the characteristics of Ti alloys because these affect their behavior which can dictate the limiting conditions under which they can be efficiently used, and a discussion of the range of properties that can be obtained by changes in processing.

Ti alloys have evolved over the past 35–40 years from a rare laboratory curiosity into a commodity. Today most Ti alloys are sold as much on the basis of price as they are on the basis of variations in performance. Nevertheless, the production methods for Ti alloys have a major impact on both cost and quality. These methods also have evolved over the years with several important breakthroughs. An awareness of this information is as important to design engineers as it is to materials engineers because it bears directly on the subject of structural reliability. Thus a brief account of Ti production will be included with emphasis on the impact of production methods on product quality. Included will be some comments on how Ti alloy production technology may further evolve during the next decade. Similarly, the manufacturing methods used for making structural components from Ti alloys will be briefly described, including some remarks on the future trends and apparent opportunities in this area. A recurring point regarding the behavior of Ti alloys is the intrinsic reactivity of Ti when it comes in contact with air or other solids when it is a very hot solid or in molten form. The basis for this characteristic is the large free energy of formation of a wide variety of compounds that can form with Ti. It also is based on the large heat of solution when many solutes dissolve into Ti.

There are numerous, detailed articles on the production of Ti alloys and the manufacture of Ti structural components, just as there are on the mechanical properties and metallurgy of Ti alloys. A representative (but not exhaustive) selection of these articles is listed under Further Reading. This chapter has a different focus: it is intended to fill the gap which separates these somewhat disconnected aspects of Ti applications. Of necessity, the level of detail regarding any particular aspect will be limited; it is hoped that the supplemental references found under Further Reading will fulfill most readers' needs for more detail. Thus this chapter could be sub-titled 'Ti alloys – a primer for design engineers'. This represents an unfilled niche in the technical information available for Ti and literally all high performance alloys. It is hoped that this chapter will help repair the gap that is currently present in the literature on structural materials.

3.2 BRIEF SUMMARY OF THE METALLURGY OF CONVENTIONAL TI ALLOYS

This section addresses some of the more common metallurgical jargon encountered in the literature on conventional Ti alloys, but it is not intended to convert design engineers into Ti metallurgists! The intent is 'de-mystifying' the metallurgical jargon for the benefit of non-metallurgists. If this effort is considered useful by readers of this chapter, the author hopes that design engineers who write articles on structural design and stress analysis will reciprocate, because much of the literature on these subjects is equally mysterious to metallurgists who try to read it.

Today, most of the work on Ti alloys is directed at quality improvement or cost reduction. There is also a large, world-wide effort to reduce the so-called Ti aluminides to practice. This latter effort comprises most of the development effort at present. Ti aluminides are alloys based on the intermetallic compounds Ti_3Al or TiAl. The Ti aluminides represent a different 'class' of structural material because of their quite different characteristics. Ti aluminides are emerging as new Ti base materials and it is useful to distinguish between these and the more mature alloys by calling these latter alloys, of which Ti–6–4 is an example, conventional alloys. The two material systems based on Ti_3Al and TiAl have the potential for extending the current use temperature of Ti based materials by as much as ≈ 225 °C. At present, however, there are *no* production applications of these materials. This is because of unresolved technical questions including their substantially lower ductility and toughness at or near room temperature, the general evolutionary nature of these materials at present and a host of production-related questions for the aluminides ranging from melting to cost. This section discusses current commercial or conventional Ti alloys. A brief section on the characteristics and status of Ti aluminides is included near the end of this chapter.

3.2.1 Crystallography and anisotropy of Ti

Pure Ti has two crystalline (allotropic) forms, a low temperature hexagonal close packed (h.c.p.) α-phase and an elevated temperature body centered cubic (b.c.c.) β-phase. The transition from α to β occurs at ≈ 882 °C in pure Ti. The h.c.p structure of the α-phase causes it to have a pronounced anisotropy of physical properties, of which the elastic constants are, practically, the most important. It also causes the plastic deformation behavior of this phase to be complex as well as anisotropic, as compared to cubic materials such as Al and Ni base alloys. The large volume fraction of the α-phase present in most commercially important Ti alloys causes this anisotropy to have significant implications for their mechanical behavior. Apart from the Zr alloys used in the nuclear industry, Ti alloys are the only metallic structural materials used in any quantity which have the degree of anisotropy that the h.c.p. α-phase imparts. Thus a brief discussion of the consequences of anisotropy may be useful to consider here.

The elastic constant of the h.c.p. α-phase in the close-packed directions perpendicular to the 'c' axis (E_{11}) is ≈ 100 GPa, whereas in the direction parallel to the 'c' axis (E_{33}), it is ≈ 145 GPa. This elastic anisotropy has a number of effects that have practical importance. First, non-random crystallographic alignment of the α grains creates a directional dependence of the elastic modulus in the material. Most stress analysis computer codes do not account for this. Even in those codes in which anisotropy is accounted for in principle, the actual degree of anisotropy must be measured to provide the necessary input data required to quantitatively deal with it. Second, the velocity of a sound wave in a solid is proportional to the elastic modulus of the solid. If regions of different modulus exist within a piece of Ti or a part, this will result in scattering of an ultrasonic beam that is used to inspect the material for defects. As a minimum, this scattering creates 'noise' in the reflected ultrasonic signal; this noise makes detection of true defects more difficult because elastic anomalies caused by texture respond in a similar manner during ultrasonic inspection.

The degree of crystallographic alignment of individual microstructural units in a material is called preferred orientation, crystallographic texture, or just 'texture'. The existence of texture in wrought Ti products depends on the processing history. This history is expressed in terms of the amount of plastic deformation that the material has undergone and of the temperature at which this deformation has been done. There are elaborate, quantitative methods for representing texture, both mathematically and graphically. The latter is done by means of a pole figure, which depicts the spatial distribution of normal directions (poles) of a particular crystallographic plane, e.g. the basal plane in α Ti. These pole figures are determined experimentally using X-ray diffraction methods. Even though major advances in the speed and reductions in the labor intensity of pole figure determination have been made, this procedure is still relatively expensive. The result is that texture is relatively uncontrolled in production grades of Ti alloys. The factors that affect and control texture formation such as the relationship between type and intensity of texture and deformation history (method of deformation, amount of plastic strain and deformation temperature) have been empirically characterized for commonly used alloys such as Ti–6–4. In principle, it is possible to specify a processing schedule that will yield a particular texture. In practice, it is difficult to faithfully reproduce this schedule for each work piece. Thus, while control of texture is possible, better routine control over processing will be necessary before materials with controlled texture can be employed routinely and affordably in practice. Even if material with reproducible, controlled texture were available today, the design methods for dealing with anisotropic materials also are required.

3.2.2 Alloying behavior of Ti

The technologically important forms of Ti all contain deliberate alloying additions. These additions affect the phase equilibria and the microstructure by altering the relative thermodynamic stability of the two allotropic forms, α and β. It has become customary to classify alloying elements according to the effect they have on the β transition temperature (also known as the β transus temperature or just β transus). Accordingly, there are three types of alloying elements for Ti alloys: α and β stabilizers and neutral elements. The most commonly used α stabilizers are Al, O, N and C. The most common β stabilizers are Mo, V, Cr, Nb, Cu and Fe. The most common neutral elements are Zr and Sn. The α and β stabilizing elements tend to be preferentially concentrated in either the α or β phases, respectively. This is known as solute partitioning. These elements stabilize either the α or β phases, leading to an increase in the volume fraction of the more stable phase or constituent in the product alloy.

It also is common and convenient to categorize conventional Ti alloys according to their constitution, i.e. according to which phases are present at room temperature under normal conditions. Following this convention, there are α alloys, which are predominantly α-phase; β alloys and α + β alloys. While each of these alloy types has a common set of applications, the high strength alloys are the latter two types. Moreover, with the exception of welding, α + β and β alloys are easier to shape and work. In fact the addition of β stabilizing elements at first was done to improve the workability of Ti alloys during the early days of higher strength Ti alloy development. Today, the most commonly used alloys (by far) are the α + β alloys, with Ti-6Al–4V comprising a significant portion of all Ti alloy usage.

While the range of elements that act as β stabilizers is extensive, only a few (V, Mo, Nb and Cr) are used to any significant extent in commercial Ti alloys produced in large quantities. In addition to basic market factors such as demand, inventory and availability, this is because of the economic benefit of commonalty of reusable scrap (also called revert). Production heats of commonly used Ti alloys can contain as much as ≈ 80% revert. The use of revert helps reduce the cost of Ti alloy mill products. The principal reason that the commonly used β stabilizers have been selected and that the number of these is so few is their effect on producibility. Many other, less expensive β stabilizers, such as Fe, Ni, Mn, and Cu exist, but are not extensively used since they cause large melting temperature reductions. These create attendant production difficulties such as chemical inhomogeneities due to freezing segregation in production sized ingots. Cr also has this effect, but to a much lesser degree. The Fe, Ni, Cu or Mn rich regions in the ingot cause a local reduction in β transus which leads to non-uniform microstructures in the final products. Such regions are devoid of primary α-phase in α + β worked products and, accordingly, are known as β flecks. These regions are viewed as microstructural anomalies (defects) because they are particularly deleterious to some properties such as low cycle fatigue.

They thus are clearly undesirable. They can be eliminated or minimized through careful control of melting practice, but the possibility of β flecks has curtailed the use of some of the least expensive β stabilizing elements because of producibility questions.

It may be helpful to use an example to discuss the role of alloying additions in α + β Ti alloys. Let us examine the composition of a common α + β alloy such as Ti–6Al–2Sn–4Zr–2Mo (Ti–6–2–4–2) in light of the previous discussion of alloying. In Ti–6–2–4–2, the Al increases the β transus and strengthens the α phase; Mo stabilizes some β phase and somewhat offsets the effect of Al on the β transus. Sn and Zr strengthen both the α- and β-phases. This alloy contains a small amount (< 0.1 wt%) Si added for creep strength and the Zr also makes this addition more effective.

The strength and other properties of Ti alloys represent a remarkable example of synergism, because the strength of either of the individual constituents (the α- and β-phases) is considerably less than that of the two phase mixture in α + β alloys, even in the annealed condition. Without this synergism, the technical interest in using Ti alloys for lightweight structures would be substantially reduced.

There has been an interest in β alloys for many years. Ten years ago, these alloys were thought to be the next generation of structural Ti alloys, but realization of this prediction has been slow. There is a variety of reasons for this, ranging from producibility to changes in design philosophy with respect to the relative importance of strength and other properties such as fracture toughness and fatigue crack propagation resistance. Today, β alloys are beginning to be more widely used in a variety of products ranging from aircraft to petrochemical equipment, but the total volume is small compared to α + β alloys.

There are several inconsistent uses of terminology in the Ti literature, which should be identified here. First, β alloys are usually used in the aged condition, and in these cases contain significant amounts of α-phase which is present as a strengthening precipitate. Second, α alloys such as commercially pure (CP) Ti or Ti–5Al–2.5Sn contain small amounts of β-phase. The β-phase is stabilized by the small concentration of Fe which is present as an impurity which is introduced during production of the sponge. This β-phase acts as a grain refining constituent and increases the hydrogen tolerance of these alloys so is generally beneficial.

Thermomechanical processing (TMP) is the combination of mechanical working an alloy in a controlled manner while simultaneously maintaining a controlled thermal history. TMP permits a range of microstructures to be attained in a single alloy that would not be possible through heat treatment alone. The use of TMP to enhance properties of structural materials has received much attention with the advent of the high strength low alloy (HSLA) steels. The properties of Ti alloys have

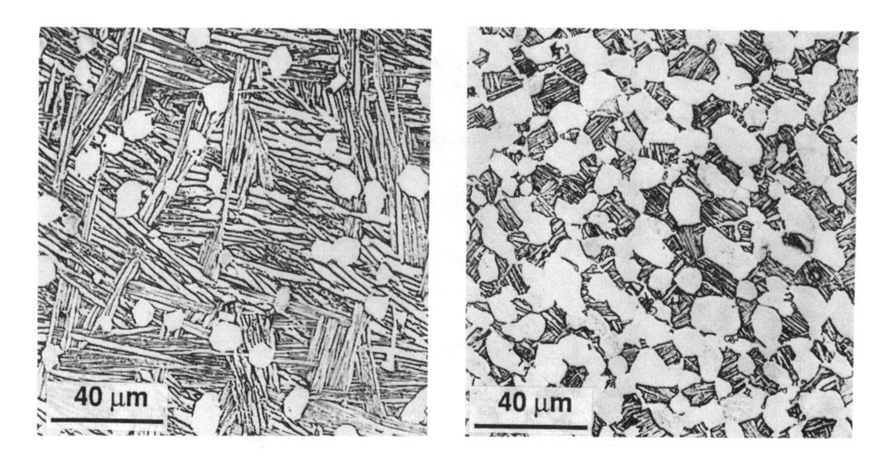

Figure 3.1 Microstructure of Ti–6Al–4V that has been $\alpha + \beta$ processed and solution heat treated to produce two different volume fractions of equiaxed primary α.

been enhanced through TMP for many years (long before micro-alloyed steels). In the case of Ti alloys, TMP has been used to control the microstructure of the alloys for the purpose of balancing strength and ductility. From a processing terminology standpoint, the basic distinction in processing is between $\alpha + \beta$ and β working of $\alpha + \beta$ alloys. These terms refer to the temperature range at which the working operation (forging or rolling) is completed, including the introduction of reasonable amounts of plastic strain. Working in the $\alpha + \beta$ temperature range (i.e. below the β transus) results in α-phase with an equiaxed or globular shape (morphology). This is known as primary $\alpha(\alpha_p)$. The volume fraction of α_p is controlled by the final working temperature and/or the subsequent solution heat treatment temperature. The microstructures of $\alpha + \beta$ processed Ti–6Al–4V with two volume fractions of primary α are shown in Figure 3.1. The structure of this same alloy which has been β processed is shown in Figure 3.2. This latter structure consists of colonies of α plates, the size of which depend on the details of the working and thermal history. The other significant microstructural feature in this condition is the presence (or absence) of a layer of α-phase along the prior β grain boundaries. This feature is called grain boundary α and can be controlled by the working history. Recent advances in hot die forging technology now make achievement of the β processed or colony microstructure possible without the concomitant presence of grain boundary α. The implications of these microstructures on the mechanical properties of these alloys will be discussed in some detail in a later section.

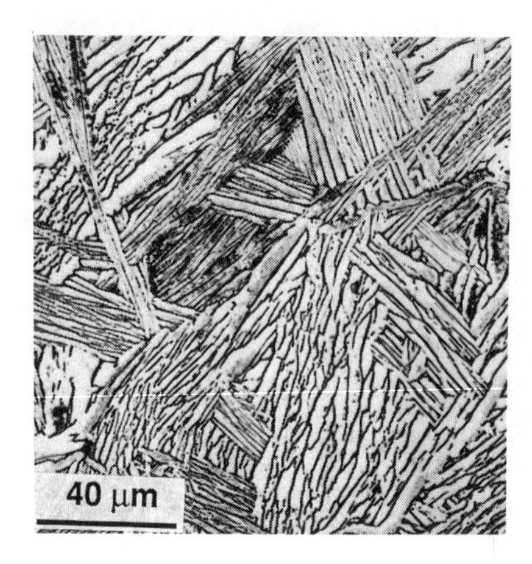

Figure 3.2 Colony microstructure in β processed Ti–6Al–4V.

3.3 THE PRODUCTION OF Ti ALLOYS AND Ti ALLOY COMPONENTS

As mentioned earlier, Ti metal is extremely reactive with many of the elements found in the atmosphere, particularly oxygen, nitrogen, hydrogen and carbon. While Ti comprises a significant fraction of the earth's crust, it always occurs in combination with other elements. The two most common forms are Rutile (TiO_2) and Ilmenite ($TiFeO_3$). These two forms are considered the ore from which Ti is extracted, but the large free energy of formation of the oxides make production and/or refinement of metallic Ti very energy intensive. That is, considerable energy is required to separate Ti ions from oxygen or other elements with which they are combined in nature. This section summarizes the main manufacturing operations that are involved in the production of Ti alloys and components. Included are the making of metallic Ti (sponge), ingots (melting), semi-finished products, forgings, sheet metal parts, castings and welds.

3.3.1 Ti metal (sponge) making

There are basically two methods of producing Ti metal in significant quantity from Rutile and Ilmenite ores today: metallic reduction and electrolytic reduction. Further, there are two variations of metallic reduction depending on which reactive metal is used to reduce the Ti ion. Both begin with Ti tetrachloride ($TiCl_4$), and either Na or Mg is reacted at elevated temperature with the $TiCl_4$ to produce Ti metal and a salt, either NaCl or $MgCl_2$. After the reaction is complete, the metal and salt mixture is cooled. The salt then is leached out of the metallic Ti and salt mixture leaving porous metallic Ti product, aptly called sponge. Sponge is the starting point for making Ti alloys. The reason to include this discussion is that the quality of Ti alloy products depends on the quality of sponge because of the possibility of melt related inclusions that can be carried over from the sponge. The actual reduction of Ti is relatively straightforward, but execution of a large quantity of sponge making process requires care in order to exclude reaction with the atmosphere, not only during the reduction phase, but afterward when the metallic reaction product is being extracted from the reaction vessel. Failure to do so can result in 'burned sponge' which is Ti sponge which has partially recombined with N_2 from the atmosphere while the sponge is still hot. Burned sponge also can result if the reaction vessel is opened to the atmosphere without an inert cover gas because the finest particles of Ti are pyrophorric and these can generate enough heat to 'burn' the bulk sponge.

(a) Melting

Up until now, most Ti alloys have been produced by consumable vacuum arc melting and remelting because molten Ti is too reactive to be compatible with the refractory lined vacuum induction melting furnaces used in the production of Ni and Fe base specialty metals. The first melt electrode of a Ti alloy heat is made by blending sponge particles typically smaller than 10 mm with master alloy granules of the same size. Master alloy is a pre-prepared mixture that contains two or more of the major alloying elements in the correct proportions but in much higher concentration than the final alloy composition. The use of master alloy instead of individual elemental additions helps achieve the correct alloy chemistry and alloy homogeneity. For example, alloys containing high melting point, refractory elements such as Mo would be very difficult to prepare without having some remaining unmelted Mo if master alloys were not available. The portion of master alloy and sponge are weighed to assure that the desired alloy composition is achieved. The blended sponge and master alloy is mechanically pressed into briquettes and these are then welded together using any tungsten free heat source (plasma torch, etc.). This is how the first melt electrode is formed. Here again, the welding must be done under a protective atmosphere in a dry box to prevent reaction between the Ti and the atmosphere. This electrode is consumably arc melted into a water cooled copper crucible in a vacuum arc remelt (VAR) furnace. It is important that no arc interruptions, vacuum leaks or water leaks occur during this process. Inadequate process control can lead to N_2, O_2 and/or C enriched regions that are hard and brittle and which are not completely dissolved during the melting operation. These regions are variously called melt related inclusions, hard α or Type I defects. The as-solidified first melt ingot then becomes the starting point for remelting operations.

Depending on the grade of alloy being made, the material is vacuum arc remelted once or twice more or is remelted in a cold hearth furnace using either a plasma torch or electron beam heat source. Over the years it has become desirable to make larger and larger ingots of Ti alloys. Large ingots are more cost effective and permit the production of larger diameter forging billet which has been given sufficient work to produce the desired billet macro and micro structure. Large billet is required for the larger Ti parts being used in newer aircraft and engines. The technology for making heavy (5000 Kg), large diameter (90 cm) ingots of Ti alloys which are homogeneous and free from hard alpha is both a demanding task and one that is still as much the result of knowhow as it is a technological achievment. The successful application of high performance Ti alloys depends on a carefully controlled, highly reproducible ingot making process. Such a process yields ingot that can be converted into segregation free and hard alpha free material that is compositionally correct and free from other ingot imperfections such as shrinkage pipe at the top or last portion of the ingot to solidify.

In recent years, there has been a major effort to replace one or more of the VAR steps by an alternative process known as cold hearth melting. Cold hearth melting has been described in detail elsewhere, but essentially it is a melting technique which uses a high energy heat source; either plasma torches or electron beam guns to melt Ti which is contained in a water cooled copper hearth. By balancing the heat input from the guns or torches with the heat extracted from the hearth by cooling, the molten Ti alloy pool can be held in the hearth for any period of time in contact with only a solid layer of Ti alloy (called a skull) separating it from the copper hearth. This melting method has the advantage of allowing the alloy to be held in a molten state for longer periods of time than can be accomplished readily with VAR. As a result, the melt related inclusions which are soluble in molten Ti can be eliminated form the ingot by dissolution. Hearth melting is currently followed by a final VAR step to achieve better chemical homogeneity. Ti alloys made by this so-called hearth + VAR process have proven to be essentially free of melt related inclusions. Eventually, the final VAR step may be eliminated for cost and quality reasons and the goal of hard α free Ti alloy ingot will be realized. There are no major technological barriers to reaching this objective, but the current downturn in the world aerospace markets provides few if any incentives to invest the time and money necessary to qualify 'hearth only' ingot. This situation is aggravated by the current relatively low use levels of Ti alloys which makes it difficult to 'work off' the substantial amount of material produced during qualification of a new process such as hearth only melting.

3.3.2 Semi-finished products

Once the melting process is completed, the resulting ingot is hot worked into semi-finished products such as billet for forgings, rod or bar, sheet bar or finished flat rolled products including plate, sheet and strip. Among these, the closest to a commodity is strip. However, in contrast to the true commodity metals, production of Ti strip is largely confined to unalloyed grades for subsequent fabrication of welded tubing but Ti strip still is a high value product.

The details of the hot working process depend on the particular alloy and even on the intended use of this alloy which are determined by the customer requirements and are stated in the applicable material or process specification. For example, forging billet used to make rotating parts for aircraft engines frequently has macrostructure requirements because experience has shown that the most reliable way to control the microstructure of the finished forgings is to limit the variation in the material which is used in the forgings. Specifying billet macrostructure accomplishes this objective without unduly constraining the mill practice used to produce the billet. Billet macrostructure also qualitatively corresponds to the noise and attenuation of the ultrasonic waves used to inspect billets before they are made into forgings. Ultrasonic inspection techniques have evolved as have Ti production methods. Today much smaller discontinuities can be routinely found using this technique and work continues to further improve the capability of these methods as will be described in a later section of this chapter.

In other product forms (bar, rod, plate and sheet), the hot working operation has a direct effect on the ability to create the desired final microstructure in the product. The microstructure at this stage also has a major effect on the response of the material to secondary working operations such as sheet forming or pinch and roll forming operations. In addition to the microstructure requirements, sheet and plate have flatness and gage requirements which must be satisfied. The working characteristics of Ti alloys vary with alloy composition, but generally Ti alloys are considered more difficult to work than steels or Al alloy. This is because they retain a greater proportion of their room temperature strength at higher temperatures than steel or aluminum alloys. This is probably related to the more limited high temperature ductility which in turn leads to surface cracking during working. The surface cracks must be conditioned by grinding which affects both yield and cost. The hot workability varies significantly between alloys, but the more heavily β stabilized alloys such as Ti–6Al–4V tend to be more workable than the so-called near β alloys such as Ti–8Al–1Mo–1V. Alloys which contain refractory elements (e.g. Nb or Ta) as β stabilizers are also harder to work because these elements tend to increase the elevated temperature flow stress due to their lower diffusivities.

3.3.3 Forgings

Ti forgings, mainly for aerospace applications, have been a major volume item in the annual consumption of Ti alloys for a number of years. In fact, when the aerospace markets are down, the business impact on the producers of Ti alloys and Ti forgings is quite severe. The sophistication and size of the forgings continues to increase with each new generation of aircraft or aircraft engine. This is particularly true for the most common alloys such as Ti–6Al–4V because the forging is used both as a shaping operation *and* as a means of controlling microstructure through TMP. This is in contrast to practice for most steels and most Al alloys where forging is more commonly used as a shaping operation and a means of producing desired macroscopic characteristics such as grain flow.

There are numerous variations on forging practice, far too many to comprehensively review here. However, the principles of using forging practice to control microstructure and simultaneously to create near net shape articles is central to much of the successful application of Ti alloys in high performance structures. Thus these aspects of forging practice will be summarized here.

It has already been mentioned that the working temperature (relative to the β transus), the amount of plastic strain and the plastic strain rate affect the microstructure of forgings of $\alpha + \beta$ Ti alloys. This is true both for the as-forged articles and for the microstructural changes that occur during post-forging heat treatments. These factors also affect the microstructure of β alloys, but the relationship is more complicated. Consequently, the majority of the following remarks pertain much more explicity to $\alpha + \beta$ alloys.

The first order distinction is whether the working is performed above or below the β transus. Working above the transus results in a microstructure consisting of packets or colonies of plates of α similar to that shown in Figure 3.2. This is known as a colony microstructure. The scale of the colonies and of the individual plates within the colonies depends on the cooling rate after forging is completed and on the alloy composition. When cooled at a constant rate alloys which contain higher concentrations of β stabilizing elements will have smaller colonies and plates. In β forged alloys, there can be a continuous layer of α phase present at the prior β grain boundaries. This is called grain boundary α and is generally considered to be deleterious to mechanical properties.

Figure 3.3 Grain boundary α along prior β grain boundaries in β processed Ti–6Al–4V.

An example of grain boundary α is shown in Figure 3.3. Grain boundary α is viewed as undesirable by many, if not all, users of Ti forgings. This fact is reflected in the microstructural section of most users' specifications. The details of the effects of microstructure on properties will be summarized later, so here it is sufficient to qualitatively describe the effect of β forging on the properties of Ti–6–4, as is done in Table 3.1. The elimination of grain boundary α can be accomplished by working continuously through the β transus temperature. Such a working schedule results in continuous recrystallization of the β phase and little or no grain boundary α formation.

Forging under conditions of decreasing temperature starting below the β transus and finishing well into the two phase α + β region, is known as α + β working or forging. This results in an equiaxed microstructure

Table 3.1 Qualitative effects of β forging on key properties of Ti–6–4

Property	*Higher/lower with β forging*
Strength	Lower
Ductility	Lower
Fracture toughness	Higher
Fatigue crack initiation (fatigue life)	Lower
Fatigue crack growth rate (FCGR)	Lower
Creep strength	Higher
Aqueous stress corrosion cracking resistance	Higher
Hot salt stress corrosion cracking resistance	Lower

similar to that shown earlier in Figure 3.1. The detailed microstructure response to $\alpha + \beta$ working depends on the amount of work done at various temperatures below the β transus and on the rate of working (plastic strain rate). This interdependence suggests that the ability to achieve uniform microstructures in $\alpha + \beta$ worked forgings depends on a uniform distribution of plastic strain and strain rate. Until recently, this was achieved through the use of trial and error and experience based forging die and working schedule design. Today, determination of the die geometry and forging schedule for a new part is guided by computer modeling of the forging process; the result is that the desired shape and microstructures are achieved more efficiently and in many fewer iterations. Not only does this virtually eliminate expensive die modification, it also greatly reduces the number of trial forgings which must be made and cut up to verify that the die and forging process produce parts with the desired microstructure. Computer modeling also permits more uniform microstructure to be achieved throughout the forged part.

Other relatively recent advances in Ti forging technology is the use of hot die isothermal forging. These techniques permit much better control of microstructure and shape because the forging operation is performed under more controlled conditions. Isothermal forging is done at a constant temperature where the flow stress is constant and well known. It also is done in a temperature regime where the kinetics of the metallurgical reactions which control microstructural development are well understood. The combination of these factors results in forgings that have more uniform microstructures and less property variation with location in the forging. The current limitations of these techniques are the maximum size forging which can be made and the cost of the dies used to make the forgings. Hot die forging of Ti–6–4 typically is performed at 900–1000 °C inside a closed chamber designed to exclude the environment. The chamber is either evacuated or filled with inert gas such as Ar. The main reason for the chamber is the strength requirements of the die materials if they are actually operating at 900–1000 °C. The most commonly used materials are TZM, a Mo based alloy. This alloy oxidizes almost catastrophically at these temperatures and thus requires a protective chamber. The size limitation thus arises as a result of chamber and load lock size limitations, *not* because of press capacity.

The production of very large (> 1000 kg billet weight) forgings is now practiced routinely for the aerospace industry. In this regard, the high flow stress of Ti alloys at the forging finishing temperatures imposes limits on product sizes due to limitations in the force capability of hot working equipment. There are several very large capacity forging presses in the US which were originally purchased with the benefit of federal assistance. The Former Soviet Union (FSU) is believed to have the largest capacity press in existence today. Generally speaking, forging press capacity has not been a major constraint on the use of very large Ti alloy forgings. A bigger and more inherent issue is the heat transfer and flow characteristics of Ti alloys which produce inherent limitations to the ability to control the microstructure of the forgings. The growing use of modeling has been very beneficial in overcoming this limitation as has the use of more heavily alloyed grades of Ti such as Ti–6Al–2Sn–4Zr–6Mo (Ti–6–2–4–6) and Ti–5Al–2Sn–2Zr–4Cr–4Mo (Ti–17) which develop full strength in thicker sections.

3.3.4 Sheet gage parts

Production of Ti alloy parts from sheet has been a consistent challenge for the aerospace industry. This is due to the formability characteristics of Ti alloys. These characteristics are partially because of the mechanical properties of Ti alloys and partially because Ti tends to react with air at elevated temperatures (> 650 °C), resulting in oxygen enriched surfaces.

Exposure of Ti alloys such as Ti–6–4 to air above ≈ 650 °C results in an oxygen rich layer that has low ductility. This layer contains a higher volume fraction of α phase that is stabilized by the oxygen and is called α case. The presence of α case is known to promote crack initiation, especially during cyclic loading, and thus leads to substantial reductions in fatigue life.

In contrast to other commonly used sheet materials such as 300 series stainless steel, the relatively high flow stress and low modulus of Ti alloys imparts a greater proportion of elastic strain to the material during sheet forming. This strain is recovered on unloading and is often called springback. Springback makes shape and dimensional control more difficult. Ironically, this same characteristic is what makes some Ti alloys attractive for use in springs. Hot forming operations help ameliorate springback by lowering the flow stress, but the reactivity of Ti alloys requires that a coating be used if the forming is done above ≈ 625 °C. Alternatively, forming can be done in air at higher temperature, followed by a surface treatment to remove the affected layer of material. In practice this layer must be removed prior to using the part because of the deleterious effects it has on fatigue life. In sheet metal parts and castings, this is most commonly accomplished by chemical milling. In forgings, final machining normally accomplishes this.

In recent years the use of superplastic forming of thin gage Ti alloy parts has become commercially feasible. Superplasticity is a phenomenon exhibited by materials which have a stable, fine grained equiaxed structure. Superplasticity permits large plastic strains with uniform reduction of cross sectional area to be achieved using forming operations that require relatively low forces (Chapter 9). The original superplastic alloys were Al–Zn alloys, but it was soon realized that α + β Ti alloys had similar microstructures and also exhibited superplastic behavior. Superplastic sheet forming is now commonly used for the production of intricate shaped parts of Ti–6–4. Other alloys can also be superplastically formed, but there is more experience with producing Ti–6–4 sheet having the required microstructure for good superplastic response. The forming operation is typically performed with a single die which contains the desired part features. The sheet and die are heated to ≈ 850–950 °C and about 2 atmospheres of Ar gas pressure is used to press the sheet against the die. The forming rates are relatively slow compared to press or hammer forming, but the simplicity of the apparatus and the absence of the usual secondary straightening or hot sizing operations compensate for this. The dimensional preciseness of the final shape of superplastically formed parts is excellent. For some applications this preciseness is a key factor in the selection of superplastic forming for some complex geometry components.

At room temperature, Ti and its alloys have a thin, protective layer of TiO_2 on their surface. This imparts excellent chemical inertness to Ti, making it suitable for chemical and petrochemical applications. At > 625 °C Ti metal reacts with this oxide film and it dissolves into the Ti, leaving a clean metallic surface. In the absence of air or other reactive gases, this surface readily bonds to any mating Ti surface. This bonding is known as diffusion bonding and can be an effective joining method for Ti alloys. When combined with superplastic forming, this can be a powerful technique for fabricating complex components from sheet gage Ti alloys. This technique is known as superplastic forming/diffusion bonding (SPF/DB) (Chapters 9 and 10). At present it is used for making aircraft parts and hollow air foils for gas turbine engines. Today, SPF/DB is still expensive to practice, partly because the inspection of the bonds in the final part is time consuming. To a degree, it is also because there are few production volume applications for components made by this process. Still, SPF/DB permits the fabrication of intricate, structurally efficient components the likes of which have not been demonstrated with any other manufacturing technique.

3.3.5 Castings

During the past 10 years there has been an enormous growth in the use of Ti alloy castings for structural parts in aircraft and aircraft engines. This is in part due to advances in capability for making more complex shapes and larger castings. It also is in part due to the demonstrated improvements in properties (especially fatigue) which result from hot isostatic pressing (HIP) after casting. The use of HIP for castings closes and heals internal pores and voids that remain as a result of shrinkage during solidification of the casting. Removal of these pores, which act as fatigue initiation sites, greatly improves the fatigue performance of the castings. The use of castings in place of forgings or fabricated sheet metal components has become commonplace in the aircraft engine industry and is growing in importance in the aircraft industry. Castings are also growing in importance in the chemical industry for pump impellers and large piping fixtures (elbows, tees, etc.).

The principal advantage of Ti castings is the cost of the finished component. The properties of cast components of alloys such as Ti–6–4 are not strictly equivalent to those of a forging part or one machined from plate or bar (often called a hog-out). Thus a modest weight penalty is sometimes incurred as the result of lower design properties for a casting as compared to a wrought equivalent. Except in the most weight critical applications, this penalty is overridden by the substantial cost saving. These cost savings are the result of the nearer net shape of a casting when compared to a forging. In many cases, castings are chemical milled to remove α case, selectively machined at flanges and other mating surfaces and used. Forgings, on the other hand are typically machined all over before use. The cost per pound of metal removal by machining equals or exceeds the per pound cost of the as-forged shape itself. Therefore, the near shape characteristics of castings become a significant economic benefit. In many applications of castings the design limiting property is stiffness, so the reduced design properties do not come into play and the weight penalty mentioned earlier is not realized.

In the as-cast condition Ti–6–4 castings have a colony microstructure similar to that shown in Figure 3.2. This microstructural condition has lower ductility and fatigue strength compared to an equiaxed $\alpha + \beta$ structure. There have been several efforts to refine and modify the microstructure to ameliorate the effects of this microstructure. Included in these efforts to produce an equiaxed microstructure in situ and without the aid of mechanical working are the use of thermal cycling near the β transus and by introducing hydrogen as a temporary alloying addition. Between these, the hydrogen method appears to be the most promising because it produces a fine, equiaxed microstructure. The technique consists of hydrogenating the casting in a purged and hydrogen filled retort and then subsequently de-hydrogenating it under controlled conditions in a vacuum furnace. The detailed metallurgy of this process is not pertinent here, but essentially, the hydrogen stabilizes the β-phase. The vacuum treatment permits the α phase to reform with a fine, equiaxed microstructure. The principal limitations on the use of this technique are the inherent cost of the hydrogenation/de-hydrogenation treatment and the possibility of dimensional distortion due to the volume change that accompanies hydrogen charging and subsequent removal. Since the hydrogen is introduced into the bulk by solid state diffusion, there are additional practical section size limitations imposed by diffusion distances which can be realized in a reasonable time. The cost issue and the fact that most castings used today are stiffness limited, has been a significant barrier to extensive use of castings that have been treated by this temporary alloying method. That is, the properties associated with the colony microstructure have not been limiting, so there has been little incentive to employ the microstructural modification methods, especially at increased cost. The availability of this technique will some day be important when Ti users contemplate the direct substitution of casting for forgings in critical applications.

The majority of Ti castings used today are either commercially pure Ti for chemical applications or Ti–6–4 for structural applications. Only limited use of other alloys in castings has been made because the higher strength alloys such as Ti–17 or Ti–6–2–4–6 are more difficult to cast with acceptable chemical homogeneity and because, as stated earlier, the most common design limiting property for castings is stiffness, not strength. There is growing interest in casting of other alloys, e.g., Ti–8Al–1Mo–1V (Ti–8–1–1), which has a higher modulus and the β alloy Ti–3Al–8V–6Cr–4Mo–4Zr (beta C), which has the capability of much higher yield strength and thus better fatigue strength. At this point the interest in these alloys is largely at the evaluation stage.

3.3.6 Welding

The use of fusion welding as a fabrication process for Ti alloys is well established for a variety of applications. Because of the reactivity of Ti, all welding processes require protection from atmospheric contamination. This means that gas tungsten arc welding (GTAW), gas metal arc welding (GMAW), plasma arc welding (PAW) or electron beam welding (EBW) in vacuum are the only common methods practiced today. There is some use of electroslag welding for very heavy sections, but the purity requirements for the flux and the relatively few heavy section applications make this technique uncommon today. Weldability of unalloyed Ti is excellent and GTAW is used routinely as a fabrication method for manufacturing tubing for heat exchangers and other chemical plant applications. The weldability of Ti–6–4 is also very good and a variety of high strength components such as aircraft wing boxes are manufactured with the help of fusion welding.

The welding heat source obviously causes localized melting and re-solidification of the fusion zone, but the adjacent unmelted region also is heated to quite high temperatures creating a heat affected zone. The heat affected zone temperatures range from well above the β transus to ambient. This creates a microstructural gradient and also basically removes any effect of prior thermomechanical processing on the base metal microstructure. Further, the fusion zone is surrounded by a considerable mass of cooler material so the cooling rates in the weld and heat affected zone are rapid. The actual cooling rates vary with the weld process used, but even for high, less localized heat input such as PAW, the cooling rates are quite high. The resulting microstructures are metastable and will decompose on subsequent exposure to elevated temperature. As a result, a post weld stabilization is typically used to re-establish partial equilibrium in the weld and heat affected zone.

The weldability of higher strength alloys such as Ti–6–2–4–6, Ti–17 and Ti–6Al–6V–2Sn (Ti–6–6–2) is not as good as Ti–6–4. This is because of the tendency of the latter two of these alloys to hot crack. This is because some of the β stabilizing elements added to these alloys (Fe, Cu and Cr) also lower the melting point and extend the freezing range of the alloy. (This is related to the tendency for solute segregation in ingots.) Thermal stresses created during cooling cause the partially solidified weld to separate. This type of cracking is called liquation cracking and is different in origin from the high temperature solid state cracking observed in other high alloy systems such as steel. Another reason that the more highly alloyed grades of Ti are not considered as weldable is because the weld fusion and heat affected zones can have very low ductility. In Mo rich alloys such as Ti–6–2–4–6, this often is due to formation of a metastable phase known as orthorhombic martensite (α''). The α'' is not brittle, *per se*, but it rapidly decomposes during re-heating even to quite low temperatures, for example during a stress relieving operation at 650 °C. The decomposed α'' is very strong (> 1400 MPa yield strength), and the associated low ductility follows quite naturally. In other alloys such as Ti–17, the β phase can be retained in the fusion and heat affected zones and also can decompose during re-heating to create a very strong region containing either fine α or ω precipitates. These strong regions also have very low ductility.

All of these alloys can be welded if significant precautionary measures are employed. Such precautionary measures include pre-heating the work piece to control the thermal stress formation and the types of decomposition products that form during cooling. Complete re-heat treatment is of course possible, but this is seldom practical due to the likelihood of distortion, surface contamination by oxygen and due to the sheer size of many welded components. In the more difficult to weld alloys, it is common to use a starter and run out tabs to ensure that the weld contained in the actual part is uniform. These tabs are subsequently removed from the part. These techniques have been successfully employed in a variety of applications where the added cost is offset by the benefit obtained from the use of welding to make large or complex parts. For very large parts, welding may be the only feasible alternative to mechanically fastened construction and welded structures are considerably lighter and usually lower cost.

Welding is also used as a repair method for reworking defects that occur in Ti castings during the manufacturing operation. The ability to weld repair castings is an important factor in the economics of casting production, since small, surface connected defects are commonplace. Without weld repairs, casting yields would be much lower and the cost benefit of castings would be much less significant than is the case today. The need to easily and effectively weld repair castings also places some limitations on which alloys are commonly used for castings. This is because the more difficult to cast alloys are often also more difficult to weld. Moreover, the manufacture of weld filler wire is reasonably difficult for all alloyed grades of Ti, even Ti–6–4. It is considerably more difficult for less workable alloys such as Ti–8–1–1. Accordingly, the cost of weld repair is higher for both those alloys that are hard to draw into filler wire and those that require special weld processes because they are difficult to weld.

3.4 THE MECHANICAL BEHAVIOR AND PROPERTIES OF COMMON Ti ALLOYS

It has been briefly illustrated (Table 3.1) and frequently implied that the mechanical behavior of Ti alloys is strongly dependent on microstructure, which in turn depends on processing history. One of the unusual aspects of Ti alloy metallurgy is the ability to alter and control the microstructure on two size scales more or less independently. The 10–250 μm sized features affect the fracture and fatigue behavior, whereas the sub-micron scale features control the strength. The 10–250 μm scale microstructure features are mainly affected by processing, whereas the sub-micron features are mainly affected by heat treatment. At strength levels up to ≈ 1200 MPa, these features can be altered almost independently of each other. Above ≈ 1200 MPa the volume fraction of fine structure required to achieve this strength level approaches 100% and the coupling between strength and structure is no longer independent of the coarser microstructural features. At lower strengths, this aspect of Ti metallurgy can be viewed as an opportunity to tailor the strength and other properties of Ti alloys. It can also be viewed as an additional degree of complexity, which if not monitored and controlled, can lead to larger variations in properties than observed for simpler alloy systems.

In reality, microstructural engineering of high performance alloys is one of the last areas of significant opportunity for improving the performance of metallic materials. If this opportunity is left unexploited, the erosion of the metallic materials application domain by composites and intermetallics will accelerate. In the view of the author, the wealth of experience in designing with metals that exists should not be allowed to become obsolete. This obsolescence can be prevented through more innovative application of improved metals, thereby keeping the metals component designers 'in the game'. There are bound to be 'surprises' associated with the introduction of these new materials which may lead to set backs. The ongoing availability of metallic materials design and application capability is needed at least as a back-up until the merits and structural capability of composites and intermetallics is fully understood.

Further, too many new applications of composites and intermetallics may tax the ability of the design and materials groups within the producers of high performance structures beyond their collective ability to do the best possible job of 'digesting' the application technology for these new materials. Thus slow, incremental evolution away from metals is advocated and encouraged here.

This section will attempt to summarize the major effects of processing and chemistry on the behavior of a limited number of the most commonly used Ti alloys. Again, this is not intended to be an exhaustive or scholarly review. Rather, it is intended to illustrate the more important factors that govern the behavior of Ti alloys under the more common application circumstances. This section will commence with a short summary of the principles that govern the relations between microstructure, composition and properties. It will next summarize selected data for some alloys to show how these principles apply to real alloys. The reasoning behind this approach is that, practically speaking, it is impossible and probably not useful to summarize all the microstructural conditions encountered throughout the Ti alloy user community. However, elucidation of some of the underlying principles governing properties and mechanical behavior should assist in the efficient use of a wide range of Ti alloys.

3.4.1 Tensile properties and fracture toughness

Ti alloys are strengthened by a combination of solid solution strengthening, boundary strengthening and precipitation hardening. In $\alpha + \beta$ alloys, solid solution and boundary strengthening are the more important because there is only limited opportunity to utilize precipitation hardening. In β alloys precipitation hardening is very important because the metastable β phase can decompose to form a uniform distribution of either α phase or ω phase precipitates. The details of this strengthening are discussed later.

(a) $\alpha + \beta$ *alloys:*

The microstructural scale of the $\alpha + \beta$ alloys is consistently fine because these alloys have a duplex microstructure in the equiaxed or $\alpha + \beta$ worked condition. This is similar to the duplex or dual phase steels. The partitioning of alloying elements between the α and β phases causes the fine two phase mixture to be quite stable up to temperatures within ≈ 250 °C of the β transus. This aspect of the metallurgy of these alloys causes the *annealed* yield strength to be a stronger function of alloy composition than of detailed heat treatment history. That is, the relative constancy of microstructural scale essentially fixes the boundary strengthening contribution. The result of this is that the greatest variation of strength between alloys in the annealed condition is derived from the solute strengthening of the individual constituents.

Included in the term solute, as used here, are the interstitial elements carbon, nitrogen and oxygen, the most important of which is oxygen. While these elements are nominally present as impurities, they have a very potent strengthening effect at temperatures below ≈ 550 °C. The strengthening effect of oxygen is illustrated in Table 3.2. These data pertain to three different lots of material, containing 800, 1200 and 1900 p.p.m. oxygen by weight. These data also illustrate the complexity of the interaction between processing history, chemistry and properties. Note that the ultimate strength monotonically increases with increasing oxygen, but that the yield strength does not show the same trend. The microstructure of these alloys was relatively constant, so another explanation must be sought for the inconsistent variation in yield strength with oxygen content. The only other major factor that affects strength is crystallographic texture. The effect of texture is important, not only for strength, but for other properties such as stiffness, as has been discussed earlier. The implications of texture on mechanical behavior of Ti alloys can be significant and will be mentioned as appropriate throughout this section. Here it is sufficient to mention that texture can be responsible for as much as 70–150 MPa difference in yield strength in the longitudinal and transverse directions in unidirectional worked products such as plate and sheet. Texture is probably responsible for the variation in

Table 3.2 Effect of oxygen content on the tensile properties and fracture toughness of annealed Ti–6–4 [1]

Heat treatment condition – oxygen level	*Yield strength (Mpa)*	*Ultimate strength (MPa)*	*Elongation (%)*	*R of A (%)*	K_{IC} *(MPam$^{1/2}$)*
Anneal – low oxygen	735	820	21.3	40	≈ 124*
Anneal – medium oxygen	707	873	12.4	36	≈ 120
Anneal – high oxygen	855	920	12.3	24.9	≈ 68

* The approximate symbol (≈) indicates that the K_{IC} tests did not meet thickness requirements for valid K_{IC} values to be reported.

yield strength shown in Table 3.2, although no texture data was obtained to support this point.

In fact, oxygen is a sufficiently powerful solid solution strengthening element in the α-phase that there are two grades of Ti–6–4, a special grade known by the designation extra low interstitial (ELI) and standard grade, which is always implied by the designation Ti–6–4 without ELI. These grades differ mainly in oxygen content and the ELI grade typically has oxygen concentration that is lower by a factor of ≈ 2. Representative oxygen concentrations are 1500–2000 p.p.m. by weight vs. < 800 ppm for ELI. The ELI grade is most commonly used for cryogenic tanks and fixtures where the very high low temperature strength of the higher oxygen grades would compromise the desired toughness levels, especially in welded structures.

The yield strength of more highly alloyed grades of Ti can be varied significantly by heat treatment. In this context, the concentration of β stabilizer alloying additions is most important because these control the heat treatment response as described below. Among the common α + β alloys, Ti–6–4 is representative of the leanest alloy composition that is heat treatable. Here the term 'heat treatable' implies that the yield stress can be varied by at least ≈ 20% by heat treatment. The object of heat treatment is to use known phase transitions to decrease the microstructural scale. This is made possible by the formation of larger volume fractions of β phase on heating to higher temperatures in the 2 phase α + β region. On cooling, especially rapid cooling, this β phase becomes thermodynamically unstable and decomposes either by a diffusional nucleation and growth process to form a fine α + β mixture or by a diffusionless nucleation and shear process to form a fine, acicular product known as martensite. This is illustrated in Figure 3.4,

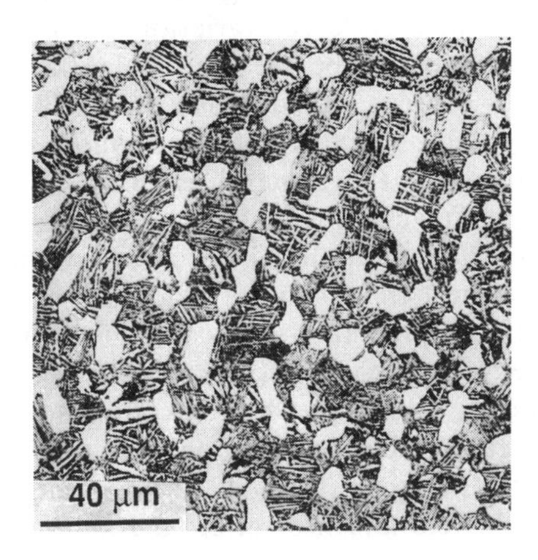

Figure 3.4 Duplex microstructure in solution treated and aged Ti–6Al–4V consisting of equiaxed primary α and fine basket weave mixture of α and β.

Table 3.3 Effect of heat treatment on tensile properties and fracture toughness of Ti–6–4 [1]

Condition No.	*Heat treatment condition*	*Yield strength (MPa)*	*Ultimate strength (MPa)*	*Elongation (%)*	*R of A (%)*	K_{IC} *(MPa m$^{1/2}$)*
1m	Anneal – medium oxygen	707	873	12.4	36	≈ 120*
3	Dupl. anneal	857	908	15.3	47	≈ 110
6	Solution treat and age	877	939	15.2	34	≈ 84
7	Solution treat and over-age	909	982	15.5	47	75
4	β anneal/forge	774	852	11.2	23	≈ 120
5	β quench	862	931	5.9	6	100

* The approximate symbol (≈) indicates that the K_{IC} bars did not meet thickness requirements for valid K_{IC} values to be reported.

which should be compared to Figure 3.1(a). The refined scale of these microstructural constituents contributes extensively to boundary strengthening, resulting in significant increases in yield strength. Some representative strengths are listed in Table 3.3.

From these data it can be seen that the yield strength of Ti–6–4 can be varied by ≈ 30% by heat treatment. In higher oxygen material the percentage variation can be even greater. The data in Table 3.3 are representative of medium oxygen (≈ 1300 p.p.m. by weight) material. It can be seen that large variations in toughness also occur as a function of microstructure, but that even larger variations in toughness occur with oxygen concentration, as can be seen in Table 3.2. The sensitivity of toughness to oxygen content also depends on microstructure, with the acicular and colony microstructures being the most sensitive. Table 3.4 shows property data for the same three lots of material that the properties shown in Table 3.2 pertain to, but these data are for an acicular microstructural condition. These data show that the combined effect of oxygen and heat treatment are indeed large. In addition, there is essentially no texture present in acicular microstructure material, so the strengthening effect of oxygen is more clearly seen in the Table 3.4 data than for those in Table 3.2, where texture also contributes to the strengthening. The rapid reduction of toughness with increasing oxygen content also illustrates why ELI grade Ti–6–4 is preferred for welded structures, where the weld and heat affected zone have acicular microstructures.

The data in Table 3.4 pertain to material with an acicular microstructure, as is shown in Figure 3.5. This microstructure is obtained in Ti–6–4

Table 3.4 Effect of oxygen content on the tensile properties and fracture toughness of heat treated Ti–6–4 [1]

Heat treatment condition – oxygen level	*Yield strength (MPa)*	*Ultimate strength (MPa)*	*Elongation (%)*	*R of A (%)*	K_{IC} *(MPa m^{1/2})*
Acicular – low oxygen	735	931	21.3	40	≈ 110*
Acicular – medium oxygen	862	931	5.9	6.0	≈ 100
Acicular – high oxygen	1000	1085	5.1	7.9	≈ 68

* The approximate symbol (≈) indicates that the K_{IC} tests did not meet thickness requirements for valid K_{IC} values to be reported.

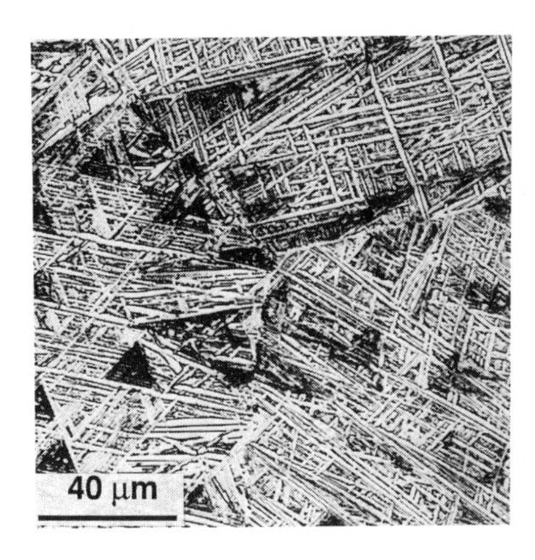

Figure 3.5 Fine acicular α + β mixture in β quenched Ti–6Al–4V

by rapidly cooling from the β phase field. In large section sizes, a water quench is required to create this microstructure in Ti–6–4. Water quenching of heavy section Ti–6–4 parts creates a surface to center microstructure gradient as shown in Figure 3.6. This gradient causes the properties of heavy section parts to vary with location within the part. This may or may not be desirable, but, as a minimum, it must be dealt with from a design standpoint if it exists. In alloys containing higher concentrations of β stabilizer, e.g. Ti–6–2–4–6, similar acicular microstructures can be achieved by rapid air cooling. In these alloys the section size dependence of microstructure is much less pronounced, which is one of the reasons that these alloys are used for heavy section applications.

The other aspects of the data shown in Table 3.3 also illustrate some general characteristics of the relation between microstructure and properties. The material used to obtain these data was from a single heat,

(a) (b)

Figure 3.6 Section size induced variation in microstructural scale of the acicular $\alpha + \beta$ mixture in a 20 cm section of β quenched Ti–6Al–4V: (a) surface; (b) center.

was forged axisymmetrically and very little texture was present. Thus the property variations represent the effect of microstructure *per se*. First the effect of boundary strengthening due to microstructural scale is clearly illustrated by the data for the annealed, duplex annealed, solution treated plus aged and β annealed conditions. The finest scale microstructure is the β quenched condition (condition 5) which also is the strongest. The next strongest materials are conditions 6 and 7 followed by 3 and then by 1. The strengths qualitatively follow these microstructural variations quite well. The other characteristic is the inverse correlation between tensile ductility, particularly as measured by elongation to fracture, and fracture toughness. For example, the condition with the lowest ductility (condition 4) also exhibits the highest toughness. This is a well known characteristic of $\alpha + \beta$ Ti alloys, but this inverse relation disappears at the very high strength levels (> 1500 MPa a yield) achievable in the β alloys. This matter has been studied extensively and is qualitatively understood. The root cause of the inverse relation between ductility and toughness in $\alpha + \beta$ alloys is related to a change in fracture mode in colony and acicular microstructures. This change affects crack initiation and crack extension differently. A tensile test combines crack initiation strain and crack extension strain whereas toughness only measures crack extension energy. Thus, the apparent reversal of these two properties is not the paradox it first appears to be.

The data presented and discussed above, taken unto themselves, scarcely merit the space devoted to them. However, they serve to illustrate the interaction between alloy chemistry, microstructure, texture and properties. These interactions emphasize the pitfalls associated with drawing hasty cause and effect conclusions regarding property variations in those instances where more than one variable may be changing. Texture is perhaps the most elusive variable because variations in texture cannot be observed metallographically and because it is more difficult and therefore less common to measure it experimentally at present.

Table 3.5 Effect of processing and heat treatment on the tensile properties and fracture toughness of Ti–6–2–4–6 [4]

Condition No	*Process. and heat treatment condition*	*Yield strength (MPa)*	*Ultimate strength (MPa)*	*Elongation (%)*	*R of A (%)*	K_{IC} *(MPa m$^{1/2}$)*
A	Acicular – STA	1048	1200	7	13	57
B	10% α_p – STA	1117	1213	13	37	34
C	50% α_p – STA	1151	1241	13	34	26
D	50% α_p – Anneal	1062	1131	13	34	26

STA = solution treat and age.

Based on the above discussion, some data for a higher strength alloy will be used to illustrate the effects of microstructure on properties in alloys heat treated to higher strength levels. The properties of Ti–6–2–4–6 in three microstructural conditions are shown in Table 3.5. All of this material contained ≈ 1200 p.p.m. oxygen by weight. These data show the same trends as Ti–6–4, except the reversal of tensile ductility and toughness between equiaxed and acicular microstructures is even more pronounced. The effect of the volume fraction of fine transformed β phase on strength also is clearly demonstrated by these data. Finally, the effect of heat treatment on strength and toughness in the equiaxed, 50% α_p microstructures (conditions C and D). Heat treating to reduce the strength, while maintaining a constant volume fraction α_p might be thought to improve the ductility and toughness. In practice, it does reduce the strength, but actually reduces the ductility somewhat and leaves the toughness unchanged. Again, these data underscore the importance of using data to make decisions regarding adjustments in heat treatment to achieve desired changes (trades) in critical properties. This becomes even more important in cases where processing history changes are also involved because these changes can also alter the type or nature of texture and lead to further unanticipated variations in property response.

There is also a good correlation between fracture toughness and work hardening rate in this and other higher strength alloys. The acicular microstructure (condition A) has the highest work hardening rate and by far the highest toughness. As in the case of Ti–6–4, there is a major change in fracture path between the acicular and equiaxed microstructures, the acicular microstructure having a much more tortuous fracture path. In fact, the fracture surface of the low toughness, high strength condition of this alloy was so macroscopically smooth that a brittle fracture mode such as cleavage might be suspected. Fractographic examination, however, showed that the fracture occurred by a ductile mode, i.e. microvoid formation and coalescence. The microstructurally controlled fracture path and attendant change in roughness is the qualitative change in local fracture behavior. Coupled with the observed increase in work hardening rate, these changes seem to adequately account for the variation in toughness and ductility.

(b) β *Alloys*

The strength levels that can be achieved in the β alloys are considerably higher than those characteristic of even high strength $\alpha + \beta$ alloys. This is because the β alloys can be precipitation hardened. In the β solution treated condition, β alloys contain 100% β phase. This phase is metastable and decomposes during subsequent aging leaving the solute enriched parent β phase and one or more of several decomposition products. The details of β decomposition and β decomposition products has been reviewed elsewhere [5]. It is not relevant to discuss the decomposition of β phase in detail here, but it is important to note several points. First, the kinetics of β decomposition depends on the alloy composition, since this fixes the degree of metastability. Second, all the β phase decomposition products are formed during aging at temperatures ≤ 550 °C. As a result they are sub-micron sized and can be uniformly distributed. The most important of the β decomposition products is the equilibrium α phase. In more solute rich β alloys, such as Ti–3Al–8V–6Cr–4Mo–4Zr (known also as beta-C or Ti–3–8–6–4–4), the lower degree of β phase metastability makes formation of the α phase during aging more difficult. Formation of these fine α precipitates is easier at preferred sites such as grain boundaries, dislocations and sub-boundaries. In such alloys, the α precipitate distribution can be non-uniform if the parent β phase is partially recrystallized during the solution treatment. Thus working the alloy in the β phase field followed by rapid cooling to prevent recrystallization, leaving a residual high dislocation density results in a more uniform density of sites for α phase formation. In the more heavily stabilized alloys such as Ti–3–8–6–4–4 a very high density of α phase nucleation sites can be achieved by solution treating and extensive cold working. If the density of α precipitates is high enough, strengths in excess of 1400 MPa can be achieved in β alloys. At these strength levels, the ductility is typically very low (< 1% elongation) so the range of applications for these very high strength alloys is very limited. Two examples are worth noting, however. These are high strength fasteners that are loaded in shear and springs, since these also are loaded in torsion (shear). There has been limited use of high strength β alloys for both of these applications.

The toughness of β alloys is not as high as for α + β alloys, largely because of the higher strength levels typically under consideration. Table 3.6 shows some typical properties of the alloy Ti–10–2–3.

Table 3.6 Tensile properties and fracture of Ti–10–2–3 in different microstructural conditions and at different strength levels [6]

Micro condition	*Yield strength (MPa)*	*Ultimate strength (MPa)*	*Elongation (%)*	*R of A (%)*	K_{IC} $(MPa\,m^{1/2})$
β processed	1262	1337	10	20	46
α + β processed	1220	1269	11	31	56
α processed	1117	1158	13	39	64
α + β processed	1083	1158	14	38	62

These data show that very attractive combinations of strength and toughness can be achieved in β alloys if care is paid to selection of the processing and heat treat schedules. The data also show that β processing does not improve toughness of this alloy at higher strength levels, in contrast to the trends shown earlier for α + β alloys such as Ti–6–2–4–6. This is a general characteristic of β alloys and the reasons for this are qualitatively understood. It is sufficient here to mention that variations in toughness with strength level are related to the local mode of crack extension but these details are not relevant here.

3.4.2 Fatigue properties

Fatigue behavior of structural alloys can be separated into fatigue life and fatigue crack growth rate (FCGR) behavior. The latter can be further subdivided into so-called short and long crack behavior, although this distinction will not be extensively discussed in this article. The fatigue behavior of Ti alloys is intermediate between high strength steel and high strength Al alloys when compared on a density normalized basis. Moreover, at least for α + β Ti alloys, there is a greater latitude for tailoring the balance between fatigue life and crack growth behavior. This is because of the wider range of microstructural options that are available in Ti alloys through the combined use of TMP and heat treatment, as described earlier.

(a) Fatigue life behavior

The effect of microstructure on fatigue life in Ti alloys is two-fold: the effect on strength and the effect on crack initiation behavior. The effects of microstructure on strength have been discussed earlier. The crack initiation behavior is controlled by the deformation behavior of the alloy. In turn, this is significantly affected by the microstructure. Deformation behavior, as used here, refers to whether plastic deformation is accommodated by twinning or slip and to the nature of the slip bands. In summary, $\alpha + \beta$ Ti alloys containing $\geqslant 4$ wt% (Al% by weight) and aged β alloys exhibit twinning so seldom that it can be ignored as a deformation mode in high strength alloys.

$\alpha + \beta$ *alloys*

The concentration of Al controls whether slip in the α-phase is planar or wavy, however, because at Al contents $\geqslant 5$ wt%, planar slip occurs in localized bands. Higher oxygen contents tend to cause the planar slip to intensify and become localized. The mechanistic factors that cause this slip transition have been widely studied and are described in detail in the literature [7]. Here it is useful only to mention the occurrence of planar, localized slip because it is a major factor in the fatigue behavior of the material. The other factor is the length of the slip band, which is directly related to the scale and nature of the microstructure. For example, in Ti–6–4 with a colony microstructure, the slip is planar because of the Al content and the slip length is controlled by the colony size, since all plates in a colony are crystallographically aligned. Thus, $\alpha + \beta$ alloys containing $\geqslant 5$ wt% Al, in this microstructural condition, exhibit long, planar slip bands. Essentially all the plastic deformation occurs in these bands; therefore these are closely related to fatigue crack initiation sites. It also follows that $\alpha + \beta$ alloys with colony microstructures have the lowest fatigue strengths of any microstructural condition, because this condition is a low strength condition and exhibits a propensity to form long, planar slip bands. That is, the stress to initiate plastic deformation is low and the onset of plasticity is concentrated in the planar slip bands, leading to early crack initiation. There is contradictory evidence regarding the effect of oxygen on fatigue strength. This is because oxygen promotes planar, localized slip, but also increases the yield stress. This point has not been as systematically studied, but it is likely that the oxygen effect depends on the microstructure, much as it does in the strength/ductility relationship outlined earlier. Data for the fatigue strength of Ti alloys is plentiful, but much of it is very empirical in nature. Often it is reported as cycles to failure as a function of stress, total strain or, less frequently, plastic strain. One of the difficulties in developing a useful data base for the fatigue strength of even a single alloy in a single microstructural condition is the number of variables involved. For example, there is stress or strain range, mean stress or strain, loading frequency, loading mode (bending or axial), test environment and specimen surface condition (finish, stress, etc.). In the end, the data must be useful to the designer and must represent the circumstances of use of the alloy. It has become common practice to use vacuum or inert gas environments to eliminate the effect of environment on fatigue behavior when the effects of microstructure or surface condition are being studied. While this is useful, it is important to avoid using such data for selecting microstructures or stress levels for real applications that will *not* operate in a vacuum or inert gas environment.

In spite of the disclaimers outlined above, it is useful to indicate some trends for the effect of microstructure on fatigue life of Ti–6–4. These are summarized in Table 3.7.

Table 3.7 Ranking of fatigue strength of Ti–6–4 with various microstructures

Microstructure	*Rank – 10^3 cycles*	*Rank – 10^7 cycles*	*Reference*
Anneal – coarse equiaxed $\alpha + \beta$	6	6	Figure 3.1
β processed – colony structure	5	5	Figure 3.2
β quench – acicular structure	3	1	Figure 3.6
Duplex Anneal – bimodal	4	4	Figure 3.4
Solution treat and age	1	2	Figure 3.5
Solution treat and over-age	2	3	Figure 3.5

These rankings are based on data and show that those microstructures that contain high volume fractions of fine, acicular structure such as solution treat and age or β quench have the best high cycle fatigue (HCF) properties. Those microstructures that also exhibit better tensile ductility and good fatigue behavior, such as solution treat and age, also tend to have somewhat better low cycle fatigue (LCF) properties as would be predicted by the well-known Coffin Manson Law. This latter point is illustrated by the reversal of ranking between low and high cycle fatigue properties shown in Table 3.7. All of the materials used to generate these data also were carefully processed to establish uniform microstructures, absent of unwanted features such as blocky α in $\alpha + \beta$ processed conditions (Figure 3.7) or extensive, coarse grain boundary α (Figure 3.3) in the case of β processed conditions. Blocky α or grain boundary α most markedly affect fatigue crack initiation behavior in the higher cyclic stress or strain ranges. Therefore, these features are most detrimental to LCF behavior and must be most closely monitored for applications where LCF is the design limiting parameter. These features also result in increased scatter in LCF life in laboratory tests and thus affect the position of the minimum life LCF curves in design data bases.

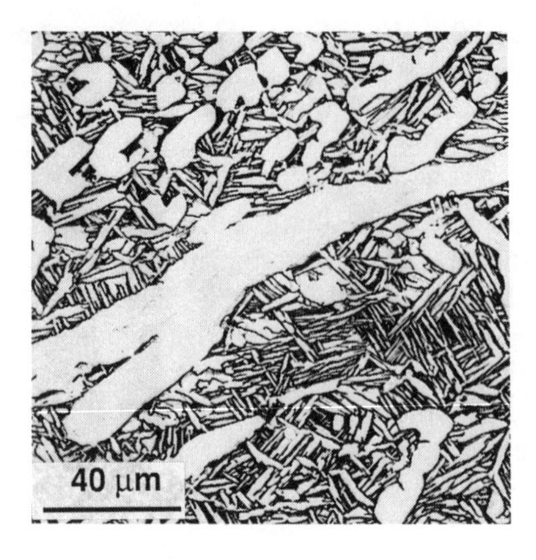

Figure 3.7 Blocky primary α in Ti–6Al–4V that has been improperly solution treated.

As a result, it is important to establish processing practice that eliminates blocky or grain boundary α. It also is essential to verify that this practice does lead to uniform microstructures throughout the entire forging and then adhere to this practice during subsequent production of all forgings. The inherent danger of designing to fatigue curves generated for uniform microstructures is that deviation from the established forging practice may allow the occurrence of unwanted microstructural features in production forgings. If this happens, the design becomes non-conservative and may lead to premature failures in service. Therefore, it has become common practice to monitor and reverify the properties of critical forgings at regular intervals, e.g. for every 250th part, to be certain that no slow, time-based deviations from the established practice have occurred. It also is common practice to completely requalify a new source of the same forging, should a change of vendors occur. As mentioned earlier, the growing use of computer based forging modeling has reduced the time and cost of such requalification procedures, but this procedure still is expensive.

The practical reality regarding the fatigue life of α + β Ti alloys is that residual compressive stress is widely used as a means of deferring crack initiation and extending fatigue life of fatigue limited Ti components. The most common and easiest to control method of introducing these residual stresses is shot peening. While shot peening is potentially a complete subject unto itself, a few comments are warranted here. First, shot peening tends to overwhelm the microstructure effects discussed earlier. Thus in practical applications, the microstructural effects become of secondary importance from a fatigue crack initiation standpoint. It is important to obtain uniform microstructures, however, since shot peening suppresses surface crack initiation but does not significantly affect internal crack initiation. Thus non-uniform microstructures, particularly those with coarse regions can lead to early sub-surface crack initiation and eliminate or reduce the benefit of surface residual compressive stresses. Second, the details of peening method are important, since it is possible to 'over peen' and to introduce surface damage instead of beneficial residual stress. Much progress has been made in automating peening to impart uniform coverage and intensity of peening, even in complex shaped parts.

β *alloys*

The high cycle fatigue strength of β alloys is generally quite good because of the high yield strengths that can be obtained. For example, if the alloy Ti–10–2–3 is heat treated to a yield strength of $\approx$ 1200 MPa, the fatigue strength at 10^6 cycles is $\approx$ 345 MPa compared to a strength of $\approx$ 250 MPa for Ti–6–4 in the STA condition. At higher cyclic stress amplitudes, e.g. 525 MPa, the low cycle fatigue strength of the Ti–6–4 begins to exceed that of Ti–10–2–3, presumably because of the superior tensile ductility of the somewhat lower strength $\alpha + \beta$ alloy. Much less attention has been paid to shot peening effects in β alloys, but qualitative improvements similar to those observed for $\alpha + \beta$ alloys are reasonable to expect.

(b) Crack growth behavior

Ti alloys also exhibit a significant microstructural dependence of FCGR. This dependence is most pronounced in $\alpha + \beta$ alloys at yield strengths < 1200 MPa because of the ability to vary micron sized microstructural features more or less independently of strength.

$\alpha + \beta$ *alloys*

The strongest microstructural dependence of FCGR in $\alpha + \beta$ alloys is found in the near threshold range of crack growth rates ($da/dn \leqslant 5 \times 10^{-5}$ mm/cycle). Some microstructural dependence also is observed in the Paris law range where the crack growth rate or crack extension per cycle (da/dn) is proportional to ΔK^n. The exponent, n, typically lies in the range of 4–6 for Ti alloys. The corresponding da/dn ranges for the Paris law regime for Ti–6–4 are $\approx 5 \times 10^{-5}$ to $\approx 1 \times 10^{-3}$. Above this range the divergence of the da/dn vs. ΔK curves reflects the microstructural dependence of K_{IC}. All of these factors are schematically illustrated in Figure 3.8.

There is such a wide range of microstructures in use for $\alpha + \beta$ Ti alloys, that it would be difficult, if not impossible, to comprehensively survey

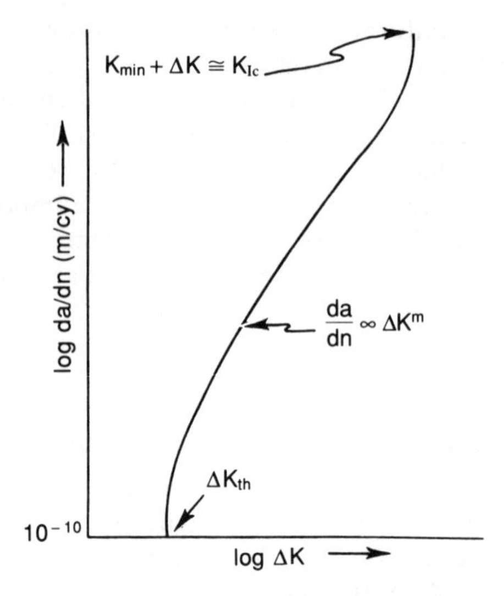

Figure 3.8 Schematic da/dn vs ΔK curve showing the near threshold, Paris law and end of life regions.

the crack growth data. Instead, a representative set of microstructures will be ranked in order of decreasing ΔK at constant da/dn in the most microstructurally sensitive near threshold region. That is, the microstructure with the best (slowest) FCGR characteristics is ranked #1. These data, shown in Table 3.8, are similar to those for fatigue life and pertain to the same microstructures that are ranked in Table 3.8.

While there are some reversals in rankings between the high and low crack growth rate regions, the high crack growth rate region generally is not of much interest because the crack is growing so rapidly at this point that very little life remains in the component. This region usually is not considered to be significant in life management calculations of fatigue crack growth rate limited structures.

Table 3.8 Decreasing ΔK ranking of fatigue crack growth rate of Ti–6–4 with various microstructures

Microstructure	*Rank – da/n ≈ 1×10^{-9} m/cycle*	*Rank – da/dn ≈ 1×10^{-4} m/cycle*	*Reference*
Anneal – coarse equiaxed α + β	3	2	Figure 3.1
β processed – colony structure	1	1	Figure 3.2
β quench – acicular structure	2	4	Figure 3.6
Duplex Anneal – bimodal	4	3	Figure 3.4
Solution treat and age	6	6	Figure 3.5
Solution treat and over-age	5	5	Figure 3.5

Much of the microstructural dependence of FCGR in $\alpha + \beta$ Ti alloys disappears at high mean stresses. This is because the effects are generally agreed to have their origin in a phenomenon called roughness induced crack closure, as described later. Different microstructures of $\alpha + \beta$ alloys induce variations in fatigue fracture surface roughness on the 10 – 100 µm scale. At low mean stresses, the mating fracture surfaces interfere during unloading, leaving a residual compressive stress field in and around the crack tip region. Upon reloading, this stress field must first be counteracted by the applied load before the crack can be opened during the next load cycle. The net result of this effect is that the applied load required to produce any particular crack tip tensile stress field to drive the crack forward is increased by the load necessary to offset the residual compressive stress. This incremental load is called the closure load, which can be converted to a gross area stress in the same way the applied load is converted for calculating cyclic stress intensity, ΔK. The well known formula for ΔK is:

$$\Delta K = \alpha \Delta \sigma_g \sqrt{\boldsymbol{a}}$$

where α is a crack front geometry correction factor, $\Delta\sigma_g$ is the gross area cyclic stress amplitude and a is the crack length. The effect of closure is to introduce an additional term to correct the applied ΔK. This is expressed as the following:

$$\Delta K_{\mathbf{eff}} = \Delta K_{\mathbf{app}} - \Delta K_{\mathbf{cl}}$$

where the subscripts, **eff, app** and **cl** refer to effective, applied and closure, respectively. From this, it can be seen that higher closure loads reduce the driving force for crack extension, $\Delta K_{\mathbf{eff}}$. These loads vary significantly between microstructures in Ti alloys such as Ti–6–4. This is the origin of the unusually large microstructural dependence of FCGR in $\alpha + \beta$ Ti alloys.

The corollary to the benefit of roughness closure is that, as mentioned earlier, this effect disappears at high mean stresses or load ratios (R). This is because, at higher R values, the crack tip remains open even at the minimum load and thus roughness induced closure does not come into play. This point has been shown quite conclusively by measuring $\Delta K_{\mathbf{cl}}$ and replotting da/dn vs. $\Delta K_{\mathbf{eff}}$ instead of $\Delta K_{\mathbf{app}}$. When this is done the crack growth curves for all microstructures tend to become coincident, suggesting that roughness induced closure *does* account for the observed variation in FCGR with microstructure. The data available for $\Delta K_{\mathbf{eff}}$ suggests that the crack remains open at $R > 0.3$. Above these load ratios, $\Delta K_{\mathbf{cl}}$ becomes 0 and the closure effects disappear. The net result of this is that microstructure can only be used to significantly enhance the crack growth portion of fatigue life in components where the significant cycles occur at low R values. The effect of load ratio on crack growth curves is schematically shown in Figure 3.9.

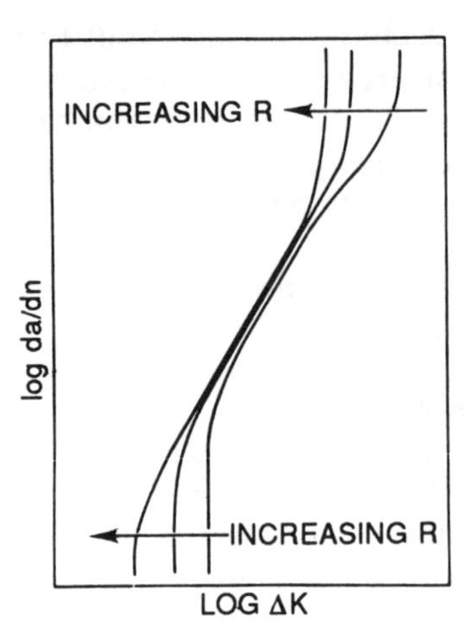

Figure 3.9 Schematic da/dn vs ΔK curves showing the effect of increasing mean stress on the threshold ΔK value (ΔK_{th}).

There are other mechanisms that lead to closure effects during fatigue crack growth, such as plasticity and oxide formation in the crack tip region. The former of these does not seem to be important in Ti alloys over a moderate range of strengths and the latter is not important at or near room temperature where the rate of oxide formation is low. Thus the major effect in Ti alloys near room temperature appears to be roughness induced closure.

At higher temperatures, $\geqslant$ 500 °C, oxide does form on the fracture surface in the crack tip region, but oxygen also affects the *intrinsic* crack growth resistance. Separation of these two factors is difficult. However, it is worth noting that, during crack growth tests conducted in air at $\approx$ 520 °C, artificially high ΔK_{th} values appeared during decreasing ΔK (load shedding) tests [8]. This effect was not investigated in detail, but the most logical explanation for the observation is oxide induced closure. Thus the *quantitative* influence of oxide formation on the FCGR behavior of Ti alloys at high temperatures appears to be an unresolved issue. Again, this effect would be expected to disappear at higher R values. Moreover, at higher R, the crack tip remains open and the penetration of oxygen should be more pronounced. This is particularly important for the higher temperature capability, creep resistant alloys such as Ti–6Al–2Sn–4Zr–2Mo(+ Si) {Ti–6–2–4–2S}, Ti–6Al–2.5Sn–4Zr–.5Mo–.45Si (Ti–1100) or Ti–6Al–4Sn–4Zr–.7Nb–.5Mo–.35Si(+ C) {IMI–834}.

For Ti–6–4, the maximum use temperature is limited by the creep strength of the alloy to $\approx$ 350 °C. At these temperatures, no significant oxide thickness develops during the interval of one cycle during normal crack growth testing at 20 Hz. In these cases the intrinsic effect of oxygen on crack growth at least can be qualitatively assessed. Tests using Ti–6–4, done at 20, 120 and 315 °C, showed that there was essentially no effect of temperature at $R = 0.3$, but that a pronounced effect was observed at $R = 0.7$. In the higher R tests, the FCGR decreased between 20 and 120 °C, then increased substantially at 315 °C, in comparison to either of the other temperatures. The cause of the intermediate temperature FCGR reduction is not clear, but the acceleration at 315 °C is the combined result of the lower yield strength at this temperature and an intrinsic effect of oxygen on the FCGR of Ti. Other environments, such as aqueous NaCl solutions, also can increase FCGR. The effect of environment is largest at intermediate ΔK levels, as shown in Figure 3.10. The 'hump backed' da/dn vs. ΔK curves have been widely studied and analyzed.

For the purposes of this chapter, it is sufficient to note that the greatest effect of environment occurs at ΔK values which have corresponding K_{max} values just above the threshold for sustained load cracking. Thus at very low ΔK levels, no effect is observed because the material is immune to the environment. At higher ΔK levels the crack growth rate is large enough due to applied load normal driving force, ΔK_{app}, that the incremental effect of environment is small at normal cyclic loading frequencies, e.g. 20 Hz. At lower frequencies, or if the peak load is sustained for even a matter of seconds, as in a square wave load with dwell at maximum load, the shape of the crack growth curve can shift dramatically. The general principles outlined above still pertain, but the

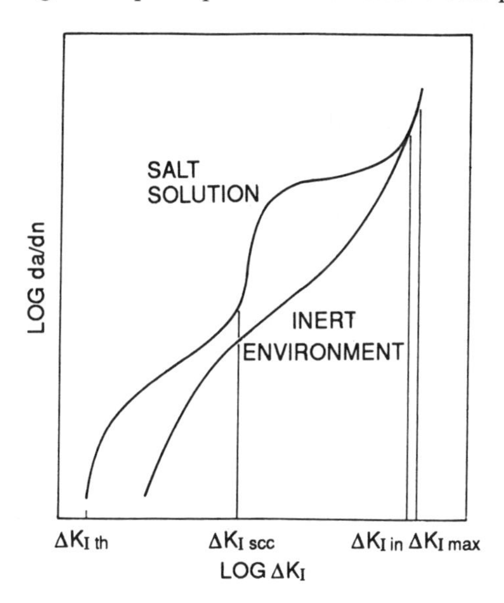

Figure 3.10 Schematic da/dn vs ΔK curves showing the acceleration of crack growth at intermediate ΔK levels.

details depend on the environment, exact loading scheme and the intrinsic characteristics of the alloy. It has been shown that high temperature Ti alloys such as Ti–6–2–4–2 can exhibit accelerated FCGR if the peak stress is sustained during the loading cycle. The details of this hold time effect are fairly complex, but have been shown to be temperature dependent. Currently, the best explanation for this effect is stress assisted migration of hydrogen to the crack tip region with resulting local reduction of crack extension resistance. Hold time allows hydrogen enrichment of the crack tip region because hydrogen moves to sites of tension stress.

The intended message here is that designing to account for environmentally accelerated fatigue crack is a complex process and requires detailed knowledge of the actual operating parameters of the component. Under such circumstances, it is relatively easy to end up with non-conservative designs with undesirable life characteristics. The most common way to avoid this is to count on design margin (alternatively known as safety factors) to mask the lack of detailed knowledge of the operating environment. In those instances where there is incentive to reduce the design margin for weight or other reasons, the risk of incurring non-conservative designs is greater where environmental effects are involved.

β *alloys:*

The fine microstructure of β alloys that imparts the high strength levels, also leads to a relative microstructural insensitivity of crack growth behavior. The factors that lead to roughness induced closure are not generally relevant to alloys such as Ti–10–2–3 at ultimate strengths ⩾ 1250 MPa, i.e. in the strength range where there currently is the greatest interest in these alloys. The only exception is the ω-phase strengthened condition which leads to intense strain localization with an attendant increase in fracture surface roughness in fatigue fractures. This localization also leads to very low ductility so this exception has little practical value.

In fact, the FCGR of Ti–10–2–3 in the α strengthened condition at ⩾ 1250 MPa strength levels is typically equivalent to the least crack growth resistant microstructures of Ti–6–4. Thus for crack growth limited structures, the FCGR of alloys like Ti–10–2–3 becomes a limiting factor. Considerable work has been done to attempt to improve the crack growth characteristics of Ti–10–2–3, but very little improvement has been observed.

(c) Summary

The intent of this section has been to illustrate with examples that the number of variables and factors that affect the fatigue behavior of Ti alloys is considerably greater than for static properties (strength, ductility and toughness). There are useful rules to help overcome this increased degree of complexity. These rules work reasonably well when supplemented by conservative design assumptions. The difficulties which can arise are usually associated with introduction of unrecognized variables which can create unanticipated, non-conservative circumstances. Where highest structural efficiency is a prime consideration, the risk of inadvertently creating components that do not meet the design intent in terms of life is increased. In such cases, the underlying design philosophy can make or break a design. One example is the use of shot peening but designing to properties of unpeened material. This creates a significant margin which can offset the unexpected or unforeseen introduction of non-conservative factors. If the decision is taken to remove this margin to improve structural efficiency, then the associated risk is increased considerably unless the details of all the operating conditions (stresses, temperature and environment) are known with considerable precision. Reduction of this margin also increases the importance of using material that is produced under tight process control so that the desired microstructure and properties are faithfully reproduced and that no unwanted microstructural anomalies of the type described earlier are present in the material.

3.4.3 Time dependent properties

Time dependent properties, as the term is used here, refers to time dependent plastic deformation (creep), time dependent crack initiation (hydrogen induced cracking) and time dependent crack extension (stress corrosion cracking). Among these, creep is the most important, but it seems appropriate to also include a few comments about the other properties.

(a) Creep

The creep behavior of Ti alloys is the primary factor that limits the maximum use temperature of Ti alloys. There also is an issue with room temperature creep in Ti alloys and while this is seldom a limiting factor, it can pose a problem for highly stressed parts where very tight dimensional tolerances are required. In the case of room temperature creep, it has been shown quite clearly that small, time dependent plastic strains can develop at room temperature. It also has been shown that the amount of strain reaches a constant value in a relatively short time (tens of hours) after which no further strain is developed. This is called exhaustive creep, since the creep rate tends to zero with increasing time. The propensity for exhaustive creep in Ti alloys depends on the microstructure, in particular the residual dislocation density that remains after processing and heat treatment. It appears that the mobile dislocations glide at relatively high stresses until they become pinned or locked in the general substructure. Since the temperature is too low to permit recovery, this dislocation structure is stable and no further dislocation motion occurs. At this point the creep rate becomes zero and the maximum achievable creep strain has occurred. As stated earlier, this phenomenon is only important under limited circumstances and can be avoided by sustained loading prior to final machining if the small creep strain is troublesome from a dimensional control standpoint.

Elevated temperature creep of Ti alloys can be separated into several discrete contributions. Phenomenologically, these are primary creep strain and secondary creep strain, but the latter has two distinct components. Mechanistically, primary creep appears to be closely related to the exhaustive creep described above, except that recovery allows it to continue for a longer time and the total strain is larger. The secondary creep is divided into boundary sliding creep and steady dislocation creep. The microstructural and compositional factors that affect these two are quite different. The addition of small amounts of Si has been recognized as beneficial to creep strength for many years. This recognition was initially based on empirical observations. The beneficial role of Si on creep resistance has more recently been shown to promote dynamic strain aging of dislocations in the α phase. It also is possible that Si strengthens the β phase and reduces α/β boundary sliding by increasing the strength of the β film along these boundaries. Early data obtained for Ti–6–2–4–2 showed that $\approx 0.08 - 0.1$ wt% wt% Si was adequate to impart significant improvement in creep strength to this alloy. Other high temperature alloys such as Ti–1100 and IMI 834 exhibit excellent creep resistance; these alloys have higher concentrations of Si, typically 0.25–0.5 wt%. At Si concentrations $\geqslant 0.15$ wt%, silicide particles are typically present, especially in alloys that contain 2–4 wt% Zr (Zr is common in high temperature alloys). These particles affect the processing response of the alloy, but do not seem to contribute to creep strength in a first order way. The other contribution to creep is boundary sliding. This appears to be a diffusion controlled process since the biggest factors are the composition of the β phase and the shape of the α laths. Microstructures consisting entirely of elongated plates of α phase have the highest creep strength, presumably because these plates cannot slide over each other as easily as equiaxed grains. Secondly, the composition and volume fraction of the β phase appears to be another important factor affecting creep. In this regard, low residual Fe ($\leqslant 0.5$ wt%) has been shown to improve the creep strength when all other factors are held constant. It is well established that Fe is a potent β stabilizer, partitions strongly to the β phase and diffuses rapidly in the β phase. Variations of only a few hundredths of a per cent of Fe in the overall alloy composition can increase the Fe content β phase dramatically by alloy partitioning. Increased Fe also increases the β phase volume fraction and simultaneously decreases its resistance to diffusional flow.

Thus, it could be concluded that β processed microstructures are the best for creep limited applications. If the creep requirement is by far the most demanding, this is the right solution. However, this is seldom the case, and other constraints such as LCF or strength also are secondary limiting factors. Recall that the β processed colony microstructure has lower strength and poorer LCF; these may become limitations even though the creep limitation is addressed. Until fairly recently, the designer was forced to compromise by balancing creep strength against yield strength and LCF life due to the occurrence of strain localization. There is now an alloy, IMI 834, which eases this compromise. This alloy contains a small amount of carbon, present as a deliberate addition. The effect of carbon is to permit processing IMI 834 very close to the β transus without producing the colony microstructure. This alloy appears to combine the LCF and yield strength benefit of α + β processing with very good creep strength. The creep strength would appear to be derived from the presence of 0.35 Si and ≈ 0.7 wt% Nb coupled with adherence to very low Fe concentrations. Nb is a low diffusivity element and retards the diffusional flow of the β phase.

3.4.4 Stress corrosion cracking

The environmental cracking resistance of Ti alloys when exposed to chlorides is divided into two service regimes: low temperature, aqueous environments and elevated temperatures where moisture is basically absent.

Ti alloys were originally thought to be immune to aqueous stress corrosion cracking (SCC) because they did not exhibit any tendency to fail when exposed to aqueous NaCl solutions using the types of sustained load tests that had been developed to evaluate high strength steels and aluminum alloys. It wasn't until the mid-1960s that it was discovered that many α + β Ti alloys, including Ti–6–4, were susceptible to stress corrosion cracking in aqueous solutions containing 3.5% NaCl if they contained a sharp stress concentration such as a fatigue pre-crack. This discovery stirred a large research and evaluation effort, because, among other things, it affected the choice of material for the US supersonic transport. There are several excellent reviews of the phenomemology and proposed mechanisms of SCC of Ti alloys [9], so this subject will be treated only superficially here.

The application of Ti alloys can, under some circumstances, be limited by the susceptibility of these alloys to SCC. This phenomenon causes the load bearing capability of an alloy such as Ti–8–1–1 to be reduced by

50% or more if the design limiting factor is SCC failure. Unlike high strength steel and aluminum alloys, SCC of Ti alloys usually requires a crack. This allows the susceptibility to be described in terms of the threshold value of the mode I stress intensity factor (T_I) for crack extension. In an aqueous NaCl solution, this threshold K_I value is designated K_{ISCC}. Using this parameter, alloys and microstructures within a single alloy can be rank ordered with regard to their susceptibility to SCC. More recently, K_I–crack velocity curves have been shown to be a more discriminating ranking method, but the use of K_{ISCC} is still a qualitatively helpful parameter that is widely cited. Using K_{ISCC} as a measure, several susceptibility trends can be summarized as follows.

- Alloys containing higher Al or oxygen tend to have lower K_{ISCC} values.
- The presence of grain boundary α tends to lower K_{ISCC}.
- Microstructures that have continuous α-phase are more susceptible and have lower values of K_{ISCC}.
- The β phase in $\alpha + \beta$ alloys is immune and acts to arrest and blunt stress corrosion cracks.
- Alloys with texture have far more directionality in K_{ISCC}, with the lowest value corresponding to loading along the direction with the maximum density of basal plane poles.

Although laboratory tests clearly demonstrate that Ti alloys exhibit time dependent crack extension in aqueous NaCl environments, there have not been well documented stress corrosion field failures in Ti alloys, unlike the situation for high strength steel and aluminum alloys. The reasons for this are not entirely clear, but the requirement for a sharp pre-crack in addition to stress and environment may be sufficiently restrictive that all of these conditions have not been widely met in actual hardware.

At elevated temperatures, chloride induced cracking occurs in Ti alloys. This is known as hot salt stress corrosion cracking (HSSCC). HSSCC appears to be quite different form aqueous cracking described earlier. First, cracking can occur in the absence of a notch or pre-crack. Second, the simultaneous exposure of Ti alloys to salt and temperature appears to introduce reversible damage in the material. That is, elevated temperature exposure to NaCl either unstressed or at stresses too low to cause cracking, causes the material to exhibit a ductility loss during tensile testing. If exposure is followed by removal of the salt and vacuum annealing, complete recovery of the material properties occurs. This reversibility suggests that hydrogen is involved, since hydrogen is the only known deleterious species that can be introduced into and removed from Ti alloys. The suggested mechanism for hydrogen introduction is a hydrolysis reaction between NaCl and moisture in the air. There have not been documented examples of HSSCC in actual service, and it has been suggested that the cyclic nature of service temperatures may account for this fact. Nevertheless, laboratory tests show unambiguously that HSSCC is a real phenomenon and there is no reason to believe that it cannot occur in service under the right combination of temperature, stress and exposure time.

Similar embrittling behavior is observed when Ti alloys are exposed to Ag at elevated temperatures. This also appears to be related to hydrogen which forms during exposure to NaCl. In the case of silver, AgCl forms first followed by the hydrolysis reaction. Silver plated fasteners are often the source of this reaction, and there have been numerous examples of HSSCC due to silver in real hardware made of Ti alloys. The elimination of contact with silver in elevated temperature service Ti structure is the surest way to avoid Ag induced HSSCC. This has not always proved to be easy in practice and service failures continue to appear periodically.

3.5 Ti-BASED INTERMETALLIC COMPOUNDS

There is growing interest in the class of materials based on two of the intermetallic compounds that form between Ti and Al, Ti_3Al (known as α_2) and TiAl (known as γ). These two alloy classes or material systems have different characteristics, yet they have been collectively given the label 'titanium aluminides'. This inappropriately vague terminology only tends to introduce further confusion into discussion of these materials. As was the case with many of the peripheral subjects discussed earlier, a detailed treatment of this subject could comprise an entire chapter. Given the relative lack of maturity or readiness of these materials for application in production components or structures, only a summary will be included here.

The situation regarding these materials is changing rapidly because of the large effort currently underway in many laboratories around the world. If this effort is successful, the situation with regard to real applications could change rapidly and in 18 months a very different picture may exist.

The attractiveness of these materials will be described later in this section. The author believes that, eventually, these materials will be widely used as lightweight substitutes for Ni based alloys and to extend the temperature capability of Ti-based alloys. Before this occurs there are numerous technical barriers to overcome and there is no real production base for these alloys at present. The current level of maturity is approximately equivalent to that of Ti alloys in the mid-1960s. The changing military-industrial base suggests that, in the US at least, there will be fewer public funds to advance the implementation of α_2 and γ alloys. This will make it more difficult to reduce these materials to practice and bring them to full production status. Focused investment can make niche applications a reality fairly quickly. In the current economic environment, some significant volume niche applications are needed badly. Such applications can help induce the capital investment needed to bring these alloys to the degree of maturity that permits a broader application range and that approaches the maturity exhibited by conventional Ti alloys today. Absent such application opportunities, and these materials may languish for a long time as a laboratory curiosity.

3.5.1 General characteristics of α_2 and γ alloys

The principal attraction of both of these alloy systems when compared to conventional Ti alloys is lower density, higher modulus, better oxidation resistance and retention of strength to higher temperatures. The general characteristics of the two material classes, α_2 and γ, as represented by the binary Ti–Al compounds, are shown in Table 3.9. From these data it is clear that the major benefits are strength and stiffness and that the major drawbacks are lower room temperature ductility and toughness, particularly for the γ alloys, lower strength up to ≈ 425 °C. Both of these characteristics are well known in many other intermetallic compounds, so they come as no surprise. The potential difference between the Ti-based intermetallics and most others is the attractive combinations of strength, stiffness and density. These characteristics make the ongoing R&D investment directed toward ameliorating the ductility and strength issues a sensible one, provided it is clearly understood to be a long term, high risk / high pay-off proposition. There are also a host of unanswered questions regarding the metallurgy and production of these materials, especially as the compositions become more complex as a result of the development efforts aimed at finding improved alloys.

These two compounds can be and have been alloyed with a variety of β stabilizers to alter their characteristics, particularly room temperature ductility. As in most situations, alloying produces a trade-off between properties and seldom does an alloying addition of 'universal goodness'

Table 3.9 General characteristics of two families of Ti intermetallic compounds

Property	*Conventional Ti*	Ti_3Al	*TiAl*	*Ni-base*
Density (g/cm^3)	4.54	4.28	3.83	8.45
Stiffness (GPa)	110	145	175	206
Max temp./creep (°C)	535	815	900	1095
Max temp./oxidation (°C)	595	650	955	1095
Room temp. ductility (%)	≈ 20	2–4	1–3	3–10
Operating temp. ductility (%)	High	5–12	5–12	10–20

emerge. One result of these alloy development programs is that a variety of 'preferred' compositions have emerged over the past 10–15 years. Thus, while the properties in Table 3.9 are illustrative of the attractions of these two material classes in their binary form, they cannot be used as binary compounds. Clearly, specific properties must be evaluated for each alloy composition. For example, the most promising α_2 alloys have as much as 24 wt% Nb in them for improvement in ductility, but such additions considerably increase the density and lower the modulus of these alloys. Another attractive aspect of these materials is that the high Al concentration, particularly in γ alloys, results in the formation of Al_2O_3 surface oxide. This is much more protective than the TiO_2 that forms on conventional Ti alloys. The effect of this is seen in the significant increase in maximum use temperature for oxidation limited applications, particularly for the γ alloys.

α_2 *Alloys*

There are now several generations of α_2 alloys under study and evaluation around the world. The time-based succession of representative compositions for each generation is (approximately) as follows, with all compositions stated in atom percent:

- Ti–25Al–5Nb (Ti–25–5)
- Ti–24Al–11Nb (Ti–24–11)
- Ti–25Al–10Nb–3V–1Mo (Ti–25–10–3–1)
- Ti 24.5Al–12.5Nb–1.5Mo (Ti–25–13–2)

These particular alloys are ones that are being actively evaluated in the US. Other compositions are being evaluated in other countries. The above list is only intended to illustrate the evolutionary compositional trends, rather than to 'tout' any particular alloy.

It is interesting to note that the second generation alloy contains significantly more Nb, which was added for the express intent of improving the room temperature ductility of Ti–25–5. It has been shown that the low temperature brittleness of the binary compounds occurs because they fracture by cleavage. This is much like mild steel at temperatures below the ductile brittle transition temperature. The microstructural factors which affect the occurrence of cleavage in mild steel have been studied in detail for many years. The results of these studies, coupled with observations of the effect of Nb on microstructure, have been used to guide the modification of Ti–25–5 toward higher Nb content. Nb is a stronger β stabilizer in α_2 alloys than it is in conventional Ti alloys. The effect of β stabilizer additions such as Nb is to permit the microstructure of α_2 alloys to be more widely manipulated by processing and heat treatment, particularly the latter. In particular, Nb additions permit the formation of a fine basket weave, or Widmanstatten structure, during cooling from above the β transus. The scale of this structure depends on cooling rate but, even at relatively slow rates, the microstructural unit (α_2 plates) size is reduced by more than ten-fold. With 11 atom% Nb, Ti–24–11 also contains a small volume fraction of residual β phase. The β phase is much more ductile than the α_2 and behaves as a ductile inclusion. In combination, the reduced microstructural scale and the presence of β phase produce room temperature ductility in the range of 4%, which is within the lower limit of acceptability to most designers.

The subsequent generations of α_2 alloys have been developed to retain the benefits realized in Ti–24–11 but to permit the strength to be improved and the density to be reduced by substituting V, Mo and other lower density, less expensive β stabilizers for Nb. These efforts have been reasonably successful as can be seen from the data for Ti–25–10–3–1 shown in Table 3.10. These data show that the strength of the third generation alloy is better and that the fracture related properties have not been degraded or compromised. The Ti–25–10–3–1 alloy also has a somewhat lower β transus, which makes it more amenable to forging. Compared to the older Ti–25–5 alloy, all the subsequent generation alloys have proven to be more workable, consistent with the higher volume fraction of ductile β phase that is present at the working temperature. At present it has been shown that all product forms can be made from Ti–24–11. However, the yields are low compared to conventional Ti alloys and the amount of conditioning required during the production of plate and sheet is excessive from an economic standpoint. The current situation is that plate, sheet and even foil can be made from current generation α_2 alloys, but the economics of these operations are unattractive. This further increases the pay-off requirement for an attractive application. In turn, the increased pay-off requirement tends

Table 3.10 Room Temperature Properties of Newer Generation α_2 Alloys

Alloy/Micro	σ_y *(MPa)*	σ_u *(MPa)*	*Elong (%)*	*R of A (%)*	K_{IC} *(MPa m$^{1/2}$)*
Ti–24–11 – β processed – fine micro	611	703	1.3	2.6	≈ 20
Ti–25–10–3–1 – β processed – coarse micro	556	745	3.5	6.7	21
Ti–25–10–3–1 – α + β processed – fine micro – 10% α_p	753	967	3.5	6.7	20
Ti–25–10–3–1 – α + β processed – coarse micro – 10% α_p	581	821	6.5	8.7	21
Ti–25–10–3–1 – α + β processed – coarse micro – 25% α_p	563	806	7.0	9.9	21

to underscore the need for a stronger alloy in order to increase the potential for weight savings. The best solution to this matter is the use of the higher strength, next generation alloys. In this regard, the Ti–25–13–2 alloy is stronger than the earlier generation and has better creep resistance because Mo has replaced V in Ti–25–10–3–1. Alloys such as Ti–25–13–2 also have considerably better environmental resistance because they do not contain V.

Finally, it has recently become clear that Ti–25–10–3–1 and Ti–24–11 both exhibit ductility losses when tensile tested in air at temperatures above ≈ 650 °C. This effect has been shown to be related to oxygen interaction with the specimen surface during the test. The seriousness of this discovery is still under examination, but it may turn out that any of the α_2 alloys known today will require coating for elevated temperature service, at least above ≈ 650 °C, where the greatest pay-off is for α_2 alloys. If this turns out to be the case, it will be another economic deterrent to the use of these alloys.

γ Alloys

The γ alloys are the more attractive among the two alloy classes because they have higher temperature capability, lower density and higher modulus. These alloys also are the higher risk and have proven to be less amenable to improvement through alloying. Nevertheless, significant progress has been made. As in the case of the α_2 alloys, multiple generations of alloy compositions exist today. Again, in rough chronological order, these include the following, where, again, all compositions are in atom%:

- Ti–48Al–1V (Ti–48–1)
- Ti–48Al–2Mn (Ti–48–2)
- Ti–48Al–2Cr–2Nb (Ti–48–2–2)

The Ti–48–2–2 alloy consistently exhibits > 2% room temperature ductility, which makes it borderline from a design standpoint. However, this alloy has a very flat temperature dependence of strength and appears to be an attractive alternative to wrought Ni base superalloys at temperatures around 700 °C. The room temperature strength is relatively low and this limits the types of application which can be considered for this alloy. The alloy has been shown to be readily castable into relatively complex shapes. It also has been forged, but this requires considerable care. One benefit of the forged (wrought) version of this alloy is that the finer grain size in the wrought materials provides a significant strengthening contribution while maintaining or increasing the ductility of the alloy. On the other side, wrought, fine grained Ti–48–2–2 has lower creep strength, which also limits the application opportunities. In those applications where low temperature strength is important and minimum sustained high temperature loads are involved, wrought γ may be a good solution. The major limiting factor at present is the cost of γ forgings. This is related to the producibility issues, some of which will be mitigated as an increased experience base is developed with these materials. This will take time, however, and it appears that the highest probability for initial application will be cast (+HIP) γ components. Unlike the α_2 alloys, the γ alloys do not appear to require coating. This is an attractive feature and probably is the result of the nature of the adherent, protective Al_2O_3 layer that forms on the surface of these alloys.

3.6 SUMMARY

In this chapter an attempt has been made to discuss a wide range of issues related to the applications and use of Ti alloys in high strength structures. The intent is not to create specific expertise to permit the reader to become a practicing materials application engineer. Rather, it is to create an awareness of the interactions between microstructure, processing and properties. It also attempts to call attention to the various aspects of Ti production practice. Without faithful execution of production practice under carefully controlled circumstances, all the microstructure understanding available will not be sufficient to yield reproducible performance of Ti components. A wise man once said:

> Materials scientists still believe that microstructure controls properties, but a good metallurgist knows that microstructural defects really are controlling.

In high performance structures such as those designed and manufactured in the aerospace industry, this is certainly true. Progress toward higher performance structures and better structural efficiency will require better materials of construction, both in concept and in execution. It will also require an improved understanding of the operating conditions under which these components are used.

ACKNOWLEDGEMENTS

The careful reading and constructive comments of my colleagues, Walt Buttrill, Jim Chesnutt and Cliff Shamblen is gratefully acknowledged. The assistance of Brian Marquardt in providing some of the α_2 data is also acknowledged.

REFERENCES

1. Chesnutt, J. C., Thompson, A. W., and Williams, J. C., (1978) Influence of Metallurgical Factors on the fatigue Crack Growth Rate in Alpha – Beta Titanium Alloys, AFML-TR-78-68, Air Force Materials Laboratory, May.

2. Chesnutt, J. C., Thompson, A. W., and Williams, J. C., (1981) Fatigue crack propagation and fracture of titanium alloys, *Titanium '80, Science and Technology* (eds H. Kimura and O. Izumi), Vol. 3, TMS–AIME, Warrendale, Pa p. 1875.
3. Chesnutt, J. C., Rhodes, C. G., and Williams, J. C., (1976) The Relationship Between Mechanical Properties, Microstructure and Fracture Topography in $\alpha + \beta$ Titanium Alloys, ASTM STP 600, ASTM, Philadelphia, p. 99.
4. Williams, J. C., and Starke, Jr., E. A., (1984) The role of thermomechanical processing in tailoring the properties of aluminum and titanium alloys, *Deformation, Processing and Structure* (ed. George fKrauss), ASM, Metals Park, OH, p. 279.
5. Duerig, T. W. and Williams, J. C., (1984) Overview: microstructure and properties of β–Ti alloys, *Beta Titanium Alloys in the 1980s* (eds R. R. Boyer and H. W. Rosenberg), TMS-AIME, Warrendale, Pa, p. 19.
6. Kuhlman, G. W., Pishko, R., Kahrs, J. R., and Nelson, J. W., (1984) Isothermal forging of beta and near beta titanium alloys, *Beta Titanium Alloys in the 1980s* (eds. R. R. Boyer and H. W. Rosenberg), TMS-AIME, Warrendale, Pa, p. 259.
7. Williams, J. C., and Luetjering, G., (1981) The effect of slip length and slip character on the properties of titanium alloys, *Titanium '80, Science and Technology* (eds. H. Kimura and O. Izumi), Vol. 1, p. 671, TMS–AIME, Warrendale, PA.
8. Allison, J. E., (1982) PhD Thesis, Carnegie Mellon University, Pittsburgh, Pa.
9. Blackburn, M. J., Smyrl, W. H., and Feeney, J. A., (1972) Stress corrosion cracking of Ti alloys, in *Stress Corrosion Cracking in High Strength Steels, Titanium and Aluminum Alloys* (ed. B. F. Brown), Naval Research Laboratory, Washington DC, p. 246.

FURTHER READING

This is not a comprehensive list; it is a good place to look for additional information. A computer based literature search will show many more articles, many of which are useful, but may also be highly specialized.

General Ti alloy metallurgy, processing and properties

The proceedings of the qradriennial International Conferences on Ti Alloys:

1968 (London): *The Science, Technology and Application of Titanium*, (eds. R. I. Jaffee and N. E. Promisel), Pergamon Press, London, 1970.

1972 (Boston): *Titanium Science and Technology*, Vol. 1–3 (eds. R. I. Jaffe and H. M. Burte), Plenum Press, NY, 1973.

1976 (Moscow): *Titanium and Titanium Alloys, Scientific and Technological Aspects*, Vol. 1–3 (eds. J. C. Williams and A. F. Belov), Plenum Press, NY, 1982.

1980 (Kyoto): *Titanium '80, Science and Technology*, Vol. 1–4 (eds. H. Kimura and O. Izumi), TMS-AIME, Warrendale, Pa, 1981.

1984 (Munich): *Titanium, Science and Technology*, Vol. 1–3 (eds G. Lütjering, U. Zwicker, W. Bunk), Deutsche Gesellschaft für Metallkunde e.V., 1985.

1988 (Cannes): *Sixth world Conference on Titanium – Proceedings* (eds. P. Lacombe, R. Tricot and G. Beranger), les éditions de physique, 1989.

Ti alloying and phase transformation behavior

Williams, J. C., (1973), Phase transformations in titanium alloys: a review, *Titanium Science and Technology*, Vol. 3, p. 1433 (eds. R. I. Jaffe and H. M. Burte), Plenum Press, NY, 1973.

Fracture and fatigue of α + β *Ti alloys*

Chesnutt, J. C., Rhodes, C. G., and Williams, J. C., (1976) The Relationship Between Mechanical Properties, Microstructure and Fracture Topography in α + β Titanium Alloys, ASTM STP 600, ASTM, Philadelphia p. 99.

Williams, J. C., and Starke, Jr., E. A., (1984) The role of thermomechanical processing in tailoring the properties of aluminum and titanium alloys, *Deformation, Processing and Structure* (ed. George fKrauss), ASM, Metals Park, OH, p. 279.

Margolin, H. Chesnutt, J. C., Luetjering, G., and Williams, J. C., (1981) Fracture fatigue and wear: critical review, *Titanium '80, Science and Technology*, (eds. H. Kimura and O. Izumi), Vol. 1, TMS–AIME, Warrendale, Pa, p. 169.

Chesnutt, J. C., Thompson, A. W., and Williams, J. C., (1978) Influence of Metallurgical Factors on the fatigue Crack Growth Rate in Alpha – Beta Titanium Alloys, AFML–TR–78–68, Air Force Materials Laboratory, May.

β–*Ti alloys*

Beta Titanium Alloys in the 1980s (eds R. R. Boyer and H. W. Rosenberg), TMS–AIME, Warrendale, Pa, 1984.

Texture effects

Properties of Textured Ti Alloys, F. Larson and A. Zarkades, MCIC Report 74–20, June 1974.

Research Toward Developing an Understanding of Crystallographic Texture on Mechanical Properties of Titanium Alloys, A. W. Sommer and M. Creager, AFML TR 76–222, Air Force Materials Laboratory, January 1977.

4

Nickel-based alloys: recent developments for the aero-gas turbine

M. McLean

4.1 BACKGROUND

The principal requirements for materials for the hot-end of gas turbines for aerospace applications are:

- high melting point
- good oxidation/corrosion resistance
- high temperature strength
- microstructural stability at high temperature
- low density
- high stiffness
- good fabricability
- acceptable cost
- reproducible performance

Nickel-based alloys have evolved as the metallic materials with the best combination of these properties for these particular applications in gas turbines.

The evolution of nickel-based alloys for aero-engines, commonly known as superalloys, over the past 50 years [1,2] has been largely driven by the need to improve the efficiency of the gas turbine for aerospace applications. In common with all heat engines, the thermodynamic efficiency of the gas turbine is improved if the temperature and/or pressure of the working fluid can be increased. Figure 4.1 shows how the temperature capability of nickel-based superalloys has progressively increased with the year in which various developments were introduced; the measure of temperature capability is taken to be the maximum

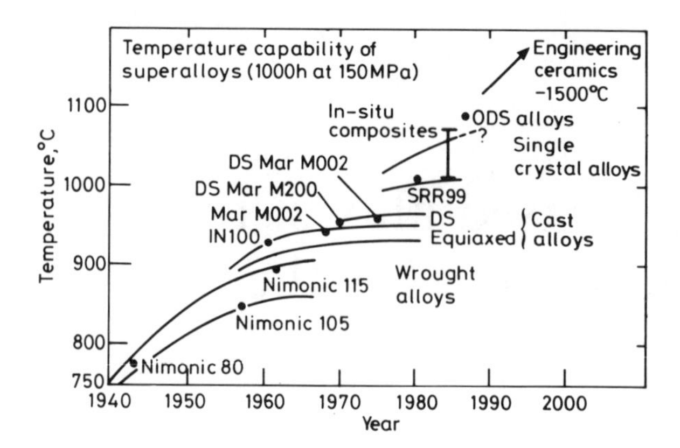

Figure 4.1 Increased temperature capability of nickel-based superalloys, shown as the temperature for 1000 h life with a stress of 150 MPa, as a function of the year of introduction into service [3].

temperature at which the alloy can achieve a life of 1000 hours with an applied stress of 150 MPa, which is a fairly typical service condition for high pressure (HP) blades in an advanced aero-engine. Several points can be drawn from Figure 4.1.

- The origin of the timescale is taken as 1940. The first gas-turbine powered flights were in 1940 and 1941 in Germany and England respectively, and the first true precipitation-strengthened superalloy Nimonic 80 was available commercially shortly afterwards. Consequently, both aero-gas-turbines and nickel-based superalloys are roughly 50 years old.
- The progressive increase in temperature capability of superalloys has occurred in waves associated with different methods of processing the alloys into shaped components, e.g. forging, investment casting, directional solidification, single crystal production. Alloys specific to each type of process have been developed with progressively increased strengths. However, the major advances have resulted from the introduction of new processes in combination with appropriate alloy development.
- There are different requirements for alloys operating in different parts of the gas turbine and in gas turbines for different applications (e.g. aero propulsing land based power generation). Consequently, superalloys with different relative combinations of high temperature strength and corrosion resistance, for example, have been developed. Thus, new alloys with apparently inferior temperature capability to some earlier materials, when measured in terms of creep rupture life, will usually have improvements in other measures of performance.
- Modern superalloys operate in service at temperatures up to – 1350 K which is within 200 K of their incipient melting temperatures. Taking the melting temperature of pure nickel (1728 K) as an

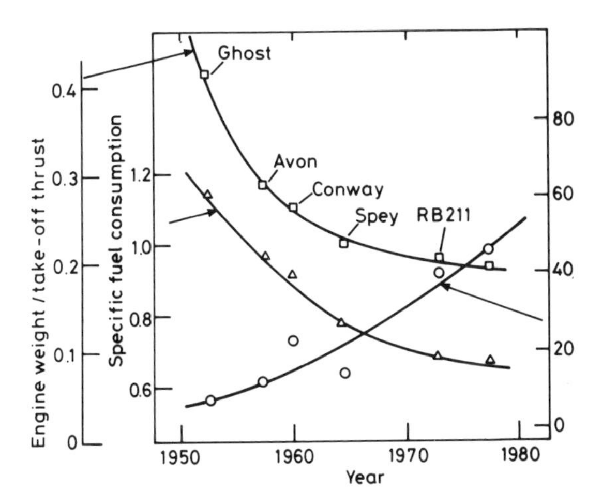

Figure 4.2 Efficiency of aero-engines manufactured by Rolls Royce, measured as the ratio of engine weight to take-off thrust, as a function of the year of introduction into service. The periods of use of wrought, investment cast, directionally solidified and single crystal blades are indicated (after Higginbotham [4]).

indicator of the absolute ceiling to the development of these materials, it is clear that there is little scope for further substantial improvements in the temperature capability of these materials. However, there continue to be significant benefits from modifying the alloys to enhance other aspects of the material performance.

- The combination of alloy development with major engineering developments have led to very large improvements in efficiency. In particular, the use of air cooling in turbine blades has enabled them to operate in streams of gas at temperatures exceeding the alloy melting temperature. Consequently, the capability of reproducibly forming intricately shaped components is crucial to superalloy development.

Figure 4.2 shows how the efficiency of the aero-engines produced by Rolls-Royce [4] has been progressively improved over the past 50 years and how these developments relate to the process developments of nickel-based superalloys. The same trend would be shown by the products of other companies.

4.2 ALLOY CONSTITUTION AND DEVELOPMENT TRENDS

The early superalloys were based on the nichrome alloys (80/20 Ni–Cr) that were being used for electric heating elements in the 1930s. Indeed the solid solution nickel alloy Nimonic 75 is a minor modification of nichrome alloy with small additions of titanium (0.4 wt%) and carbon

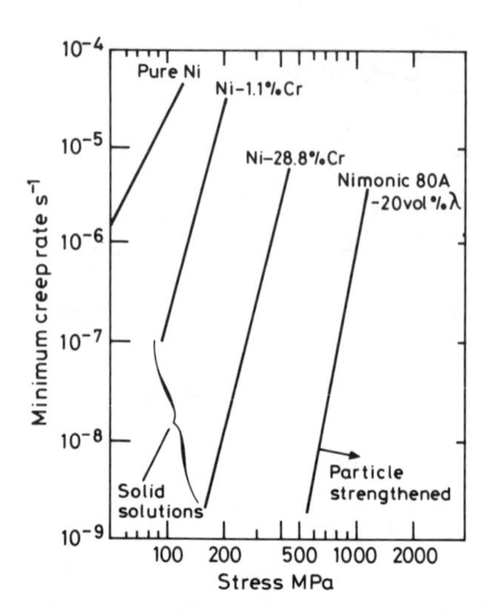

Figure 4.3 A comparison of the minimum creep rate as a function of applied stress at 600 °C for nickel, nickel–chromium solid solution alloys and Nimonic 80A.

(0.12 wt%) to produce carbides that give grain size control and some additional creep strength. However, the breakthrough in the development of nickel-base alloys came when it was recognized that relatively small additions of aluminium and titanium led to very large increases in high temperature strength. Figure 4.3 shows that the earliest of the Al, Ti modified alloys, Nimonic 80A, has creep rates of about 10^{-3} times those of solid solution (Ni, Cr), alloys such as Nimonic 75, at equivalent stress and temperature; the rupture life is inversely related to the creep rate.

As with the development of many other engineering alloys, specific compositions have been defined through a combination of intuition, empiricism or, at best, a broad impression of the likely effects of various modifications on a relevant property. Basic metallurgical principles have been used to develop multi-component alloy systems that make use of solid solution strengthening, grain boundary stabilization by carbides, oxidation/corrosion enhancement, etc. However, the key feature that imparts high temperature strength to superalloys is that the aluminium and titanium forms a high volume fraction of the intermetallic phase γ' Ni_3 (Al, Ti) that is precipitated within the γ (Ni, Cr) solid solution at relatively high temperatures. The earliest precipitation strengthened materials, such as Nimonic 80, contained about 10% by volume of the γ' phase which dissolved in the γ at temperatures above about 850 °C at which point there was a severe loss of creep strength. The dominant theme to the development of high temperature superalloys has been to increase the volume fraction of γ' and at the same time increase its

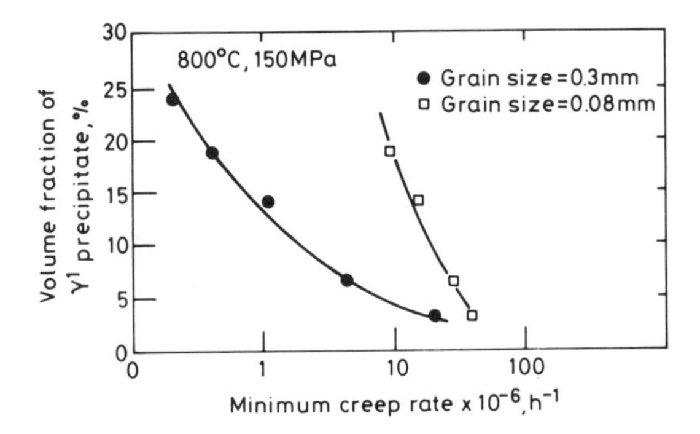

Figure 4.4 Increase in creep strength with increasing γ' volume fraction for an experimental family of nickel-based superalloys [5].

solution temperature. This has the dual advantages of increasing the level of creep strength and of maintaining it to higher temperatures. As a measure of the success of this strategy the most modern single crystal superalloys (e.g. SRR 99) have ~ 70% vol of γ' and a γ' solvus temperature of ~ 1200 °C. The extent of the dependence of creep performance on γ' volume fraction is shown in Figure 4.4 [5].

Control of grain size has also been a key factor in optimizing the high temperature mechanical behaviour of superalloys, as it has been for other engineering alloys. Figure 4.4 also shows clearly that the creep performance of superalloys is enhanced by increasing the grain size; by contrast the fatigue behaviour deteriorates as the grain size increases. Thus, it is necessary to tailor the processing conditions to give the best properties for a given component. For gas turbine blades that are relatively lightly stressed at very high temperatures, creep is the life limiting property, whereas for the disks on which the blades are attached which operate at lower temperatures and with higher stresses the fatigue behaviour is most important. Consequently blades generally have large grain sizes and disks have fine grain sizes.

Many alloys have been developed to improve properties other than the high temperature strength. In particular, oxidation of hot-salt corrosion behaviour depends on retaining high chromium contents which tend to lower the γ' solvus temperature. The highest strength superalloys have been achieved through drastic reduction in the chromium level with a consequent deterioration in environmental resistance. This is shown quite clearly in Figure 4.5 which compares the effect of chromium content of a wide range of superalloys on both corrosion resistance and stress rupture performance.

The conflicting requirements of high temperature strength and corrosion resistance cannot be separately optimized in nickel-based super-

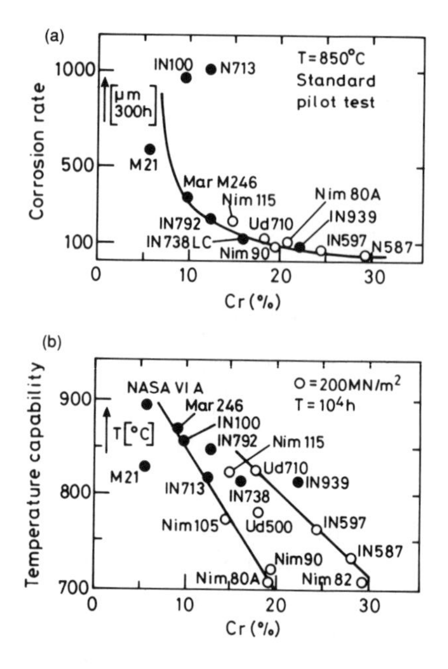

Figure 4.5 Influence of chromium content on (a) corrosion resistance; (b) creep rupture strength of nickel-based superalloys.

alloys. However, sophisticated procedures have been developed that enhance the corrosion resistance of high strength superalloys by the application of various protective coatings. It is beyond the scope of this chapter to review the science and technology of corrosion resistant coatings. However, the reader should be aware that a wide range of types of coatings have been, and continue to be developed specifically for use on superalloys for gas turbine applications. These range from surface aluminizing, deposition of relatively thick (– 1 μm) overlay coatings of a different alloy composition (e.g. CoCrAlY) by vapour deposition to covering the surface with a layer of ceramic (thermal barrier coating).

Heat treatments to control the γ' size, shape and distribution are thought to be of considerable importance in optimising the mechanical properties. Detailed heat treatment specifications, involving both solution and ageing components, are normally available for each individual alloy. It was possible to achieve complete solution of γ' in the early, low γ' volume fraction alloys so that a homogeneous distribution of γ', often bimodal, could be achieved during ageing. However, with many more advanced alloys (e.g. IN738, Mar M002) complete γ' solution could not be achieved because incipient melting occurred at a temperature below the γ' solvus. A major advantage of the most modern superalloys that have been developed specifically for use as single crystals is that the melting range has been increased by removing extraneous grain boundary strengthening elements giving a practicable heat-treatment window in which to manipulate fully the γ′ morphology. Caron and Khan [6] have clearly shown the benefits of optimizing the heat treatment of the single crystal alloy CMSX2, which appears to be associated with the regularity of distribution of the γ′ particles rather than their size. Indeed, since the ageing temperature is often similar to, or in excess of, the likely service temperature the γ′ will be considerably modified during operation; the long term properties, in particular, arc rather insensitive to initial γ′ size.

Many of the properties of nickel-based superalloys are particularly sensitive to the presence of minor concentrations of certain elements and, as a consequence, compositional control is of particular importance in the production of nickel-based superalloys. Among the detrimental species are the metalloids (Bi, Pb), sulphur and oxygen which, in extreme cases, can reduce the creep rupture life and ductility by as much as a factor of ten with concentrations of a few parts per million. Demanding specifications are placed on the alloy producers to keep these elements at very low levels and these are satisfied by using advanced processes such as primary vacuum induction melting that may be followed by a secondary remelting process such as electro-slag remelting (SR) or vacuum arc remelting.

Other elements present in minor concentrations can have very beneficial effects on a range of properties, although in most cases the mechanisms leading to the improved properties have not yet been clearly established. The addition of 'magic dust' to produce various property alloys seems to be as prevalent now as it was in the early days of alloy development. Among the most important of the minor additives in nickel-based superalloys are:

- Hafnium: This is now almost a standard additive in concentrations of ~½% to high strength cast alloys in order to increase the high temperature ductility. Its importance, first identified by Lund at Martin Marietta Corp, appears to be to remove embrittling elements from grain boundaries and to fix them in stable precipitates in the crystal.
- Yttrium and rare earths: The oxidation resistance of some materials is greatly increased by these elements at concentrations of less than 1%. The effect is not uniform to all superalloys and the principal role appears to be to enhance the adherence of the oxide layer inhibiting oxide spallation.
- Rhenium: This is the most recent additive to attract considerable attention. It has been shown to enhance considerably the high temperature strength of a range of superalloys. There is no clear evidence of the strengthening mechanism, although it is thought to be a particularly potent solid solution strengthener.

4.3 PROCESSING DEVELOPMENTS

The early superalloys were processed into useful shapes by thermo-mechanical processing – indeed forging is still the preferred method of manufacturing many components. Forging is carried out at high temperature where the alloy is relatively weak and ductile, i.e. where the γ' phase is in solution. This generally requires a working temperature range of about 200 °C between the γ' solvus and the alloy melting temperature. A range of forging techniques is currently used to produce turbine blades and disks of complex external and internal geometry.

By the early 1960s alloy compositions had been identified for which the heat treatment window was too narrow to allow industrial forging to be undertaken. This required use of an alternative method of producing shaped components. The ancient craft of lost-wax casting, which has allowed many ancient civilisations to produce artefacts of high precision, has developed into the modern investment casting process in which the microstructure and integrity of superalloy castings can be controlled. Superalloys with high γ' volume fractions can now be cast directly as complex shapes, often with complex internal cooling channels. This remains an important industrial process for a wide range of gas turbine components. However, as superalloy strength was increased even further some alloys exhibited very low ductilities in the conventionally cast form that were unacceptable for engineering applications, although their intrinsic strength and corrosion resistance were attractive. The low ductilities in cast superalloys were almost invariably associated with premature fracture at grain boundaries that lay transverse to the applied stress. Ver Snyder and Guard [7] pioneered the use of directional solidification of the castings to produce an aligned grain structure in which the intergranular fracture was suppressed. (Such directional solidification had been achieved much earlier by Northcott [8] on a range of alloys with apparently no specific application in mind.) The directionally solidified materials show much higher ductilities (~ 25%) than the conventionally cast alloys (~ 2%) and this contributes to much improved creep rupture and thermal fatigue resistance.

Although the impetus for the development of directional solidification technology has been the control of the grain morphology, the major benefit with respect to thermal fatigue resistance results from the crystallographic texture produced by the process. The cyclic stress $\Delta\sigma$ induced in a turbine blade due to differential expansion of different regions of the blade during heating and cooling cycles over a temperature range ΔT is approximately represented by:

$$\Delta\sigma = \alpha E \Delta T$$

where α is the coefficient of linear expansion and E is Young's modulus. The stress range that determines the fatigue life is therefore proportional to E which varies significantly with crystal direction. The columnar grain structure in directionally solidified superalloys results from the rapid growth of crystals in the ⟨001⟩ direction; this gives a very sharp crystallographic texture in the solidification direction. By good fortune, E has a minimum value in the ⟨001⟩ direction being about 20% smaller than in equiaxed, non-textured superalloys.

It was a relatively small technological step to produce single crystal components by producing a geometrical constraint in the early part of the directionally solidifying casting. This allows only one of the many ⟨001⟩ grains that emerge by competitive growth to develop into the main casting. An example of a single crystal together with a starting block with columnar grains and a spiral selector is shown in Figure 4.6. There is little advantage of single crystal over directional solidification for normal

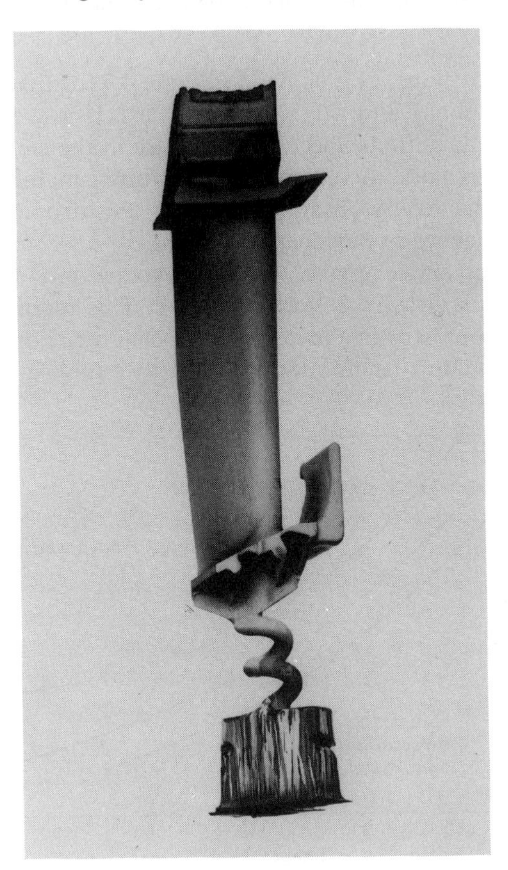

Figure 4.6 A single crystal turbine blade casting of the superalloy SRR99 showing a starter block with columnar grains produced by directional solidification and a spiral grain selector that allows an angle ⟨001⟩ oriented grain to penetrate to the main casting (courtesy of Rolls Royce plc).

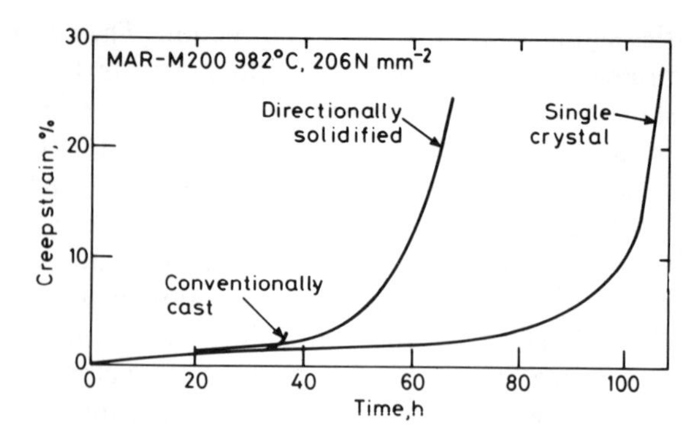

Figure 4.7 Comparison of creep curves for MarM 200 tested at 1255 K and 206 MPa in conventionally cast, directionally solidified and single crystal forms [9].

superalloy compositions, as is shown in Figure 4.7 for the alloy MarM200 in equiaxed, columnar and single crystal forms. However, a new family of superalloy compositions specifically for use in the single crystal form has led to very dramatic increases in temperature capability (Figure 4.8). This has been achieved by producing a highly anisotropic material which creates new challenges in engineering design.

The benefits of single crystal superalloys can only be used if their integrity in complex castings can be assured. This requires the use of a wide range of sophisticated inspection procedures. For example, real-time X-ray diffraction techniques are used to establish that the crystal

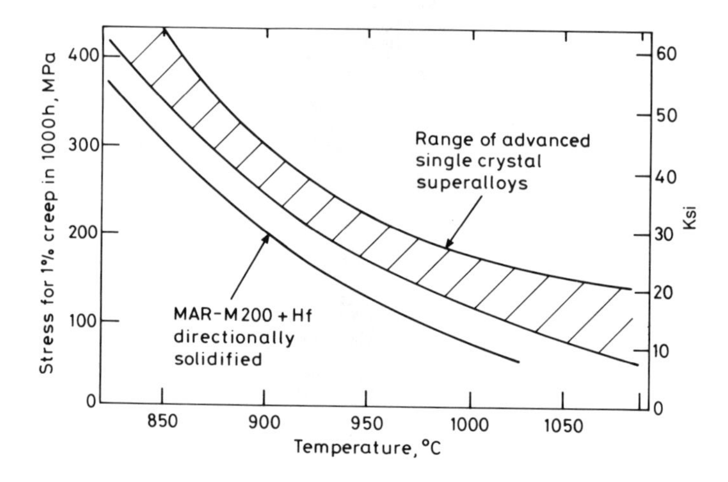

Figure 4.8 Comparison of the variation of creep strength as a function of temperature for the most advanced directionally solidified superalloys and the new generation of single crystal compositions.

orientation falls within the required range (generally within about 7 ° of ⟨001⟩). Of just as much importance is that there has been no accidental grain nucleation, particularly in some of the intricate cooling channels. Clearly, since grain boundary strengthening elements that were present in the earlier alloys have been removed for single crystal alloy compositions, these materials would be particularly weakened by such stray grains. These inspection requirements are a major factor in the relatively high cost of single crystal components.

In the more highly stressed components such as turbine disks which operate at much lower temperatures than do turbine blades, fatigue rather than creep is likely to limit the service life. This requires a small grain size so that the investment casting, directional solidification and single crystal processes are inappropriate. Rather, forging and powder metallurgy in conjunction with advanced secondary melting techniques, such as electron beam cold hearth refining, plasma melting and electroslag refining, have been developed to combine material homogeneity, grain distribution and inclusion control. The latter is proving to be of particular importance in ensuring component integrity with respect to brittle fracture.

4.4 MICROSTRUCTURE AND HIGH TEMPERATURE DEFORMATION

The unique characteristics of nickel-based superalloys largely derive from the mechanical behaviour of the γ' particles and from the very unusual crystallographic relationship between the γ and γ' phases.

Superalloys containing high volume fractions of the γ' phase, when tested at high stress levels, exhibit an increase in yield stress with increasing temperature[10] to a maximum temperature of 700 °C after which the yield stress decreases as shown in Figure 4.9. (This type of behaviour is also shown in tests on monolithic Ni_3Al, and on other ordered intermetallic compounds.) This anomalous behaviour occurs at stresses where dislocations originating in the γ phase can also shear the γ' particles. However, at lower stresses which are more appropriate to the conditions experienced in service, other deformation modes are likely to be dominant; in particular, dislocations may be restricted to the minority γ phase so that they can only move by climbing around the γ' particles.

The γ and γ' phases are both face-centred-cubic, have almost identical lattice dimensions and the two lattices are similarly oriented. Consequently, the two phases are almost totally coherent. In the γ-nickel all lattice sites are totally equivalent, the atoms constituting the solid solution being distributed randomly. However, the γ' phase is ordered with the nickel-atoms being situated in the faces and the aluminium/titanium atoms at the corners of the face centred cubic lattice.

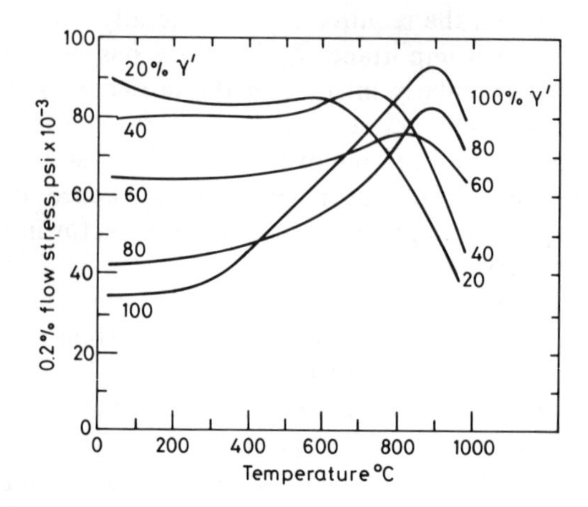

Figure 4.9 Temperature dependence of yield strength of nickel-based alloys with various volume fractions of γ' precipitate [10].

Slip in the primitive fcc lattice occurs by dislocations with a ½ $\langle\bar{1}01\rangle$ Burgers vector moving on a {111} plane. This leaves the γ lattice unaffected, but creates a fault in the stacking sequence in the LI_2 lattice of the γ' phase. This fault, known as an anti-phase-boundary (APB) is annihilated by the passage of a second similar dislocation. Thus the unit dislocation in the LI_2 superlattice has a Burgers vector of $\langle\bar{1}01\rangle$. Where deformation involves γ' cutting, the dislocations must move in pairs. This requires an additional stress to overcome the repulsion of like dislocations in γ'.

Slip with a ½$\langle\bar{1}01\rangle$ displacement can occur on the {010} plane producing a smaller APB than that associated with {111}. Consequently, there is an energy advantage in the dislocation cross-slipping from {111} to {010}. This cross-slip provides a barrier to dislocation glide on {111} and, therefore, locally increases the yield stress.

The cross-slip events are thermally activated and so their density increases with increasing temperature. Since each cross-slip increases the local yield stress, there is a progressive increase in the overall yield stress with increasing temperature. The maximum occurs when the rate of removal of pinning points by diffusional recovery exceed the rate of increase of pinning points by cross-slip.

There has been an extensive literature on the creep deformation of metals and detailed mechanisms have been proposed which generally require that a steady state dislocation substructure be established. However, the creep behaviour of superalloys differs from those of simple metals in a number of respects that suggest that the mechanism of high temperature deformation is quite different from that in pure metals. The most important differences can be summarized as follows:

- *Shape of creep curve*
  Superalloys do not exhibit an extensive steady state but have a progressively increasing creep rate over most of the life that is not associated with obvious damage, such as cavitation or cracking that are thought to cause tertiary creep in simple metals.
- *Stress σ and temperature T dependence*
  The steady state creep rate $\dot{\varepsilon}_{SS}$ of simple alloys are often represented quite well by a power law and Arrhenius equation:

$$\dot{\varepsilon}_{SS} = A\sigma^n \exp(-Q/RT)$$

  where the stress exponent $n \sim 4$ and activation energy $Q \sim$ the self diffusion coefficient Q_{SD} as predicted by various theoretical analyses. However, attempts to fit this equation to creep data for superalloys requires the use of values of n ($\geqslant 8$) and Q ($\sim 3Q_{SD}$) that are physically unrealistic.
- *Effect of prestrain*
  When a simple metal is subject to plastic deformation before creep testing, the material is strengthened (i.e. the creep rate is reduced). Prestrain weakens superalloys with respect to creep.
- *Dislocation density*
  In simple metals, the dislocation density increases during primary creep to a steady state configuration that is maintained over most of the creep life. Superalloys, however, show a progressive increase in dislocation density over all of the creep life; at all stages the dislocation densities are very low.

These, and other observations, indicate that existing models of recovery controlled creep that have been successfully applied to simple metals are inappropriate for superalloys. A model of creep of superalloys has been developed by Dyson and McLean and Barbosa *et al.* [11,12] that envisages that all dislocations are mobile but are constrained to move predominantly by climb rather than glide due to the presence of relatively impenetrable particles. The features of the creep behaviour described above are accounted for in the following way:

- The increasing creep rate with increasing creep strain is the result of generating additional mobile dislocations since $\dot{\varepsilon} = \rho b v$, where v is the dislocation velocity and b is the Burgers vector of the dislocations.
- The high stress and temperature dependence of creep rate results from dislocation velocity being determined by reaction rate kinetics which leads to an exponential rather than power law expression for creep rate.
- The effect of prestrain is to generate additional mobile dislocations which increase the creep rate.

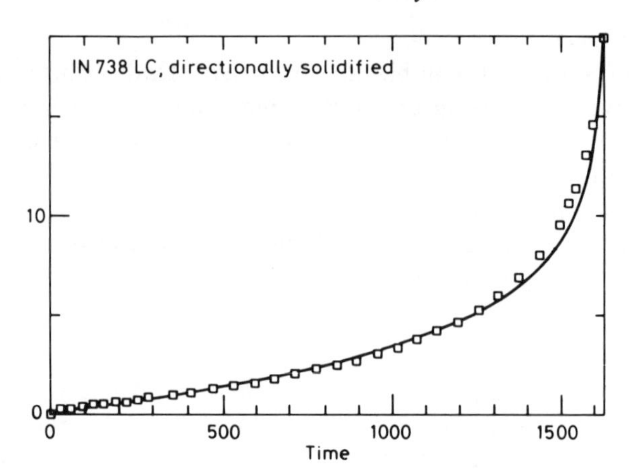

Figure 4.10 Comparison of the curve generated by the optimum fit of equation set 4.1 to experimental creep data for the directionally solidified superalloy IN738LC tested at 200 MPa and 1223 K.

The model outlined above has been used to develop physics-based constitutive laws for high temperature deformation. Here the model is expressed in terms of dimensionless state variables S and w rather than the microstructural quantities such as dislocation density that cannot be measured easily in engineering practice. For superalloys the appropriate equations are:

$$\begin{aligned} \dot{\varepsilon} &= \dot{\varepsilon}_i(1 - S)(1 + w) \\ \dot{S} &= H\dot{\varepsilon}_i(1 - S) - Rw \\ \dot{\omega} &= C\dot{\varepsilon} \end{aligned} \tag{4.1}$$

A creep curve which may consist of 100 or more data points is fully determined by the four constants ($\dot{\varepsilon}_i$, H, R, C). Figure 4.10 shows the fit of this equation set to some typical creep curves. A database of such constants as a function of stress and temperature can allow the strain evolution to be predicted in quite complex and varying stress/temperature conditions.

4.5 TURBINE DISK APPLICATIONS

The principal requirements for a turbine disk alloy can be summarized as follows:

- high tensile strength to prevent yield and fracture in overstressed conditions
- resistance to fatigue crack nucleation
- slow crack nucleation rates

- high fracture toughness
- adequate creep resistance for higher temperature operation (~ 700 °C)
- high ductility to allow relaxation at stress concentrations
- corrosion resistance

As the loading on turbine blades has increased due to the successful development of high temperature nickel-based superalloys, described above, increasing demands have been made on the turbine disks to which the blades are attached. Here the operating temperature (~ 600 °C) is much lower than for the blades, but the stresses experienced are considerably higher. For disks there is normally no significant creep problem and design is carried out in relation to the yield characteristics of the material. Consequently, the strategy in developing alloys for disk applications has been to progressively increase the yield strength. However, it is well known that there is an inverse relationship between the yield strength and the fracture toughness of materials; an unfortunate consequence of this relationship is that the fracture characteristics of disk alloys, particularly in fatigue conditions, have become progressively sensitive to the presence of unwanted defects rather than being controlled by the inherent strength of the alloy. This is potentially a very serious problem since an unexpected fracture of a gas-turbine disk in service would be catastrophic, almost inevitably leading to loss of life. By its nature it is unpredictable, being determined by the chance occurrence of a damaging defect. It is therefore essential to have reliable techniques for non-destructive valuation and quality control during disk manufacture in order to control and be aware of the defect density.

There is growing evidence that in the most advanced disk alloys, such as the wrought alloy IN718 and powder materials such as Astroloy, non-metallic inclusions that are introduced in various stages of the material processing provide a limit to their low cycle fatigue performance. This is illustrated in Figure 4.11 which shows that the low cycle fatigue performance of the disk alloys RENE95 decreases dramatically with increasing maximum inclusion size when the inclusion size exceeds 25 μm; below this level the LCF behaviour *of this material* is relatively insensitive to the presence of inclusions [13]. However, for alloys with higher yield strengths than RENE95, the critical defect size would be expected to be smaller.

Pineau [14] has proposed that the presence of inclusions obviates the need for crack initiation and that fatigue crack growth from the largest inclusion determines the life. He has developed a probabilistic approach in which the most important parameters are the statistical distribution of inclusions and the fatigue crack law. One important consequence of this approach is that large components will have inferior LCF behaviour to small components (or mechanical test specimens) because there is a

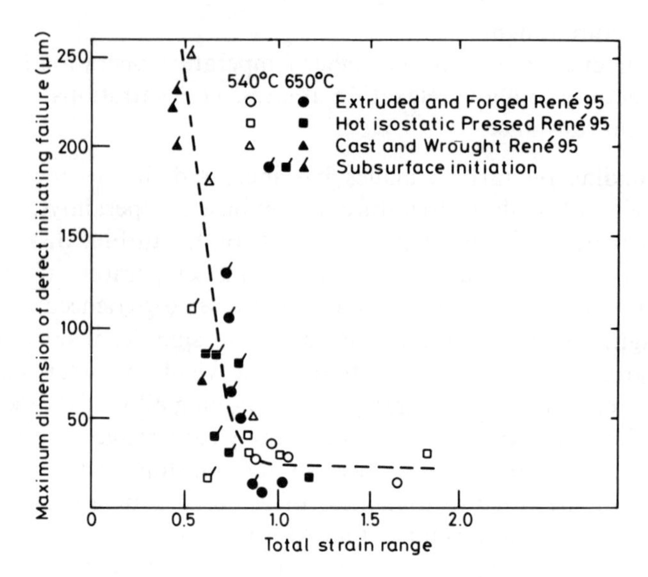

Figure 4.11 Variation in strain range to cause fracture in 10^4 cycles during low cycle fatigue of IN781 containing various sizes of non-metallic inclusions [13].

higher probability of there being large inclusions present. This is illustrated in Pineau's calculations shown in Figure 4.12.

There have been two distinct and related approaches to improving turbine disk alloys. The first has been to use 'super-clean technology' to remove, as far as possible, the detrimental non-metallic inclusions. The second has been to recognize that some low inclusion content is inevitable and to devise approaches to characterizing and controlling the level of alloy cleanness.

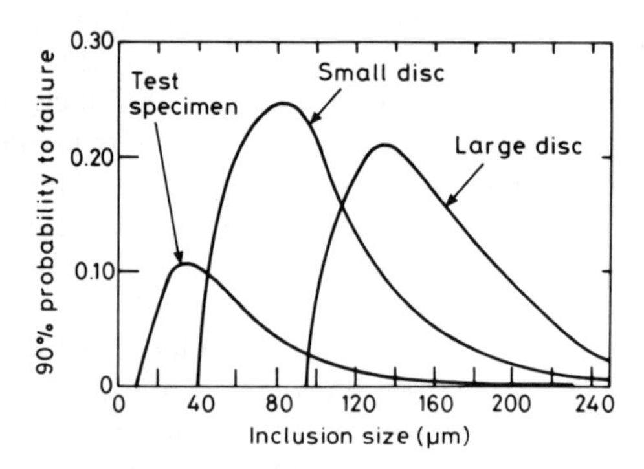

Figure 4.12 90 % probability of failure as a function of inclusion size calculated for three geometries for IN718 and a stress of 1050 MPa [14].

It is beyond the scope of this chapter to detail the philosophy and practice of superclean technology. It involves a disciplined approach to the manufacturing process to both minimize the introduction of unwanted inclusions and to remove the few that are inevitably introduced. In the production of aerospace materials, melt filtering during the primary alloy production is now used extensively. However, growing attention is being given to an additional cleaning cycle involving secondary melting. A range of secondary melting techniques is being considered including:

- vacuum arc remelting (VAR)
- electro-slag refining (ESR)
- electron beam cold hearth refining (ECBCHR)
- plasma melting with cold hearth refining

The cold hearth refining procedures are gaining increasing prominence; Figure 4.13 shows schematically the EBCHR process. The original bar stock is drip melted, by electron beam heating in a relatively high vacuum, into a flat horizontal water-cooled copper hearth from which the melt flows into a vertical mould. The aim is to let the inclusions float to the surface of the melt in the cold hearth where they can be physically separated by a system of weirs and barriers. The bar-stock produced by EBCHR can be used to produce disks directly by a forging process or as the material source for powder produced by, for example, inert gas atomization where inclusions can also lead to inferior LDF behaviour.

As the advanced disk alloys have been progressively refined, the number and size of inclusions present in them has become too low to be

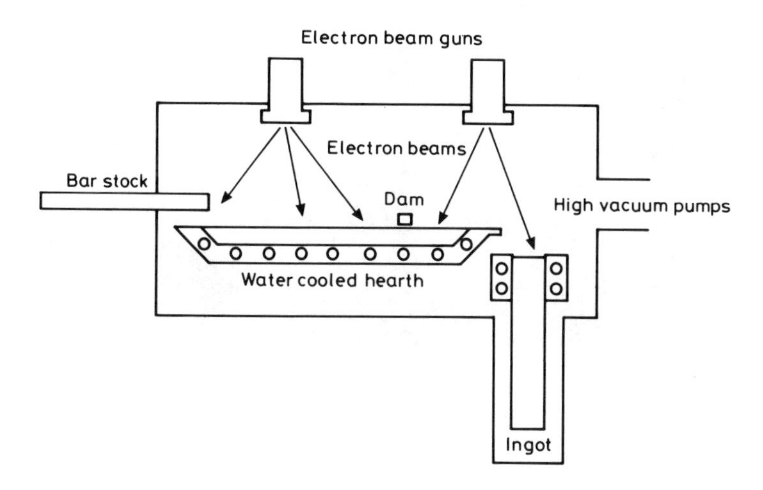

Figure 4.13 Schematic illustration of the electron beam cold hearth refining (EBCHR) process.

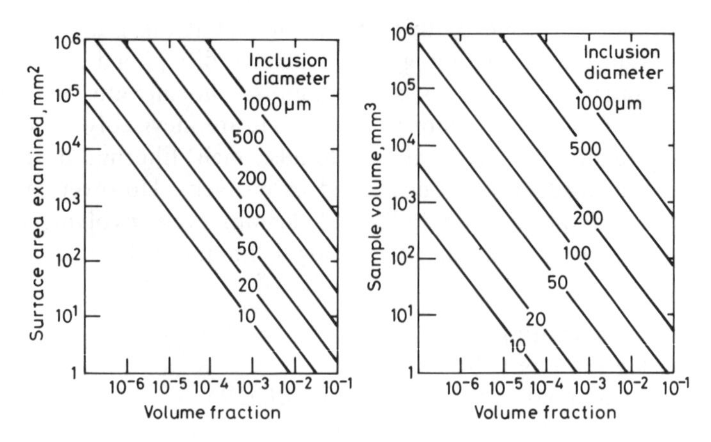

Figure 4.14 Surface area and volume of material that must be examined to give 80% confidence in measuring the volume fraction of inclusions of various diameters assuming a random distribution [15].

Figure 4.15 Typical electron beam button melting (EBBM) specimen of IN718 showing the final solidified cap in which inclusions have been concentrated [16].

detected by conventional metallographic techniques or non-destructive evaluation [15]. This is illustrated in Figure 4.14, which shows the areas and volume of material with different inclusion sizes that must be examined in order to obtain a measure of the inclusions volume fraction with an error of ± 20%. However, even at ppm levels, the defect characteristics can determine the fatigue and fracture behaviour. Various new methods of characterizing the inclusion content in superclean material have been investigated. One of the most promising has been the electron beam button melting (EBBM) technique [16]. In this approach, about 1 kg of the alloy is drip melted by electron beam heating into a hemispherical crucible where it is solidified in a controlled manner; the inclusions from the entire volume concentrate in the final liquid portion which solidifies as a cap where they can be examined in some detail. Figure 4.15 shows a typical EBBM specimen and Figure 4.16 indicates the level of detail in particle size and type that can be achieved.

4.6 FUTURE PROSPECTS

There will inevitably be further improvements in the chemistry and processing of nickel-based superalloys. For example, greater sophistication in the use of protective coatings will enable the stronger, but inherently

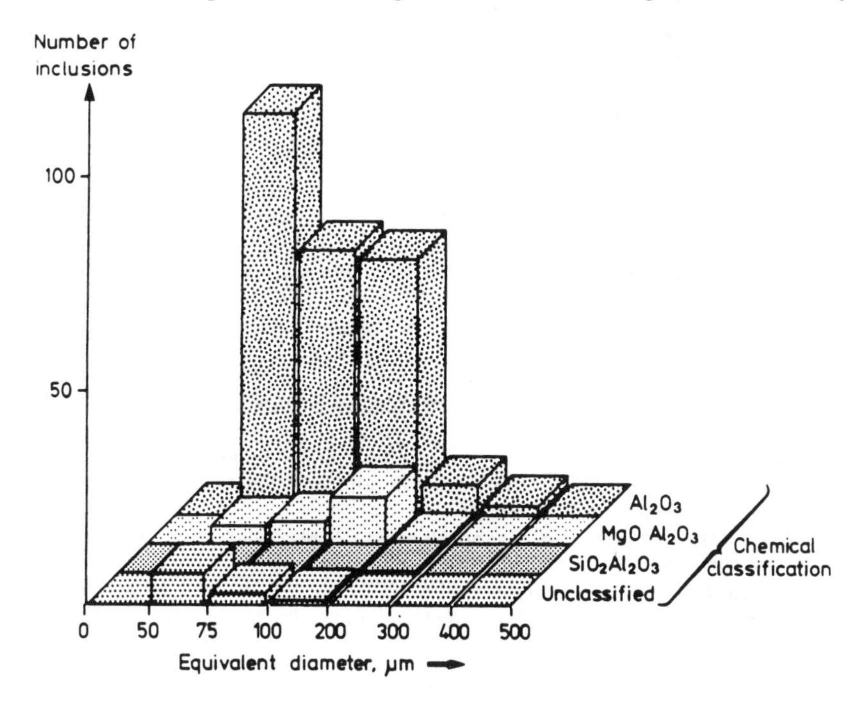

Figure 4.16 Analysis of the inclusions in the cap of an EBBM specimen of IN718 showing the distribution of sizes of inclusions of different chemistries that were originally distributed throughout 1 kg of the alloy [17].

corrosion-prone, alloys to be used in more demanding environments. (Indeed, coating technology is already critical to the use of nickel-based alloys in aerospace; space limitations have prevented this aspect being treated in this chapter). However, the melting point of nickel places an effective upper limit on the temperature capability of solid-solution based alloys.

The major thrust of recent research on nickel-based materials has been on intermetallics. The most advanced superalloys are 70 vol% γ solid solution. The matrix phase is the 30 vol% γ solid solution. The monolithic γ' material has inferior mechanical properties, as have other intermetallics such as NiAl. However, there are indications that multiple phase intermetallic systems will be identified that have the benefit of the high melting point of some intermetallics with attractive mechanical properties. The great challenge is to impart adequate toughness in intermetallics that are inherently brittle.

REFERENCES

1. Sims, C. T. and Hagel W. C. (eds) (1972) *The Superalloys*, Wiley, New York.
2. Betteridge, W. and Shaw, S. W. K. (1987) *Mater. Sci. Tech.*, ***3***, 682.
3. McLean, M. (1983) in *Directionally Solidified Materials for High Temperature Service*, Book No. 296, The Metals Society, London.
4. Higginbotham, G. J. S. (1986) *Mater. Sci. Tech.*, **2**, 442.
5. Gibbons, T. B. and Hopkins, B. E. (1984) *Metal Science*, ***18***, 273.
6. Caron, P. and Khan, T. (1983) *Mat. Sci. Eng.*, ***61***, 173.
7. Ver Snyder, F. L. and Guard, R. W. (1960) *Trans. ASM*, ***52***, 485.
8. Northcott, L. (1938) *J. Inst. Metals*, ***62***, 101.
9. Kear, B. H. and Piearcey, B. J. (1967) *Trans. Metall. Soc. AIME*, ***239***, 1209.
10. Copley, S. M. and Kear, B. H. (1967) *Trans. AIME*, ***239***, 984.
11. Dyson, B. F. and McLean, M. (1990) *ISIJ International*, ***30***, 802.
12. Barbosa, A., Taylor, N. G., Ashby, M. F. *et al.* (1988) *Proc. 5th Int. Symp. on Superalloys*, (ed. D.N. Dehl *et al.*), Metall. Soc. AIME, New York, p. 683.
13. Miner R. V. and Gayda, J. (1984) *Int. J. Fatigue*, ***6***, 189.
14. Pineau, A. (1990) in *High Temperature Materials for Power Engineering*, Part III, (ed. E. Bachelet *et al.*), Miner Academic Publishers, Dordrecht, pp. 913–4.
15. Chare, J. *International Symposium on Quantitative Metallography, Florence, 21–23 November, 1978*, Associated Metall, pp. 209–24. Ass. Italiana di Metallurgia.
16. Chakravorty, S., Peak, M. S. and Quested, P. N. (1990) *Proc. Euromat '89, 22–24 November, 1989*, DGM Informationsgesellschaft, Aachen, Germany.
17. Quested, P. N. and Chakravorty, S. (1990) *Proc. Electron Beam Melting and Refining – State of the Art 1989, Reno, Nevada, 29–31 October, 1989.*

5

Structural steels

D.P. Davies

5.1 INTRODUCTION

Although the use of steels in the aeronautical field has been gradually declining over the years, they still account for approximately 10% of the structural weight in fixed and rotary winged aircraft. Furthermore, the applications for which they are used are considered highly critical, in which failure of one of these vital components could jeopardize the safety of the aircraft.

The highest concentration of steel usage in aircraft applications appears to be (a) gearing, (b) bearings and (c) undercarriage applications, hence the chapter will specifically relate to these three areas.

5.2 GEAR STEELS

5.2.1 Helicopter transmission design

The helicopter designer faces the problem of reducing by some 100:1 the rotational speed from a power turbine to that required by the main rotor. In helicopters this is achieved by the main gearbox which reduces the high speed turbine input, turns the main drive on to a vertical axis and transmits power to the main rotor, tail drive shaft and accessory equipment (Figure 5.1).

5.2.2 Design criteria/mechanical requirements for gear materials

In helicopter transmissions it is not a simple matter to determine the critical properties in a flexible design situation. The simplest approach is to consider the common modes of failure of a given design in relation to the fundamental properties which influence that failure mode.

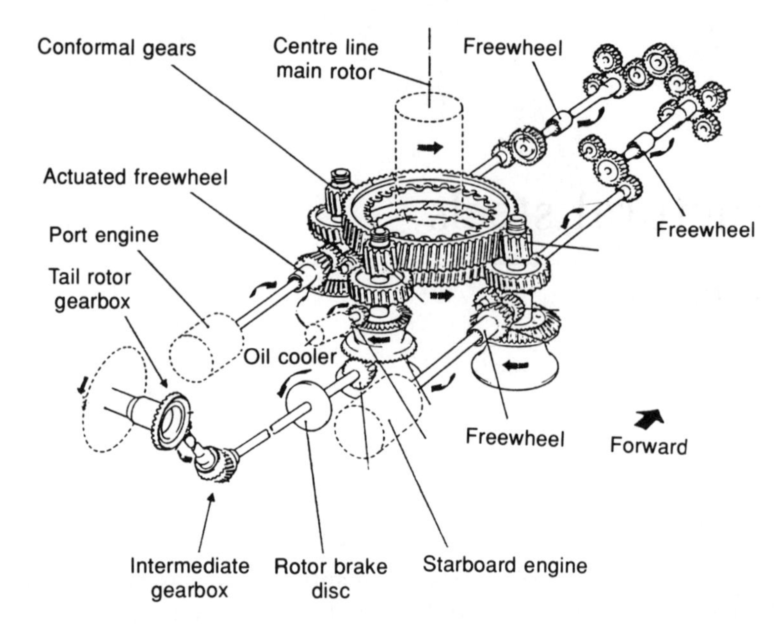

Figure 5.1 Lynx helicopter main gearbox.

In the first instance, logistic and economic requirements dictate that a single or limited number of gear steels be used. On a fundamental level, a gear steel requires:

- sufficient static and fatigue strength to withstand instantaneous and dynamic loads
- adequate toughness to withstand shock loads
- adequate resistance to deformation, i.e. pitting, scuffing and wear
- resistance to hot oil environments (90–130 °C)

5.2.3 Gear materials

In order to meet the stringent mechanical requirements, transmission components have to be case hardened by either carburizing or nitriding. As the majority of gears are carburized the chapter will concentrate on the metallurgical/mechanical properties relating to carburizing.

Table 5.1 lists the typical gear steels used in helicopter transmission applications. As can be seen, without exception low alloy steels predominate. The most commonly used is AMS 6265 whereas the UK has traditionally relied upon a higher nickel variant known as BS S156 for higher toughness. Nitriding steels also differ with the Europeans relying on chromium as the predominant nitriding forming element while the USA uses aluminium.

Table 5.1 Chemical composition of conventional and advanced gear steels

Material identification	*Nominal chemical composition* (wt%) *C*	*Mn*	*P max*	*S max*	*Si*	*Cu*	*Cr*	*Mo*	*V*	*Ni*	*W*	*Al*	*Fe*
Conventional carburizing steels													
BS S156	0.17	0.31	0.01	0.01	0.26	–	1.14	0.27	–	4.07	–	–	BAL
AMS 6265	0.16	0.46	0.01	0.01	0.22	–	1.0	–	–	3.29	–	–	BAL
Conventional nitriding steels													
Nitralloy N	0.25	0.51	0.01	0.01	0.30	0.08	1.2	0.26	–	3.46	–	1.27	BAL
32CDV13	0.32	0.48	0.01	0.01	0.33	–	2.89	0.89	0.28	0.07	–	–	BAL
Advanced carburizing steels													
Pyrowear 53	0.1	0.37	0.01	0.01	0.98	2.07	1.05	3.3	0.12	2.13	–	–	BAL
Vasco X-2M	0.14	0.30	0.03	0.03	0.9	–	5.0	4.0	0.45	–	1.35	–	BAL

5.2.4 Metallurgical factors affecting gear performance

(a) Steel cleanness

The size, shape and type of inclusions are important since they influence the stress concentration effect [1]. In terms of fatigue resistance (endurance and crack growth) and fracture toughness each steel has a critical inclusion defect size which, if exceeded, provides a potential source of failure. Furthermore, as the strength of the steel increases so the critical defect size will decrease. At the surface strengths typical of case hardening it might be expected that the critical defect size will be extremely small. However, the compressive residual stresses that are normally present in the surface layers will permit that material to tolerate a large inclusion or defect. Beneath the case, the core material is less strong but generally tough. Since the residual stresses there are tensile, the critical defect size is necessarily much smaller than in a material containing no residual macro stress. Therefore this zone beneath the case is a potential failure zone and sensitive to the presence of non metallic inclusions. For this reason gear steels are processed using VAR melting techniques.

(b) Grain size

Austenite grain size is governed by the thermomechanical history of the material, which in turn influences the size of the martensite plates. For optimum material properties it is essential that the grain size of a carburized component is both uniform and fine. An ASTM grain size of 6–8 or finer is required because as the grain size coarsens, properties such as ductility and fatigue strength will decrease [2].

(c) Core microstructure

The microstructural constituents existing within the core of a surface hardened component are determined by the transformation characteristics of the steel and by those factors influencing the rate at which the component is cooled from the austenite region. The transformation characteristics depend mainly upon the chemical composition and to a lesser extent upon the grain size of the material, whereas the cooling rate is influenced by the section size, the quenching medium and the austenitising temperature.

Helicopter gear steels are essentially deeply hardenable such that adequate case and core hardenability is achieved in sections in excess of 300 mm. Moreover, the as-quenched core must be martensitic in nature to ensure (a) the highest possible UTS to YS ratio and (b) absence of bainite and ferrite. Small amounts of ferrite are particularly harmful in the core especially close to the case–core interface where the total active stress can greatly exceed the fatigue limit of ferrite thereby significantly lowering the fatigue strength of the gear steel.

(d) Case microstructure

Retained austenite.

In steels austenite is stable at temperatures above the Ac_3 and Acm phase boundaries and although during cooling from these temperatures it becomes unstable and will transform to martensite, retention of some retained austenite in the final microstructure is extremely likely. As alloying additions, surface carbon content and quenching temperature increase, so too will the quantity of retained austenite.

Retained austenite is relatively soft (Figure 5.2) and although it is saturated with carbon its presence in coexistence with martensite will reduce the overall hardness of the structure. The failure of the austenite to transform during quenching means that the volume expansion which accompanies the austenite to martensite reaction will be reduced, giving lower residual compressive stresses. As the resistance to fatigue failure at a case hardened surface is related to the inherent strength (hardness) of the material and degree of compressive residual stress, the presence of retained austenite would appear to be undesirable. In general a trend has been established that bending fatigue limit falls by approximately 10% for each 20–30% retained austenite left untransformed in the structure.

However, gears are not carburized solely for the purpose of improving bending fatigue resistance as resistance to rolling contact fatigue is also an important consideration. Wiegand and Tolosch [3] have shown that under rolling contact conditions, as retained austenite contents were increased up to 50%, resistance to scuffing also increased. The reason appears to be that retained austenite will transform when subjected to stress, thereby the initial surface hardness will increase and any surface irregularities will be reduced giving a more favourable distribution of load.

Figure 5.2 Retained austenite.

Figure 5.3 Grain boundary carbides.

However, two additional points need to be considered in that:

- As retained austenite is unstable, transformations occurring within a helicopter gearbox would be undesirable as it would be likely to lead to dimensional instability and thereby increase the risk of meshing problems.
- Grinding has been shown to induce cracking and microstructural modifications in steels containing retained austenite contents in excess of 20%.

For these reasons a refrigeration treatment is almost inevitably employed after hardening. Its function is to transform the retained austenite after the hardening quench, to below 20%, increase surface hardness and ensure dimensional stability during gear operation. Doubt has been expressed about the desirability of this process, with some workers suggesting that it is detrimental to fatigue resistance. Trials carried out at a number of helicopter manufacturers have, however, refuted this assumption [4].

Carbides.
In controlling the case-hardening process to obviate the formation of globular carbides the three main factors affecting carbide formation are (a) the temperature above the Ac_1, (b) the carbon potential of the atmosphere and (c) the composition of the steel. Under equilibrium cooling conditions steels having a carbon content in excess of the eutectoid concentration will be rejected from solution and precipitate out as carbides. Free carbide can markedly influence the residual stress distribution, in that the matrix loses alloying elements to the carbide,

resulting in localized depleted zones which do not undergo a martensitic transformation. Furthermore, the carbides themselves cannot undergo a martensitic transformation and therefore do not suffer the expansion of the matrix associated with the transformation. With reduced residual compression stresses or in the worst case the occurrence of tensile stresses at the surface, the bending fatigue strength of a component may be reduced by 25–30% [5], when carbides are massive or decorating the grain boundaries (Figure 5.3). However, the presence of free carbides often improves wear and scuffing resistance, and as the proportion of carbides increase, so too does the value of these properties [6]. Consequently, carbides are considered acceptable providing they are fine and uniformly distributed.

Decarburization.
Decarburization can take place if the atmosphere surrounding the hot component contains molecules of the decarburizing gases (CO_2, H_2O, H_2), which combine with carbon atoms at the gas/metal interface and thereby extract them from the metal surface in an attempt to establish some measure of surface equilibrium. The way in which the surface microstructure is influenced by decarburization depends mainly on the prior carbide distribution and quenching rate. Partial decarburization has been shown to lower residual compressive stress. Moreover, depending on the degree of decarburization a non-martensitic microstructure may be formed, producing a lower than expected surface hardness. The consequence of reducing the surface hardness and of developing unfavourable surface residual stress is to impair the bending fatigue strength [7].

Internal oxidation.
When oxidation is allowed to occur in the carburizing furnace oxygen can adsorb onto the metallic surface from which it diffuses along austenite grain and sub-grain boundaries. During oxidation not only are the substitutional elements removed but the carbon level may also be reduced, thereby affecting the steels hardenability (in the oxidized layer) and martensitic behaviour. Its effect on properties will be related to depth of oxidation and microstructural features produced; however, 20–25% reductions in fatigue strength have been reported with depths in excess of 13 μm [8].

5.2.5 Mechanical factors affecting gear performance

(a) Case depth

It is not always true that deep cases will ensure maximum protection against bending fatigue, for the following reasons:

- The development of a deep case by prolonged carburizing can produce microstructural defects such as internal oxidation, excessive retained

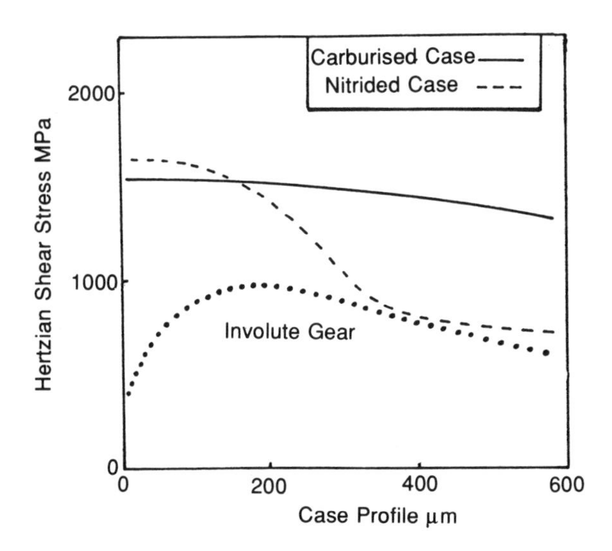

Figure 5.4 Hertzian shear stress distribution through a typical involute gear compared to the strength of a nitrided and a carburised case.

austenite and free carbides and thereby modify the residual stress distribution within the case in a manner which is not always beneficial.

- Bending fatigue strength is influenced by section size, i.e. there is an optimum value of case depth to section thickness, beyond which fatigue will fall [10].

To ensure optimum protection against case cracking (case/core separation) it is important to ensure the case depth exceeds the depth of maximum shear stress [9]. This can be seen schematically in Figure 5.4 in which the shear stress curves calculated from typical carburized and nitrided cases are compared against a typical Hertzian stress distribution for an involute gear. From this figure it can be seen that the strength of the carburized case is lower than that of the nitrided case close to the surface, conferring inferior scuffing resistance, but holds up well at greater depths, whereas the shear strength of the nitrided case declines sharply after approaching 0.3 mm. The carburized case maintains a safe working margin above the induced stresses, while the nitrided case approaches dangerously close to the strength required, inferring the possibility of case crushing. The case depth specified to achieve optimum Hertzian shear stress resistance will generally be longer than the depth required for optimum bending fatigue resistance.

(b) Surface roughness

The Hertzian stress profile discussed previously is only idealised and for many practical situations the maximum shear stress occurs at the

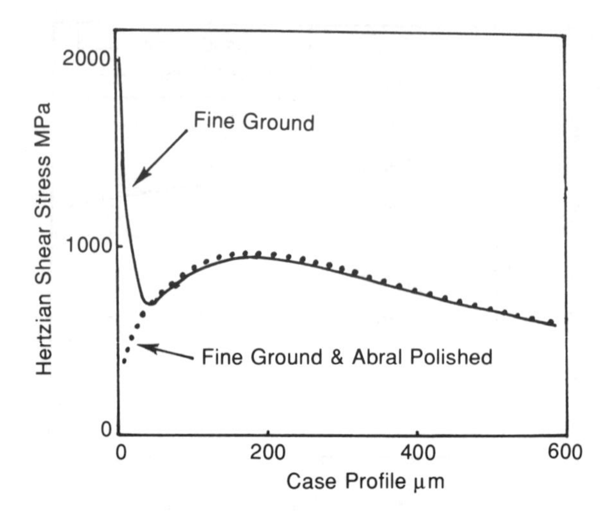

Figure 5.5 Influence of surface roughness on induced shear stress.

surface [11] (Figure 5.5). Consequently, the roughness induced near surface stresses, particularly in thin lubricating film situations, may affect the rolling contact fatigue properties associated with carburizing to a greater extent than has been realized.

(c) Surface hardness

There appears to be a relationship between surface hardness and bending fatigue resistance [12] in that as surface hardness increases up to approximately 700 HV, so does the bending fatigue limit. Above 700 HV other factors such as a core properties and case structural features become more significant. The presence of free carbides, indicative of hardnesses in excess of 800 HV, have been found to improve wear and scuffing resistance and as the proportions of carbides increase, so too does the value of these properties. For this reason surface hardness should be controlled within the range 750–800 HV.

(d) Core strength

A core of excessive strength can reduce the effectiveness of carburized case by virtue of its influence on the surface residual stresses. In addition a weak core can deform plastically under load while the case remains elastic with the result that residual compressive stresses in the case will be increased and bending fatigue strength reduced. Similarly, under heavy loading conditions, the case will not be supported by the soft core and case crushing will result.

Therefore in practice steels are selected to give sufficient core strength to resist yielding at the design loads with an adequate margin to resist the occasional overload. In general with carburized steels optimum core strength appears to vary from 1100 to 1500 MPa, depending on gear design.

5.2.6 Failure modes

As gears are structural members of a rotating mechanical system they are subjected to alternating dynamic loads and therefore can suffer fatigue failure. The two most common areas for this to occur is either at the tooth root or flank.

(a) Tooth root

Essentially the tooth root is subjected to bending or tension–compression forces from the tooth contact and therefore failure usually occurs at the surface of the root radius on the loaded side of the tooth (Figure 5.6). The origin is usually situated at the midpoint between the ends of the tooth in which the propagation direction will depend on the imposed stresses. The likelihood of bending fatigue increases with applied load.

(b) Tooth flank

Gears operating under conditions of high power, particularly at low speed, experience highly loaded contacts on the tooth flanks. It has been

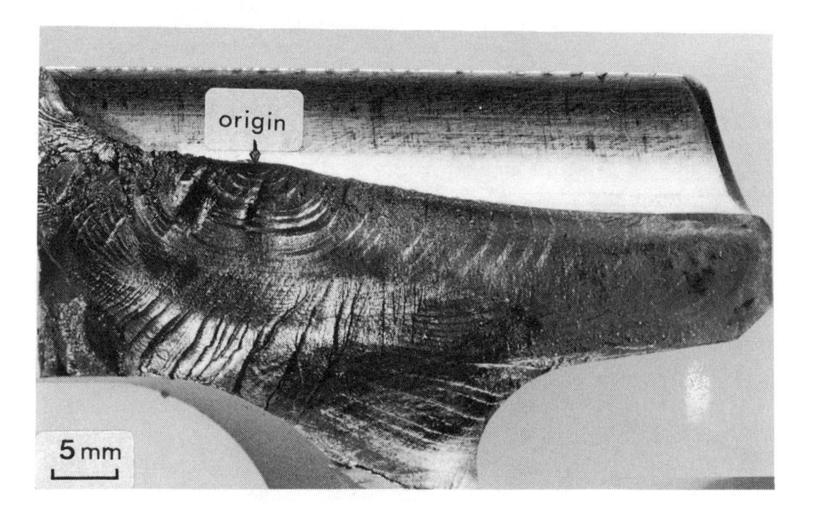

Figure 5.6 Tooth bending failure.

found that under the prevailing conditions of contact stress (i.e. rolling and sliding) a number of distinct failure modes are possible:

Scuffing (scoring).
This is the instantaneous friction welding and immediate separation of mating gear teeth. It is normally associated with either excessive contract stresses or with inadequate lubrication, but material variables have been shown to play a significant part, particularly in the absence of a good lubrication [13]. The load required to produce scuffing decreases with increasing sliding speed, although there can be a reversal at very high speed. The rapid heating and cooling of the gear surfaces in scuffing can result in a thin apparently amorphous 'white' surface layer at the gear surface.

Pitting.
This is a fatigue phenomenon where the breakdown of the tooth flank occurs by a process of progressive cracking. With VAR steels pitting originates at the surface and two basic forms are recognized:

- macro – where the pit size is comparable with the width of the localized band of contact (Figure 5.7)
- micro – where the pits are extremely small (Figure 5.8) and confined to localized irregular patches in the more severely loaded regions.

Micropitting arises as a result of inadequate boundary lubrication, although surface finish has a strong influence. Moreover, recent work

Figure 5.7 Macropitting.

Figure 5.8 Micropitting.

has suggested that micropitting [14] is not only a possible precursor to macropitting but may alternatively be followed by high rates of wear.

The onset of pitting is greatly influenced by (a) sliding in the contact, (b) surface temperature and (c) lubricant. The effect of the lubricant is particularly complex, in that fatigue life has been shown to fall off rapidly with lower film thicknesses and inferior boundary lubrication due to asperity interactions becoming more numerous and severe. Furthermore, the presence of dissolved water in oils, even in small amounts, has been shown to reduce fatigue life [15].

Wear.
This is not a serious problem in helicopter gear teeth, although there are exceptions to that [16]. Wear can be substantially influenced by hardness and surface finish.

5.2.7 Alloy developments

High performance gears in the modern helicopter are required to operate at ever increasing torque and speed. The net consequence of this is that a significant increase in temperature at the contact surface can be expected which can only be partially alleviated by improved (and heavier) lubrication cooling systems. It is likely that in future designs the main gearbox input speeds will be as high as 27 000 rpm and the nominal operating temperature currently 100–110 °C will be up to 150 °C. It is also anticipated that the next generation of military helicopter gearboxes will have to cater for extra oil loss tolerance (i.e. gear steels must have the capability to withstand once-only excursions to 300 °C as a result of battle damage or the failure of a critical component). Furthermore, although helicopter transmissions have been traditionally designed according to a 'safe life' philosophy the introduction of damage tolerant principles FAR 29.571 has meant that materials should possess a degree of slow crack growth.

Essentially conventional carburised steels are tempered in the range 140–160 °C which is considered adequate to ensure complete absence of microstructural modifications and hence dimensional instability during normal transmission operation. Recently, however, it has been shown that temperature measurements taken at the tooth flank can be as high as 200 °C, with flash temperatures even higher. Consequently, the limited temperature capability of conventional carburizing steels may influence normal operation and failure modes to a greater extent than has been realized. Unfortunately, by raising the tempering temperatures of conventional carburizing steels to achieve the desired oil loss tolerance of 300 °C, a substantially lower surface hardness, impaired pitting resistance, a reduction in favourable residual compressive stress and a corresponding drop in fatigue performance would result.

The limited elevated temperature performance of conventional carburizing steels led to the development of a number of temper resistant carburizing steels (Table 5.1). The aim of these developments had been to retain as many as possible of the advantages currently obtained from conventional carburizing steels while making sufficient improvements to meet the highest oil loss tolerance requirement of up to 300 °C.

5.3 BEARING STEELS

5.3.1 Bearing design

Simplistically, rolling element bearings provide a means of transmitting loads between a rotating shaft or gear and a stationary structure. Rolling element bearings come in several designs, the most common of which are the ball or roller bearings. Essentially, rolling element bearings consist of three distinct elements: (a) balls or rollers, (b) inner and outer raceways and (c) a non-ferrous cage.

5.3.2 Helicopter/aeroengine requirements for bearings

The high power-to-weight ratio requirements of helicopter gearboxes result in an arduous environment for the main bearings. Shaft misalignment and high vibrational levels, which are generally worse than most

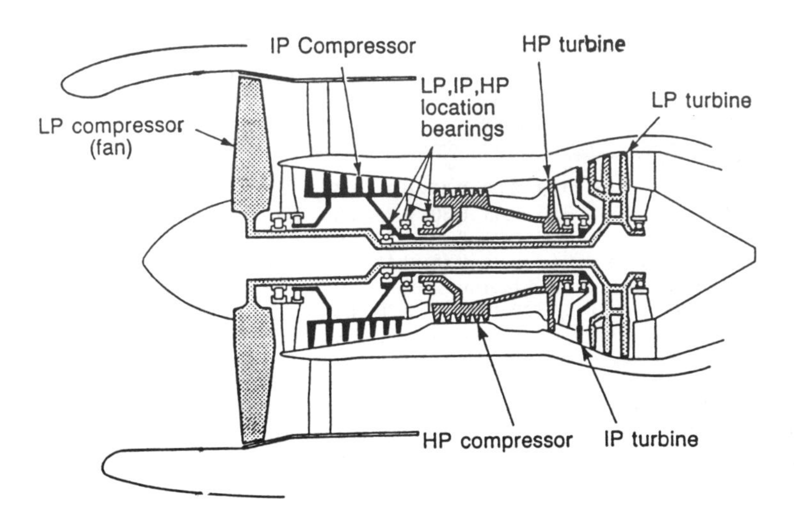

Figure 5.9 RB 211 Aeroengine.

industrial and aeroengine applications, and the saline atmosphere of seaborne operations all have a severe debilitating effect on bearing life.

With the desire to improve performance, increase fuel economy and produce a light engine weight, aeroengine bearings are subjected to possibly the most exacting service requirements of any bearing (Figure 5.9). Currently aeroengine bearings are required to operate at turbine speeds approaching 2.2 million DN (i.e. bearing bore (mm) × shaft speed (rpm)) in a hostile temperature environment of nominally 350 °C and under high hoop stresses.

5.3.3 Mechanical requirements for bearing steels

On a fundamental level a bearing steel must have:

- sufficient fatigue strength to withstand the high rotational speeds and hoop stresses
- high hardness to withstand rolling contact fatigue damage
- temperature resistance up to 400 °C
- adequate corrosion resistance

5.3.4 Bearing materials

Table 5.2 lists the typical bearing materials used in aeroengine and helicopter transmission applications. As can be seen, without exception low alloy steels predominate. Generally, the current practice is to use through hardened materials, for mainshaft applications. The most commonly used material in the USA is AISI M-50, whereas European

Table 5.2 Chemical composition of conventional and advanced bearing steels

Material identification	*Nominal chemical composition* (wt%)												
	C	*N*	*P max*	*S max*	*Mn*	*Si*	*Cr*	*V*	*W*	*Mo*	*Co*	*Ni*	*Fe*
Conventional bearing steels													
SAE 52100	1.0	–	0.03	0.03	0.35	0.3	1.45	–	–	–	–	–	BAL
AISI M50	0.8	–	0.03	0.03	0.30	0.25	4.0	1.0	–	4.25	–	–	BAL
Tl (18–4–1)	0.7	–	0.03	0.03	0.30	0.25	4.0	1.0	18.0	–	–	–	BAL
440C	1.03	–	0.02	0.01	0.48	0.41	17.3	0.14	–	0.5	–	–	BAL
Advanced carburizing/nitriding bearing steels													
M50 NiL	0.13	0.33	0.02	0.01	0.25	0.20	4.1	1.23	0.25	4.25	0.25	3.4	BAL
X30	0.33	–	–	–	0.44	0.51	16.2	0.02	–	1.11	–	0.43	BAL

gas turbine engine manufacturers have traditionally used more highly alloyed steels such as 18–4–1 (T1). Both materials contain alloying elements to promote high hardness and good hardness retention at the high operating temperatures typically experienced by the main shaft bearing.

Helicopter bearings are traditionally lower alloyed (i.e. of the 52100 through hardened or Ni–Cr–Mo low alloy carburizing types), as high operational temperatures are not experienced in helicopter gearboxes. 440C, a stainless steel, is also used in specific applications which require corrosion and oxidation resistance.

5.3.5 Metallurgical factors affecting bearing performance

The major processing/development steps taken to improve the bearing life of M50 is shown schematically in Figure 5.10 and discussed more fully in the following sections:

(a) Melting practice

One of the technical breakthroughs in materials engineering was the advent of the vacuum melting process in steel industries in the late 1950s and early 1960s. With the development of larger scale production melting furnaces, substantial improvements in micro and macroscopic homogeneity and cleanliness were realized. Such improvements have resulted in:

- improved macroscopic inhomogeneity (elimination of freckles, segregation and casting structure)
- enhanced microscopic cleanliness (reductions in inclusions and interstitial gas contents)
- enhanced carbide distribution and reduced chemical inhomogeneity.

(b) Grain flow

An important variable affecting bearing fatigue life is the forging flow orientation with respect to rolling contact fatigue surfaces. Trials carried

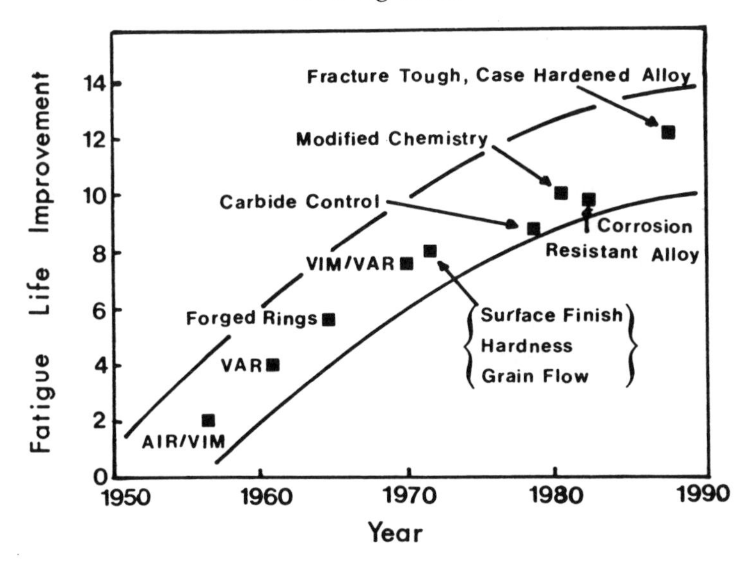

Figure 5.10 Process/development steps taken to improve the bearing life of M50.

out by Zaretsky [17] and others have shown that increasing fatigue life accompanies a transition from perpendicular to parallel forging flow.

(c) Grain size

As coarser grain sizes imply inferior fatigue properties, turbine manufactures generally specify the prior austenite grain size shall be ASTM 8 or finer, with individual grains not exceeding ASTM 5.

(d) Hardness

Control of hardness was one of the first procedures adopted in the bearing industry for maximizing fatigue life. Early studies on 52100 carried out by NASA [18] showed that fatigue life increased with hardness. Furthermore, a differential hardness concept for optimum fatigue life in 52100 was also confirmed [19] in which provided a differential hardness of 1–2 Rc exists between the race and rolling elements maximum fatigue life can be expected. For these reasons aerospace companies stipulate that rolling elements and races shall have an average hardness in the range 62–64 Rc.

(e) Microstructural features

Complete absence of ferrite and bainite is essential if the bearing steel is to achieve its desired hardness and fatigue strength. Retained austenite contents should also be kept below 3%, to prevent structural instability.

5.3.6 Lubrication factors affecting bearing performance

Bearing failure by fatigue is affected by the physical and chemical properties of the lubricant. In general, rolling element fatigue life is affected by lubricant viscosity, i.e. higher viscosity improves fatigue life. Moreover, lubricant additives can prevent or minimize wear and surface damage to bearings whose elements are in contact under very thin film or boundary lubrication conditions. These antiwear or extreme pressure additives either adsorb on to the surfaces or react with the surfaces to form protective coatings or surface films. Their use is commonplace in helicopter and aeroengine transmissions.

5.3.7 Failure modes

The most prevalent causes of damage to bearings include:

(a) Abrasive and improper assembly

If during assembly localized surface imperfections (i.e. nicks or damage marks) are caused this is likely to cause high stress concentrations, which often leads to a fatigue spall, usually characterized by a distinct arrow head formation pointing to the origin.

(b) Inadequate or improper lubrication

(c) Contamination

Foreign matter of a coarse and hard nature tend to produce pits in the rolling contact surfaces which can readily be distinguished as embedded abrasive particles over the entire rolling contact surface.

(d) Corrosion

This takes several forms, varying from pitting which can become the starting point for subsequent fatigue, to general surface corrosion which can cause progressive wear by the lapping action of the abrasive oxides. It occurs from the ingress of water or the presence of moisture from condensation. It can also be caused by acidic deposits from oxidation of the lubricant or contact with bare hands on unprotected bearings.

(e) Seizure and smearing

This can occur due to misalignment, cage breakdown, etc.

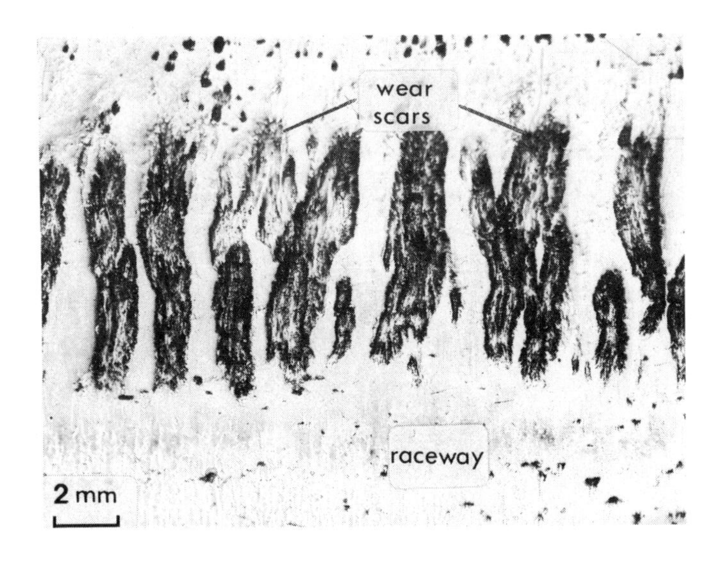

Figure 5.11 False brinelling.

(f) Stray electrical currents

The effect on the surfaces can take several forms according to the type of arcing and other conditions. Apart from pitting and other damage it tends to cause localized softening of the surface due to local high temperatures.

(g) False brinelling of raceways

This is a form of fretting corrosion caused by vibration transmitted through the rolling elements when the bearing is not rotating. In appearance it is similar to true brinelling but is characterized by pronounced polishing of the depressions made by the balls and rollers, especially at the edges. It is usually associated with a cocoa-like deposit of iron oxide which is the result of oxidation of minute particles of steel worn away by the fretting motion (Figure 5.11).

(h) Rolling contact fatigue

Contact fatigue of the rolling elements is caused by repeated stress reversals during operation. It generally results in minute fissures below the surface which propagate until they become recognizable as surface cracks, although with the use of vacuum melted steels surface initiation becomes more prevalent. Propagation eventually leads to flaking (Figure 5.12).

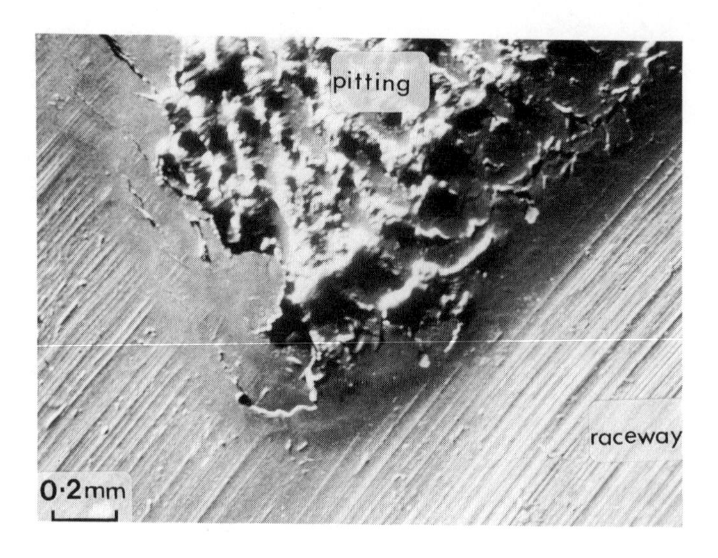

Figure 5.12 Pitting damage on bearing raceway.

(i) Material and manufacturing defects

Carbides are an essential phase in high carbon, high alloy steels used for rolling element bearings in that they provide wear resistance and inhibit grain growth. Increasing carbide size and volume fraction leads to improved wear resistance accompanied by lower toughness and ductility. In certain circumstances carbide segregation found in highly alloyed steels can cause microstructural problems. Furthermore, fatigue performance can be inferior in steels containing larger, more angular and elongated carbides than ones which are finer and more uniformly distributed.

One of the most insidious of manufacturing defects is a grinding burn on a rolling surface. Even though no grinding cracks are present the reduction in hardness and severe internal stress accompanying such a burn will drastically reduce the bearing life. Nital etch inspection has practically eliminated this type of defect getting into service.

5.3.8 Alloy developments

(a) High speed bearings

Since the late 1950s main shaft bearing speeds have increased from nominally 1.5 million DN to approximately 2.5 million DN due to requirements for higher performance, better fuel economy and a lighter engine weight. However, as can be seen in Figure 5.13, as bearing speeds increase, bearing fatigue life decreases, due to increased rate of stress

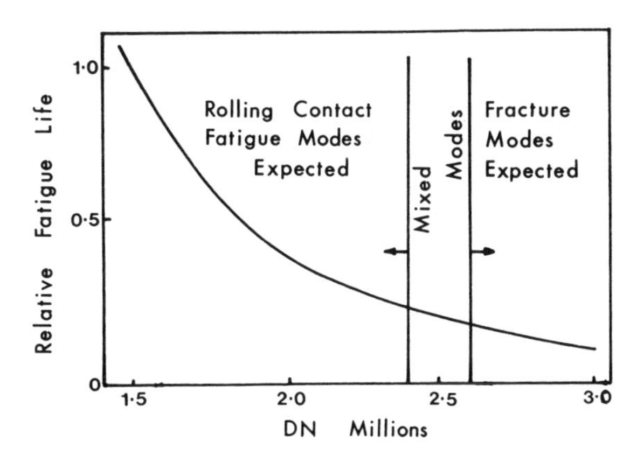

Figure 5.13 Influence of speed on bearing life/failure mode.

cycle accumulation and centrifugal effects on the rolling element. Therefore to maintain an acceptable removal and/or overhaul time, increased rolling-element fatigue lives in stress cycles or revolutions must be attained. Hence, modifications to the current alloys or the development of new steels, along with improved processing and heat treatment, is required to provide the desired fatigue life improvements. In addition, at high rotational speeds bearing races made of through hardened materials are susceptible to catastrophic fracture when exposed to the high tensile hoop stresses often present in high speed ball and roller bearing inner races. This type of fracture is a totally unacceptable mode of failure, because of its potentially serious secondary effect on engine integrity.

To overcome the technical challenges of (a) enhanced fracture toughness and (b) improved rolling contact fatigue performance, it was recognized that the traditional through hardened steels would have to be replaced by new materials and processes. Generally, carburized materials offer advantages over through hardened materials in that they possess favourable compressive residual stress patterns likely to promote improved rolling contact fatigue properties. Moreover, they possess higher toughness, due to the inverse relationship between hardness and fracture toughness. However, the deterrent to using conventional carburizing materials for aircraft engine rolling element bearings has been the relatively low operating temperature capability. Recently, however, high temper resistant carburizing steels have been developed (Table 5.2) with significantly improved temperature performance. Such materials are now entering service.

(b) Corrosion resistance

The problem of corrosion is not unique to high speed aeroengine bearings, in that general corrosion is one of the major causes of rejection at overhaul (e.g. nearly one-third of all aircraft turbine engine bearings are rejected due to corrosion). The problem is most severe in systems with long periods of inactivity.

The reduction in rejection rates due to corrosion is being tackled in three main ways:

- using corrosion resistant materials (i.e. AISI 440C, AMS 5749, X30)
- with surface modifications to the alloy steel (i.e. thin dense chromium (TDC))
- with corrosion inhibited lubricants

5.4 ULTRA HIGH STRENGTH STEELS

5.4.1 Undercarriage design

Aircraft landing gears consist mostly of highly loaded non-redundant components with severe restrictions on weight and space. For these reasons many landing gear components are made from high strength, low alloy steel, the most common of which are summarized in Table 5.3. A

Table 5.3 Chemical composition of conventional and advanced undercarriage steels

Material identification	*Nominal chemical composition* (wt%)										
	C	*Mn*	*Si*	*Cr*	*Ni*	*Mo*	*V*	*Co*	*Ti*	*Al*	*Fe*
Conventional steels											
BS S155 (300M)	0.43	0.78	1.63	0.83	1.83	0.38	0.05	–	–	–	BAL
AISI 4340	0.41	0.70	0.28	0.80	1.83	0.25	–	–	–	–	BAL
BS S99	0.40	0.58	0.23	0.65	2.55	0.55	–	–	–	0.05	BAL
35 NCD 16	0.35	0.45	0.28	1.80	3.85	0.43	–	–	–	–	BAL
DTD 5212	0.02	0.10	0.10	0.25	18.0	4.9	–	7.8	0.45	0.10	BAL
AMS 6487 (H11)	0.35	0.70	0.28	0.78	1.83	0.35	0.20	–	–	–	BAL
Advanced steel											
Aermet 100	0.23	0.1	0.1	3.1	11.5	1.2	–	13.5	–	–	BAL

typical aircraft landing gear (Figure 5.14) is, from the airworthiness aspect, considered a part of the airframe structure. It contributes nothing to the aircraft performance in its flight regimen but provides an energy absorption system during landing (of ~ 3m/s for fixed winged aircraft and up to 11 m/s for rotary winged aircraft) and a suspension system during taxiing. Furthermore, the landing gear is subjected to appreciable standing stress (i.e. up to 380 tons on an aircraft such as the Boeing 747) while the aircraft is stationary and dynamic stresses during take off, taxiing and landing.

5.4.2 Design criteria for undercarriage materials

The selection of the choice of materials will depend on a number of factors, namely:

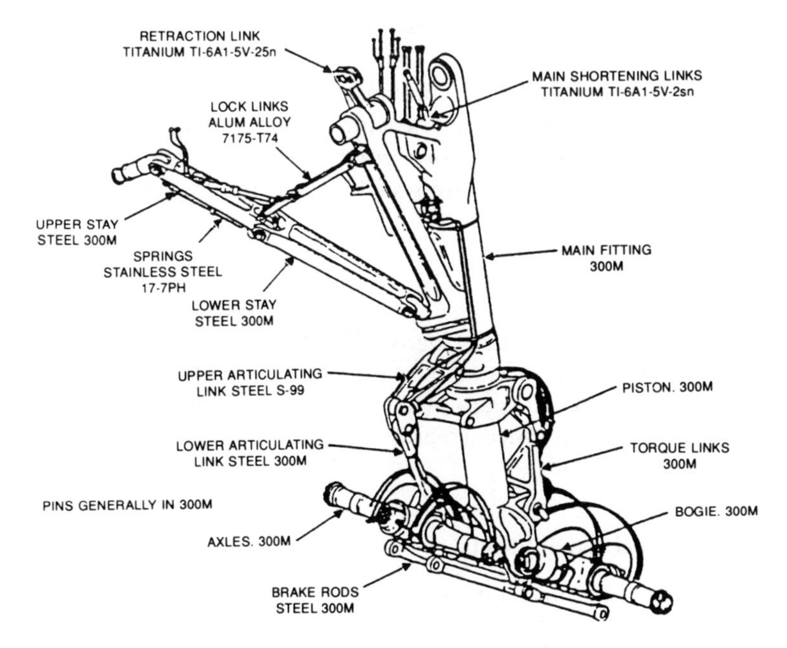

Figure 5.14 A330 undercarriage.

(a) The weight

The complete landing gear system (less wheels, tyres and brakes) represents between 2.5–4% of the AUW of the aircraft. Hence every effort must be made in keeping this to a minimum by choosing materials with high specific strengths.

(b) The geometry

In certain aircraft, space limitations within the airframe may be severely restricted and hence critical. In this case the use of steel at the highest strength level is almost inevitable because it offers the maximum strength and stiffness per unit volume.

(c) The cost

Minimum weight design may not be cost effective because of the disproportionately higher costs of the material and manufacturing process.

(d) The environment

Generally, landing gears are required to operate within the temperature range – 50 °C to + 100 °C. In certain applications (e.g. axles), however, heat resistance is also an important factor due to the localized heating due to braking (i.e. in excess of 400 °C during emergency braking). It must also be appreciated that landing gears operate in wet or marine environments, in addition to experiencing impact damage from runways. Hence resistance to stress corrosion cracking is very necessary, together with a suitable surface protection such as chromium plating to resist abrasive damage.

(e) Types of stressing

Fatigue spectrum and ultimate loads will depend on the type of operation, i.e. the landing gear of a military aircraft must withstand 5000–10 000 flights spaced over 20–30 years, while for civil aircraft a requirement of 30–60 000 flights is specified (representing of the order of 500 000 km travelled on the ground).

5.4.3 Mechanical requirements

In order to achieve a minimum weight design, each component has to be considered individually to balance the requirements of strength, stiffness and fatigue life. In addition, each component has to be considered in relation to the structure as a whole to ensure equally important secondary factors such as corrosion resistance, electrical conductivity, toughness and wear resistance.

On a fundamental level an undercarriage material requires:

- sufficient static strength to withstand the extremely high bending stresses
- high fatigue strength
- high stiffness to limit deformation
- adequate resistance to corrosion/stress corrosion
- adequate toughness to withstand shock loads
- sufficient wear resistance to resist frictional wear or bearing stress

5.4.4 Metallurgical factors affecting undercarriage performance

(a) Strength

As the majority of landing gear components are designed by their ultimate bending strength, a high ratio of 0.2% proof strength to ultimate tensile strength is therefore desirable. In the majority of cases the high strength levels are achieved through the transformation of austenite into acicular martensite. Consequently, absence of ferrite, bainite and retained austenite is critically important as these undesirable microstructural features are shown to detrimentally affect both the resultant strength level and UTS to YS ratio. The most commonly used low alloy steel in undercarriage applications is 300M as it offers the highest strength (e.g. UTS 〉 1900 MPa) to weight ratio of any commercially available alloy.

One class of steels differs from the conventional low alloy steels in that hardening is achieved not by carbon, but instead by the precipitation of intermetallic compounds. These alloys have been termed maraging steels in which the term derives from martensite age hardening and denotes age hardening of a low-carbon martensite matrix [20].

(b) Stiffness

Generally a high value of modulus to static strength is desirable, except on a few occasions where a relatively lower value can be exploited. However, without exception the stiffness of the steel cannot be altered, regardless of microstructural constituents or mechanical properties.

(c) Fatigue resistance

As a general rule, landing gears are designed to safe life principles, hence resistance to crack initiation and sub-critical crack growth are of primary importance. For a civil transport aircraft the major part of the fatigue damage to the landing gear usually occurs during the taxiing and ground manoeuvring phases, while in combat aircraft the landing phase will contribute a more nearly equal share of the fatigue damage. For either type of aircraft a high proportion of the loading cases which cause fatigue damage are in the low cycle high stress region. Furthermore as no surface can be regarded as free of stress concentrations, which may be induced either by manufacturing or by geometric effects, adequate notched fatigue properties are essential.

The most important metallurgical factors affecting the fatigue performance of high strength low alloy steels are:

Steel cleanliness
Due to the high tensile strength of undercarriage steels (> 1500 MPa) their critical defect sizes are relatively small, hence it is essential that the materials are processed using either VAR or ESR melting techniques, otherwise inferior fatigue properties will result.

Microstructural features
Complete martensite transformation is essential if steels are to achieve their required fatigue level. Furthermore, because of the problem of machining UHTS steels in the hardened condition a large proportion of the surface area of components is finish machining before the final hardening treatment. Consequently, it is important that the surfaces do not become decarburized, otherwise a reduction in fatigue life occurs.

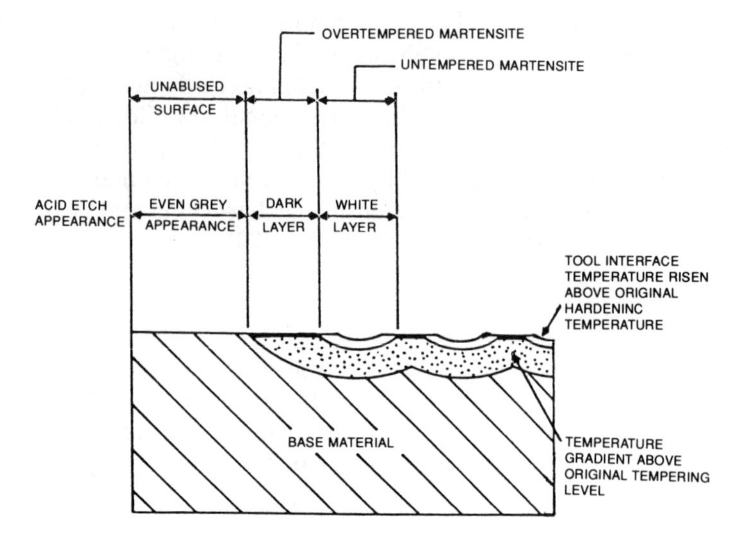

Figure 5.15 Grinding burns.

Surface Anomalies
Great care is also required in machining fully heat treated materials, in that it has been shown that machining abuse can introduce surface layer alterations, which can reduce fatigue strength by up to 30% [21]. Essentially, two basic forms are recognized (Figure 5.15):

- Untempered martensite (rehardening burns). This occurs in the chip removal and grinding operation. It essentially results from excessive overheating and sudden quenching of the surface layer by the base material or by the cutting fluid. It is brittle and often associated with the nucleation and propagation of minute cracks.
- Overtempered martensite. This invariably exists under the layers of untempered martensite on steels which have been quenched and tempered, although it can exist on its own. It is generally softer than the base material as a result of the overtempering generated from the grinding operation.

(d) Corrosion/Stress corrosion resistance

Due to the high strengths of these alloys it is vitally important that small corrosion pits on the surface of landing gear components are avoided as this can initiate cracks and thereby lead to catastrophic fracture when cracks grow beyond 2.5 mm. Moreover, due to the low alloy content of the high strength steels their corrosion resistance is fairly low, and suitable surface protection is required. The most convenient way this can be done is to apply a surface coating, in which the coating protects the base metal in two ways:

- by preventing access of the corrosive agent
- by sacrificially corroding in preference to the base metal

The most commonly used coating systems appear to be cadmium plating or chromium plating when wear in addition to corrosion is a problem. By applying either of these coating systems adequate stress relieving before plating and through de-embrittling after plating is essential. This is due to the fact that hydrogen (which is liberated during plating) has a deleterious effect in promoting premature failure in high strength alloys subjected to sustained loading.

Ultra high strength steels usually have poor stress corrosion cracking properties (K_{ISCC}), in which on average K_{ISCC} is approximately half to a third those of their corresponding fracture toughness values. In most instances these properties are too low to be used as the basis for a design in which minimum acceptable crack length is a significant parameter.

(e) Toughness

Although airworthiness requirements do not currently insist on damage tolerance analysis to undercarriage components the implementation of FAR 25.571 has meant that components should be designed according to damage tolerance principles. The major problem in trying to apply damage tolerance to landing gear components is that due to the very high strengths, the resultant fracture toughnesses are generally in the range 50–100 MPa$\sqrt{m}$, and hence the critical crack lengths are small (nominally 2–4 mm). Consequently, the time taken to develop the small critical crack lengths are essentially short. Nevertheless, higher fracture toughnesses are considered advantageous, where reductions in strength, grain size, carbide size and shape or inclusion content are all likely to enhance fracture toughness.

5.4.5 Failure modes

A world-wide aircraft accident survey by Campbell and Lahey [22] using data collected between 1927 and 1981, showed that the failure of the landing gear was the most common fatigue problem, accounting for approximately 37% of all accidents. Fortunately, however, landing gear failures seldom result in a fatal accident (i.e. approximately 1% of all landing gear failures resulted in fatalities) or severe damage to other parts of the aircraft structure.

Landing gear failures have been attributed to one of the following causes:

- manufacturing/processing deficiencies (e.g. notches, decarburization, grinding abuse, corrosion, hydrogen embrittlement)

- service damage (e.g. corrosion, nicks/dents)
- material deficiencies (e.g. inclusions)

The majority of landing gear failures can be attributable to stress corrosion or to delayed fracture, starting from areas which have been damaged [23]. The type of damage can occur mechanically or by a corrosion mechanism, in which both sources can act as a nucleus for a stress corrosion crack.

A few failures have started at internal flaws such as non-metallic inclusions in the steels (although this type of failure has been diminishing over the years with the advent of cleaner steels).

Finally, other failures have been caused by improper heat treatment, plating procedures (hydrogen embrittlement) and grinding cracks.

5.4.6 Alloy developments

The desire to reduce the weight and increase reliability of landing gear steels has culminated in a number of steel developments (Table 5.3), in which three different design philosophies are currently being evaluated:

- steels with increased mechanical strength
- steels with improved damage tolerance, without sacrificing strength
- steels with enhanced corrosion resistance

With regard to improved strength, great care must be taken to ensure that the critical crack size, which will decrease with strength, does not fall below 0.5 mm. In this size range defects cannot be avoided with certainty using conventional NDT methods and hence safety would be jeopardized.

The technical reasons for replacing low alloy steels with stainless steels is that they are likely to avoid corrosion on surfaces that are difficult to protect. Furthermore, there is an added advantage that cadmium would not be required. However, there is likely to be a cost penalty which would have to be judged against the enhanced corrosion resistance.

ACKNOWLEDGEMENTS

The author would like to thank Westland Helicopters for permission to publish this chapter and especially Messrs B.C. Gittos and N.L. Bottrell for their valued support and advice.

REFERENCES

1. Johnson, R. F. and Sewell, J. F. (1960) *Journal of the Iron and Steel Institute*, **196** (4), 414.
2. Kalner, V. D., Nikonov, N.F. and Yurasov, S. A. (1973) *Metal Science and Heat Treatment*, **15**, 752.

3. Wiegand, H. and Tolosch, G. (1967) *Harterei-Technische Mitteilungen*, **22**, 213.
4. Olver, A. V. (1985) *AGARD Conference Proceedings* No. 394, p. 10–1.
5. Kozlorskii, I. S. (1971) *Metal Science and Heat Treatment*, **13**, 650.
6. Boughman, R. A. (1960) *Journal of Basic Engineering*, **82**, 287.
7. Sagarvdze, V. S. and Malygina, L. V. (1966) *Metal Science and Heat Treatment*, p. 560.
8. Kozlorskii, I.S. Kalinin, A. T. and Novikova, A. (1967) *Metal Science and Heat Treatment*, p. 157.
9. Parish, G. American Society of Metals Catalogue Card No. 80–10679.
10. Dawes, C. and Cooksey, R.J. (1966) *Heat Treatment of Metals Special Report* 95, p. 77.
11. Sayles, R. S. and Webster, M.N. (1984) *AGARD Conference Proceedings* No. 369, p. 14.1.
12. Weigand, H. and Tolasch, G. (1967) *Harterei-Technische Mitteilungen*, **22**, 330.
13. Graham, R. C., Olver, A.V. and MacPherson, P.B. (1980) *American Society of Mechanical Engineers Conference Proceedings*, p. 80-C2/DET118.
14. Olver, A. V. and Wedden, P. R. (1978) Westland Helicopters Ltd, ML/AVO/HSH/19 006/30.24.
15. Cantley, R. E. (1977) *Transactions ASME*, **20**, 224.
16. MacPherson, P. B. and Spikes, H. A. (1980) *ASME Conference Proceedings*, p. 80-C2/DET-12.
17. Zaretsky, E. V. (1966) *Machine Design*, **38** (24), 206.
18. Irwin, A. S. (1960) *ASME Conference Proceedings*, New York.
19. Zaretsky, E. V., Anderson, W. J. and Parker, R. J. (1962) *Transactions ASME*, **5** (1), 210.
20. Floreen, S. (1978) *ASM Metals Handbook* (9th edn) **1**, 445.
21. Best, K. F. (1986) *Aircraft Engineering*, **58**, 14.
22. Campbell, G. S. and Lahey, R. T. C. (1984) *International Journal of Fatigue*, **6** (1), 25.
23. Holshouser, W. L. (1966) *Metals Engineering Quarterly*, p. 108.

6

Ceramic materials in aerospace

D.M. Dawson

6.1 INTRODUCTION

Ceramic materials, used alone or as part of an overall materials' system, have several attributes of great benefit in many aerospace applications. In simple terms they offer:

- higher temperature capability than metals – exploitable through both increased combustion temperatures and/or reduced component cooling requirements in gas turbines, and in high velocity aerodynamic control surfaces in missiles and aircraft.
- lower densities than metals – offering lower weight components, thus improving the system payload or range capability; this also offers greater flexibility in changing the centre of gravity of structures to more appropriate locations; this attribute must be assessed in terms of specific strength for a given application; often this reduces the initially perceived benefit
- higher stiffness and hardness than metals – giving improved wear capability for bearings and seals
- some application specific properties, e.g. radar, Infra-red (IR) or other electromagnetic wavelength transparency.

Unfortunately, ceramics also have some major drawbacks in that they are brittle and have relatively low thermal shock resistance. These properties, both good and bad, fundamentally result from a strong inter-atomic bond which is best achieved through the combination of the lighter elements capable of covalent bonding, i.e. Be, B, C, N, O, Mg, Al and Si. Most engineering ceramics contain one, or commonly more, of these elements.

The potential benefits are very significant, and for these reasons ceramics, especially monolithic ceramics, have been under development and evaluation for some time. With the development of hot pressed silicon nitride in the UK in the late 1960s serious work got underway for the use of high performance ceramics in gas turbine engines. The late 1970s and early 1980s saw the successful completion of major engine demonstration programs in the USA such as CATE, AGT 100, and AGT 101, to name but a few. Proof of concept was established, along with significant advances in material, fabrication, design and analysis technologies. At the same time, serious problems were identified, some very fundamental ones such as the very low levels of reliability and reproducibility, which have essentially stalled the use of these ceramics in aeroengines, at least for the time being.

Current development thrusts are therefore aimed at the development of the next generation of monolithic and toughened ceramic components with acceptable reliabilities. At the same time, a tremendous level of activity has been sustained over the last 10 or 15 years for the development of fibre reinforced composite ceramics for applications in both the aerospace and other industries.

This chapter will discuss the development and application of structural ceramics, rather than ceramic coatings which have been addressed recently, under the general headings, (1) monolithic and toughened ceramics including whisker toughening, and (2) long/near continuous ceramic fibre reinforced ceramics. The subject of fibrous ceramics will be covered in the second category.

6.2 MONOLITHIC AND TOUGHENED CERAMICS

6.2.1 Materials

The properties of some of the more important monolithic and toughened engineering ceramics are summarised in Table 6.1. Most have been under active development for 20 or 25 years and further improvements are constantly being made.

(a) Alumina based materials

Alumina is historically the most developed and understood engineering ceramic. It has been under active development for 40 to 50 years, and it remains, in terms of market volume and value, the most important engineering ceramic material. Alumina has a crucial market in the engine field as spark plugs, and other important applications include wear resistant parts, substrates and metal cutting tips. Various grades of alumina, from sapphire crystals to aluminas with significant silica contents, the debased aluminas, are available. They have very different properties and capabilities. Many alumina materials have a strength of

Table 6.1 Materials properties for a selection of monolithic and toughened ceramics

Material	*Process*	*Density* $Mg\,m^{-3}$	*Elastic modulus (GPa)*	*Flexural strength (MPa)* (25°C)	(1000°C)	(1400°C)	*Toughness* K_c ($MPa\,m^{1/2}$)
Monolithics							
99.9% alumina	Sintered	4.0	400	550	410		2–3
96% alumina	Sintered	3.7	320	350	170		2–3
Mullite	Sintered	2.8	230	180			2–3
Silicon carbide	Sintered	3.1	300	380		380	2–3
	Hot pressed	3.2	440	500		300	4–5
	Reaction bonded	3.1	410	310		190	410
Sialon	Reaction bonded	3.0	330	600			3–7
Silicon nitride	Sintered	3.1	240	420		80	5–6
	Hot pressed	3.2	310	750		300	5–6
	Reaction bonded	2.8	170	200		250	3–4
Toughened							
Partially stabilized Zirconia	Sintered	5.7	210	600	200		9 (5 @1000°C)
Al_2O_3/SiC(w)	Sintered	3.7	380	640			4–6
Si_3N_4/Sic(w)	Hot pressed			850			7–8
	Reaction bonded			620			7–8

300 MPa at ambient temperature; by 800 °C this has reduced to 220 MPa and by 1200 °C alumina deforms plastically.

The use of silicon carbide whiskers to reinforce alumina should, in theory, result in an increase in strength and toughness. Recent improvements in whisker dispersion and consolidation techniques have allowed the theoretical properties to be almost achieved in practice. Strengths of up to 600 MPa are reported together with an increase in fracture toughness from 4.5 MPa√m for plain alumina to 8.5 MPa√m for composites with 20 vol% SiC whiskers. These materials are now available as machine tool tips, deep-drawing dies and for other wear applications. These materials still do not have the reliability or properties demanded for structural aerospace applications. Indeed the theories suggest that they will not attain such levels. For this reason, and for the potential manufacturing health risk in using whiskers, research and development activity for structural use is at a low level.

(b) Zirconia based materials

Until 1975 zirconia was unusable as an engineering material. Pure zirconia undergoes a number of phase transformations; the most important occurs at 1170 °C. On cooling there is a 5% volume change as the material transforms from the tetragonal to the monoclinic crystal structure; any component fabricated in the material would have disintegrated during this transformation. The problem was circumvented in practice by the addition of cubic phase stabilizers such as MgO, CaO and Y_2O_3, thereby eliminating the phase, and hence volume, change. By controlling this phase change it has been possible to develop materials which generate compressive strain fields around crack tips; this reduces the stress concentration locally, making the crack propagation more difficult. The material is therefore considerably tougher; the fracture toughness can be effectively doubled to *ca.* 10 MPa√m. For a constant flaw size this increase in toughness corresponds to an effective doubling in strength.

This 'zirconia toughening' has been exploited in a wide range of materials. In toughened zirconia polycrystals (TZP) a single phase tetragonal fine grained material is produced. In partially stabilized zirconia (PSZ) a fine dispersion of tetragonal precipitates is induced by heat treatment in larger grains of cubic zirconia. The property of transformation toughening can be utilized in other materials; zirconia toughened alumina (ZTA) is a mixture of 20% of metastable tetragonal zirconia in a fine grained alumina matrix. The strongest reported ceramic material is a hot-isostatically pressed TZP with an addition of 20% alumina; this has a bend strength of 2.4 GPa.

These properties are impressive but they rely on the transformation mechanism. As the temperature is raised, to between 500 and 1000 °C, the thermodynamic driving force for the transformation reduces and with it the strength and toughness performance.

(c) Non-oxide based materials

The non-oxide silicon based ceramics, silicon carbide, silicon nitride and the sialons, have a lower thermal expansion coefficient than oxide materials. Thus, in high temperature applications where high thermal gradients are present or thermal shock is experienced, the thermal stresses are much smaller. Much of the recent work on the exploitation of monolithic ceramics in aerospace, in particular the gas turbine engine, has been concentrating on the newer non-oxide materials.

6.2.2 Manufacturing methods

(a) Conventional ceramic processing

Generally, a green preform of the component is shaped through the use of one of several available powder compaction routes, such as cold pressing, injection molding or slip casting. The resultant green preform is then sintered to form a dense ceramic shape.

Early methods of cold pressing ceramic powders used simple uniaxial pressing, but these were limited to simple shapes. Recent powder compaction techniques involve the use of isostatic pressing which allows greater shape flexibility and more homogeneous compaction. Variations in local packing densities in the green state can result in both large residual stress concentrations and large variations in shrinkage during sintering. In the first case stress fields can lead to component failure during manufacture or in service; in the second the component can become distorted. The technology for injection molding and slip casting has been greatly enhanced over the last several years, and these methods are very suitable for producing complex shapes with high quality.

The addition of organic binders to increase the strength of the powder compacts has resulted in the ability to machine complex shapes prior to sintering. However, binders, plasticizers and other volatile additives have to be removed from the molded part by burning and drying without generating cracks and defects. This remains one of the main processing problems, especially for thick sections, and the required technology is continuously developing. The other associated problem relates to the introduction of impurities with these admixtures which leads to defect formation during subsequent sintering.

Sintering itself poses some very basic limiting problems for the non-oxide ceramics. While necessary for densification, it occurs only at very high temperatures when the thermal energy exceeds the binding energy of the material allowing rapid atomic transport.

This does not generally occur in pure non-oxide ceramics as thermal decomposition of the material becomes dominant before significant atomic diffusion takes place. Hence, in most ceramic systems it is necessary to employ the use of impurity additions, usually various metal oxides in the case of nitride ceramics, to improve the densification kinetics. These form a low melting liquid by reaction with the surface silica (SiO_2) present on the silicon nitride grains resulting in enhanced atom transport via the liquid phase and hence densification. However, the residual intergranular glass phase which forms has been shown to severely degrade properties at elevated temperatures. Some of the best compositions are based on the sialon formulations (i.e. *s*ilicon *n*itride, with substitutions of *a*luminium and *o*xygen) sintered using yttria as the sinter additive. Yttrium aluminium garnet is subsequently crystallized at the grain boundaries and the resulting material has a strength of 500 MPa at 1400 °C. Considerable effort is also being spent in developing sintering methods without the use of sintering aids.

Silicon carbide is a rare example where higher decompositional temperatures and the use of 'surface active' impurities (B + C) permits solid state diffusional sintering to near theoretical density. Final densification requires high temperature (up to 2000 °C) and high pressure making the production of large components both difficult and expensive.

(b) Hot isostatic pressing

Some processes such as hot uniaxial pressing and hot isostatic pressing (HIP) combine the preform shape forming and the sintering steps in one, and are accomplished under high pressure of as much as 200 MPa. Hot uniaxial pressing has been successfully used to produce fully dense ceramics (e.g. the silicon nitride NC 132), but is only suitable for very simple shapes. Moreover, uniaxial pressing causes anisotropy in properties due to factors such as oriented grain growth; this can become an additional problem in a component.

The application of high isostatic pressures via gaseous medium (200 MPa) allows sintering with a great deal of control of both shape and microstructure, the use of low levels of sintering additives, increased strength and reliability. Strengths of 500–1000 MPa combined with Weibull strength distributions of 25 have been obtained at room temperature for material of near theoretical density. However, HIPing is expensive due to the extremely high pressures and temperatures used and the complexity of the equipment required to achieve these conditions (Chapter 13 and section 13.5.3).

(c) Reaction processing

Silicon nitride can be made by reaction bonding silicon metal preforms in a nitrogen atmosphere (RBSN). The metal preform can be readily machined; the fabrication route does not produce a volume change so the component can be produced to a high tolerance and requires little end machining. The strength of RBSN is only modest (200–300 MPa) because of the high retained porosity (20–30%), but this strength is retained to the decomposition temperature (1800 °C) in an inert atmosphere. Silicon carbide can also be produced using a reaction bonding route. The material, in this case, has no residual porosity (the pores are filled with silicon (*ca.* 10–20%)), and it has a typical strength of 500 MPa which is retained up to the melting point of silicon (1400 °C). The porosity and retained metallic inclusion defects are inherent in this fabrication process. Developments to reduce the defect levels and their size are ongoing.

There exists a need for new and innovative fabrication methods, several of which are beginning development. For example, the use of sol–gel, dynamic compaction, sintering without aids, and spray forming are all showing potential benefits. The goal remains the same; the production of components free of significant defects, which have an homogenous microstructure and are of stoichiometric material. Equally importantly, they need to be manufactured at a commercially acceptable cost.

6.2.3 Component design and applications

(a) General considerations

The basic framework for understanding the strength, toughness and reliability of monolithic ceramics is provided by fracture mechanics and statistics. The traditional view is that fracture occurs in a brittle manner and originates from some flaw in the material. Weibull statistical approaches have been developed with some success to predict the failure strength of ceramic components from a combination of volume stress integral of the component and statistical failure probability data of test specimens. A typical example of the way a design stress is estimated is illustrated in Figure 6.1. Considerable care is necessary in the interpretation of the test data, and the failure mechanism and initiation site, in such estimations.

The flaws in current materials are almost always associated with processing defects rather than the intrinsic microstructure. As was discussed in section 6.2.2 the elimination, or at least a reduction in the size range of these flaws, has been and continues to be addressed. Further improvements in strength are possible but are proving difficult, and the methods of achieving higher strength may not be easily applicable to large scale manufacture. In any event these materials are inherently brittle, and surface damage during machining or in service will set a lower bound to the degree of 'perfection' that can be used and hence to the allowable design stress.

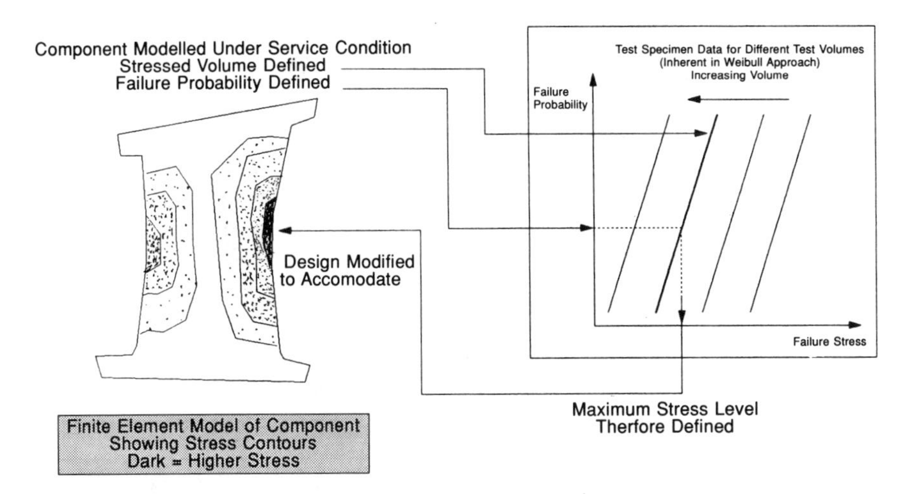

Figure 6.1 A schematic diagram of the Weibull design methodology for a monolithic ceramic component.

One consequence of the trend for increasing the strength of ceramics is that fracture, when it occurs, is increasingly catastrophic. For example, the elastic strain energy stored in a 10 mm cube of alumina stressed to 1 GPa is sufficient to fracture the material into slices only 10 µm thick. As strength capability increases it implies that the stress necessary to cause fracture will increase, but the stored energy also increases. During fracture this stored energy must be released, primarily as surface energy created. Therefore more surface must be created during failure and the component, in the limit, returns to powder. In addition, as the stress increases, fracture mechanics defines that the size of the critical flaw decreases, making its detection even more difficult.

Even when the ceramic component is designed or manufactured, to fully exploit the potential benefits of monolithic and toughened ceramics, a completely different and innovative approach to mechanical design is needed. To date the activity has been limited to the replacement of individual metal components by ceramic components. It is not widely recognized, but designing metallic style components in ceramics imposes serious additional hurdles in the use of these materials for aero engine applications. Apart from the inherent design difficulties such as ceramic to ceramic load interfaces and the susceptibility to foreign object damage (FOD) arising from the brittleness and lack of ductility in ceramics, the substitute ceramic components have to interface and survive in the surrounding metal environment and the stresses and geometries imposed. Problems arise from the low expansion coefficient of ceramics relative to the metals, and their low ductility reduces their ability to uniformly

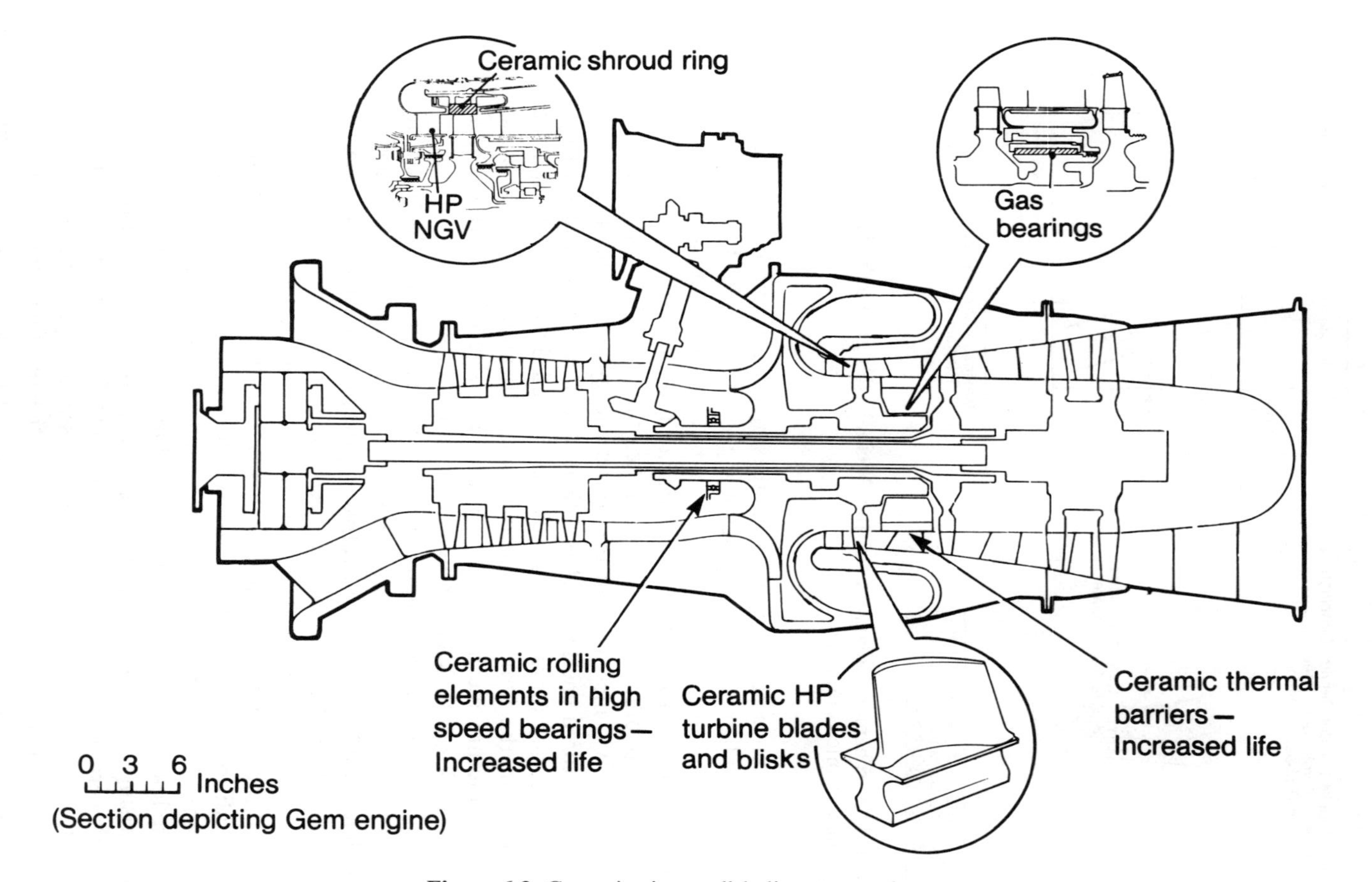

Figure 6.2 Ceramics in small helicopter engines. (Courtesy of Rolls Royce plc)

transfer loads across interfaces. These problems can be illustrated by considering a few simple components in the Rolls-Royce Advanced Mechanical Engineering Demonstrator (AMED) gas turbine engine, shown schematically in Figure 6.2.

(b) Gas turbine component programmes

The shroud ring was the first ceramic component to be investigated by the advanced engineering department at Rolls-Royce, Leavesden. It encompasses the turbine blade assembly and is a means of keeping the gas flow over the turbine blades. The smaller the gap between the blade tip and the shroud ring the greater the turbine efficiency. The ceramic material used for this component was hot pressed silicon nitride containing a sintering additive of about one percent magnesia. For the reasons described above the operational temperature was limited to about 1000 °C.

Nevertheless, this material has a dimensional stability superior to the superalloys, and remained very much more concentric; thus a smaller tip clearance was maintained, and as a result this assembly produced a significant improvement in power and specific fuel consumption. The benefits could only be achieved, however, when a complex mounting system was developed which absorbed the thermal mismatch, prevented the shroud from moving and yet did not unduly stress the shroud itself. Mechanical mounting and location of the shroud ring was achieved by four annular wire (Nimonic 90) brush seals. Apart from uniform peripheral loading these brush seals also provided a compliant mounting.

Both the higher temperatures, and greater rotational speeds of advanced gas-turbine engines, have increased the demands for more efficient bearings and lubrication systems. One option under development is a ceramic gas bearing. At high rotational speeds, the gas bearing generates its own low friction aerodynamic cushion, and it can therefore operate without conventional lubrication. This would eliminate churning of lubricant, scavenging and cooling requirements; factors that reduce the complexity, weight and cost of gas turbines. In some of the early experiments, the shells and the rotor were metallic with ceramic coated bearing surfaces. Under test conditions the coatings readily spalled. The most likely cause was probably a combination of contact stress and adverse residual stresses imparted by the coating process, with maximum effect at the coating-parent metal interface. As a result of this coating failure, the metallic coated shells were replaced by monolithic ceramics, together with an improved plasma sprayed coating (chromium oxide) on the rotor journal. Further test runs proved to be very successful. However, there was some degradation of the chromium oxide coatings, which tended to microweld on to the ceramic shells. Modification of the coating parameters improved the coating behaviour and the concept was successfully demonstrated.

The ceramic-to-ceramic tribo surfaces will, however, necessitate some surface engineering to provide low friction surfaces for the stop-start conditions, where the bearing journal will 'land and take-off' from the shells.

Depending on the operating condition, treatment of the precision bearing surfaces may have to produce coatings with certain properties:

- residual stresses in parent material or coated surface should be advantageous in nature
- good adhesion
- precise stoichiometry
- uniform thickness
- dimensionally extremely stable
- high strength
- high fracture toughness
- controlled density of structural defects
- high thermal shock resistance
- high thermal stability
- resistance to creep and wear
- resistance to oxidation and corrosion
- conformal surface topography

To produce surface treatments with these properties, a process is required which is capable of precise control and flexible in regard to material formulation. For bearing purposes, the required surface treatment, or coating, may be only of the order of a few micrometres. If this is so, then, ion assisted coating processes would be a natural route to follow.

Looking to the future development of the gas bearing: its natural mechanical neighbour could be a ceramic turbine disc with the bearing journal an integral part of the disc assembly.

The turbine blade is subjected to the most potentially life degrading conditions throughout the whole of the gas turbine. It is essential, therefore, to have a most detailed knowledge of stress and temperature distributions, and vibrational modes inherent in the design.

Early experience with monolithic ceramic turbine blades was with a hot pressed silicon nitride containing ~ 1% magnesia. This formulation produced a material with a high glass fraction, and tended to be susceptible to excessive creep at temperatures above 1000 °C. Consequently the operational temperatures for the engine trials were kept below 900 °. Considerable expertise was developed in the machining of the component to minimize surface defects and residual surface stresses. With this temperature limitation in mind, an Yttria–Si–Al–O–N material in which the minor intergranular phase was recrystallized as yttrium aluminium garnet (YAG) was examined in some detail. This material exhibited superior creep properties to the earlier hot pressed material, but was limited to temperatures below 1350 °C due to oxidative degradation with the YAG and SiO_2 (formed as Si_3N_4 oxidizes) reacting to form a liquid in this temperature region, resulting in poor creep strength. However, some turbine blades were manufactured and tested from this material using the same machining techniques. A research and development programme addressing the development of a dense silicon nitride containing a single oxide additive, capable of long term operation at 1400 °C with adequate oxidation, strength and creep resistance is underway.

Several conceptual designs over the years have addressed the fundamental weakness of monolithic ceramics by attempting to design a component system in compression. None have been developed to a satisfactory level as yet and the philosophical change makes it difficult to demonstrate such components within existing systems. It is an equally valid development route to the quest for improved materials and such studies are ongoing in various aeroengine companies.

6.3 COMPOSITE CERAMICS

6.3.1 Materials

Continuous fibre reinforced ceramics have been under intermittent research and development for over 20 years; the goal is to produce strong, stiff, tough and damage tolerant materials. Currently there is enormous interest in such materials, and major programmes are underway in the USA, Japan and Europe. Some materials have reached the stage of component demonstration but for the most part are still in the laboratory under development and assessment.

A key benefit over monolithic ceramics is the significant increases in toughness and reduction in notch sensitivity. Against these benefits have to be set the low level of materials development, design methodology and process technology maturity.

Work on continuous fibre reinforced ceramics began in the UK in the mid-1960s after the development of the early carbon fibres; indeed, much of this activity predates the use of these fibres in organic resin systems. The obvious toughening and reinforcement benefits were recognised and a wide range of matrix compositions were tried. The inherent oxidation of carbon fibres at temperatures over 400 °C was a critical limit in the fabrication and, to a greater extent, the usability of these materials. As a consequence the UK programmes ended in the early 1970s. In the United States, work continued on the development of these materials, primarily for space structures, throughout the 1970s. With the advent of the higher temperature SiC (type) fibres (such as Nicalon and Tyranno) in the late 1970s a major effort was made in the development, evaluation and exploitation of these materials. Several materials/components systems are under evaluation using both oxide and non-oxide matrices.

(a) Oxide matrix composites

Glass matrix composites, and glass-ceramic (ceramic derived from heat treatment of suitable glasses (also called silicate ceramics)) and ceramic matrix composites are a range of high temperature capability, continuously reinforced composite materials with the potential for exploitation throughout the gas turbine engine and for hot aerodynamic surfaces. Between 300 °C and 1400 °C these materials are anticipated to provide the highest specific strengths for structural applications. The major benefits are low weight (half that of titanium alloys, a third that of nickel alloys), high temperature strengths and stiffness and no fire risk.

The different chemical characteristics and physical properties of the fibre and matrix theoretically provide a stable mechanical discontinuity to produce crack deflection and promote significant toughness.

Materials with near full density are produced by hot-pressing or liquid infiltration of ceramic fibre fabrics or aligned fibre preforms with glasses and glass-ceramics, or by directed oxide growth from a molten metal. Excellent mechanical properties are claimed, flexural strengths of up to 1.25 GPa, works of fracture of 140 $kJ\,m^{-2}$ and fracture toughness as high as 26 $MPa\sqrt{m}$ for 1-D materials, and these properties can be maintained at up to 1100 °C.

With the development of more refractory fibres, currently at laboratory scale production, these classes of material can be expected to be applicable at up to 1400–1500 °C.

(b) Non-oxide matrix composites

Carbon/carbon and SiC/SiC have also had a long history of development. SEP has been associated with the SiC/SiC in particular and there is a range of commercially available materials/components.

Although very high bending strength (above 1000 MPa) was achieved on unidirectional material in the development phase, the final goal was the development of two-dimensional composites with a reliable strength at both room temperature and high temperature. Subsequent investigations produced 2-D silicon carbide matrix analogs of carbon fabric reinforced C–C composites. This material, CERASEP, had much higher strengths than expected, especially at temperature, at up to around 1200 °C. Both carbon and ceramic (SiC) fibre reinforcements were evaluated, and the composites found to be very tough and exhibit very high fracture energies due to fibre pull-out.

Figure 6.3 shows a typical stress-elongation curve for tensile specimens of C–SiC obtained at room temperature. Three zones can be observed on

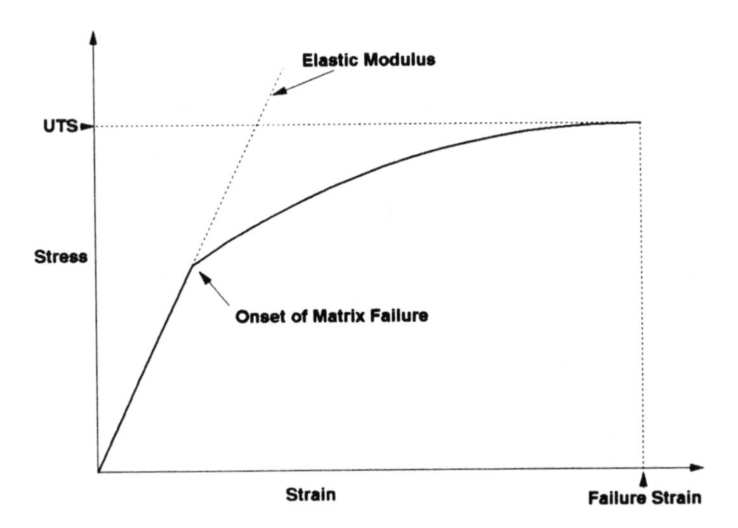

Figure 6.3 Generic stress/strain curve for axial tensile testing of 2-D woven C reinforced SiC.

this curve. In the first zone, the behaviour of the material is linear with modulus of 210 GPa up to an elongation of 0.035%. Then, the material becomes non-linear in zone 2, where there is believed to be fracture of the matrix and a transfer of the load to the fibres. At about 0.2% elongation, there is a change of the direction of the curve (zone 3) up to the ultimate strength around 190 GPa for an elongation of 0.6%.

Mechanical shock resistance using the IZOD method (ASTM 256) on a notched specimen shows the CERASEP material to be ten times more resistant to mechanical shock than monolithic silicon carbide. Water quench testing from temperatures at 200 ° to 1000 °C followed by residual 4-point bend testing shows that neither C–SiC or SiC–SiC CERASEP are greatly affected by thermal shock.

(c) Fibrous ceramics

Low density, part sintered compacts of ceramic fibres have been developed for the niche market of thermal protection. Most notably, Lockheed is associated with the development of 'Reusable Surface Insulation' materials for the Space Shuttle. Various combinations of silica and alumina fibres are sintered together, often with additions of titania (to modify the dielectric properties), boron nitride and silicon carbide, to form rigid panels.

High purity fibres confer high temperature capability; shrinkage is minimal at up to 1400 °C for extended exposure. The physical and mechanical properties are dominated by the materials density; strengths in the range 2–10 MPa are associated with materials of density $0.1 - 1\,\mathrm{Mg\,m^{-3}}$.

6.3.2 Manufacturing methods

Ceramic matrix composite materials are both hard and costly to machine; near net dimension fabrication and processing of components is necessary. Depending upon design requirements, a selection of fabrication techniques may be available, i.e. fabric lay-up, fabric moulding, tape winding, filament winding or combinations of these systems.

More recently, 3-dimensional braiding and weaving have been successfully used to obtain less anisotropic properties in the configured component. The finer fibre diameter carbon fibres and polymer precursor silicon carbide fibres, such as Nicalon, have been the fibre of choice for fabrication ease, even with moduli of 552–690 GPa. With the exception of the Nicalon type fibre, which has a 10–12 μm diameter, ceramic fibres have been found to be difficult to fabricate except for the most simple shapes. Given this preform there are several routes to introduce the matrix phase.

(a) Oxide matrix composites

Historically, the first and most widely used process is similar to resin composite manufacture. A preform of the ceramic fibres impregnated with matrix phase powder fixed by organic binder addition is made. The preforms are stacked in graphite tooling and hotpressed to a temperature where the matrix particles sinter forming a low porosity body. The process is shown schematically in Figure 6.4. Considerable efforts are underway to optimise the processing conditions to ensure a consistent preform, an optimised interface between the fibres and matrix and minimised fibre degradation. Both sol–gel derived matrices and liquid phase infiltration have also been examined to differing degrees.

Recent studies have examined the directed metal oxide (DIMOX) process developed by Lanxide Corporation. Here an oxide matrix, typically Al_2O_3, is produced, by the controlled oxidation of specific alloy compositions, directly through a preform. This process is innovative and, potentially, cheap to productionize to give near net shape components. There are problems associated with the highly reactive process degrading the reinforcing fibres and precoating the preforms, and modifications to the oxidized alloy compositions are being investigated.

(b) Non-oxide composites

Non-oxide matrices are introduced by either a gas phase or via a polymer precursor. Chemical vapour infiltration (CVI) allows the most refractory materials, such as carbon/carbon and silicon carbide/silicon carbide, to be composited directly into components. The fabrication rate for the

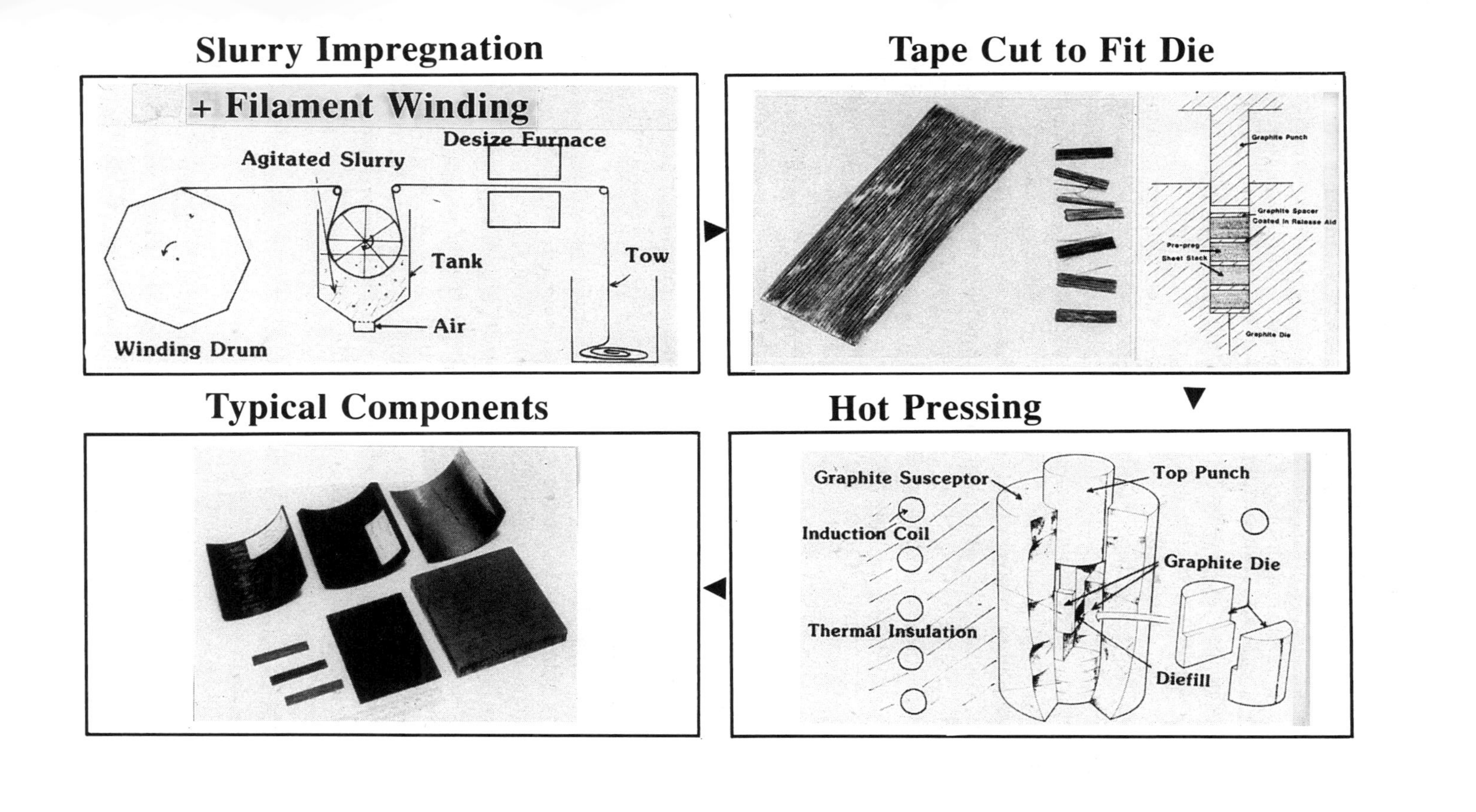
Slurry Impregnation
+ Filament Winding
Desize Furnace
Agitated Slurry
Tank
Tow
Air
Winding Drum
Tape Cut to Fit Die
Graphite Punch
Graphite Spacer
Coated in Release Aid
Pre-preg
Sheet Stack
Graphite Die
Typical Components
Hot Pressing
Graphite Susceptor
Top Punch
Induction Coil
Graphite Die
Thermal Insulation
Diefill

initially developed, isothermal processes was extremely slow, and hence very expensive. More recently a forced flow CVI under a controlled temperature gradient has been developed at Oak Ridge laboratory. This offers a significant reduction (*ca.*1/5) in the processing time and encourages the reaction to occur in a step by step, localized build-up through the preform that also reduces the residual porosity. The currently achieved minimum porosites are generally *ca.*20%.

Uniform fibre spacing and an open, non-bottleneck pore structure are critical in producing high strength, tough CVI carbon and ceramic matrix composites materials. As such, preform fabrication assumes an important role in designing, selecting constituent materials and processing desired components. The fibre form; diameter, filament count, orientation and/or weave significantly influence the options open to the fabricator. Because the CVI matrix only deposits on the exposed surfaces of the fibre preform, long tunnel-like (lenticular) pores are necessary to prevent a premature sealing of the internal or surface structure as the densification proceeds. In practice the infiltration process involves periodic surface machining to attain an acceptable density throughout the component.

While pore structure is not quite so critical in polymer pyrolysis systems, variations in fibre diameter and fibre spacing have a considerable effect on the properties and performance of the composite materials and components produced. Large diameter fibres (20 – 150 μm) and high filament count fibre bundles tend to produce composite structures with large fibre crossover discontinuities. These crossover discontinuities result in matrix rich areas which, in turn, tend to respond to composite loading in much the same manner as the low fibre volume systems. Conversely, fine fibre diameters (5 – 10 μm) coupled with low to intermediate filament count (1000–4000) fibre bundles provide a much greater uniformity in fibre spacing within the composite preform and a corresponding improvement in strength and property uniformity.

The 'green' body is then heat treated to transform the matrix into the ceramic phase. The problems of control of the interface and matrix phase shrinkage need to be addressed in considerable detail before structurally acceptable components will become available.

(c) Fibrous ceramics

Rigid fibrous ceramic components are produced to near net shape by combining the fibres, fillers and binders with water as a slurry. This is then cast or vacuum formed in a contoured mold, dried, sintered and finish machined. A variety of coatings is used to provide improved handleability or specific surface characteristics.

6.3.3 Component designing and applications

Composite materials inherently incorporate a combination of constituent materials and/or phases. Components can be fabricated through a combination of fabrication methods and experience several, very different process variables. Each of these factors must be designed to meet specific operational needs as slight variations in any of these parameters produces significant changes in component reproducibility and performance.

The issues of design, materials selection, fabrication and processing are, quite correctly, application driven, but there is a lack of objective design engineering methods to justify these decisions. The determination of what issues may be controlling in any given application is often difficult. Very often a subjective ranking method is used and these are then incorporated into the constituent materials, fabrication technique and process variable selection and used to 'optimize' the specification to meet identified component operational goals.

The largest usage of continuous fibre reinforced ceramics are carbon/carbon disc brake components. The light weight and capability of maintaining efficient performance at high temperature have lead to both military and increasingly civil application. Similarly, rocket nozzles in both C–C and SiC–SiC are production components that are in regular use. The stresses experienced by such components are of low complexity and simple fibre architectures are acceptable.

Fibrous ceramics have been used predominantly for the thermal protection of the Space Shuttle during re-entry manoeuvres. They have demonstrated considerable success in allowing re-usability of the panels and the perceived problems of attachment to the airframe have been, to a great extent, overcome. The logical extension – their use for thermal protection of exhaust structures and combustor components in gas turbines – is under active development by all the major gas turbine manufacturers.

In the main, structural applications remain at a development/demonstration stage. Figure 6.5 illustrates some of the applications being addressed in the gas turbine industry. For the oxide matrix systems a range of components are under evaluation in both the US and in Europe. A low expansion glass (such as Pyrex)/Nicalon (SiC) is being evaluated for applications at up to 600 °C and both Cordierite and Anorthite/Nicalon (SiC) are being evaluated for applications at up to 1000 °C. In the longer term there are many applications for a material with 1400 °C capability. A target matrix system for this capability is mullite (type); currently no fibre system is available for reinforcement with this temperature capability. Work is underway using Nicalon reinforcement, as a model, of these high temperature systems.

For the two lower temperature windows the materials exist and efforts are concentrated on the manufacturability of generic components and the

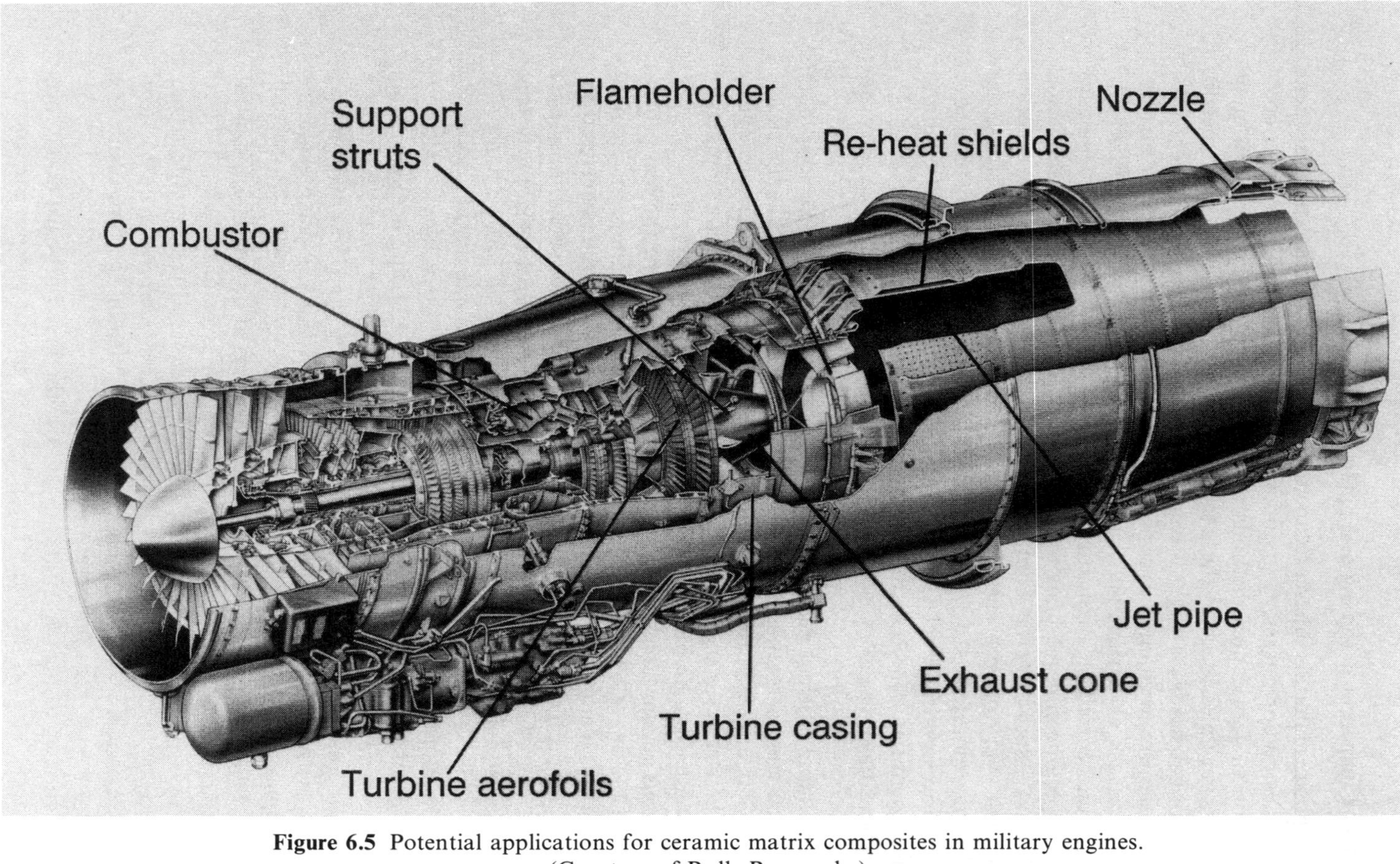

Figure 6.5 Potential applications for ceramic matrix composites in military engines. (Courtesy of Rolls Royce plc.)

development of test methods, for both coupons and sub-elements, to generate design and lifing data and methodologies. The evidence and experience to date indicates that glass and glass-ceramic matrix composites can be considered to be high temperature analogues of resin matrix systems in terms of design, processing and mechanical behaviour.

The fact that the fibres are stronger and stiffer than the matrix gives room for optimism that such an analogous approach on methodologies is appropriate. The critical issues under examination are the determination of the mechanical behaviour and failure mechanisms of these materials across the wide range of mechanical and thermal environments that will be observed in practice. The control of the fibre/matrix interface, from whence comes the advantageous properties, is a key factor in the lifing of these materials.

Considerable model development work and empirical justification is required before such materials/components will be used on a commercial scale and allow the potential benefits to be seen in practice in the aerospace industry.

7

Polymeric-based composite materials

N. Marks

7.1 INTRODUCTION

Since the chapter is dealing entirely with polymeric-based composite materials, any future reference within the chapter to composite materials will imply the use of polymeric matrices. Since my background is within the helicopter industry, many of the examples will be taken from this industry, but in all cases they are equally applicable to fixed wing components. In order to assist, a short glossary of terms is given in Table 7.1.

Composite materials are not new: they were being employed centuries ago when man was making dwellings from mud and straw but the types of composite being considered here are relatively new. These have been in use for aircraft applications over the past 30 years to some extent, but there has been a rapid increase in their use over the last 15 years.

The aircraft industry started using fibreglass components in the early 1960s but they were only used for tertiary structures such as fairings or non-structural doors. It was in the mid to late 1960s that companies started examining composites for use on structural components and this coincides with the introduction of carbon fibre which probably had some bearing on the policy change. It was realized that the fibre reinforcement would allow designers to tailor the strength of a component in the direction most needed by strategic orientation of the fibres. There was also a realization that this technology had the potential of reducing the weight of components.

During this same period, there was extensive research into improving matrix materials. The original glass fibre components were manufactured using polyester resin matrices but these had restricted properties and composites were improved by the introduction of epoxy resins. A further

Table 7.1 Glossary of terms

Aromatic	Unsaturated hydrocarbon containing at least one benzone ring in the molecule.
Curing agent	A reactive agent which when added to a resin causes polymerization to occur.
Fibre	Material which has a high length-to-thickness ratio.
Filament	A single fibre.
Interface	The boundary between two different media, i.e. between the fibre and the matrix.
Matrix	The resin which binds the fibres together.
Polymer	A high molecular weight organic compound where structure contains small units.
Polymerization	A chemical reaction in which the molecules of the monomer are linked together to form large molecules.
Precursor	The fibres, such as polyacrylonitrile, from which the carbon fibres are derived.
Specific properties	The material properties divided by the specific gravity of the material.
Void	Air or gas which has been trapped within a cured resin, laminate or adhesive.
Warp	The end running lengthwise in a woven fabric.
Weft (fill)	The end running across the width of a woven fabric.

progression was the development of preimpregnated materials using resins which cured at higher temperatures with consequent improvement of properties, particularly regarding higher working temperatures and ductility. The preimpregnates also allowed more control over processing, particularly in producing a more consistent end product.

The introduction of newer matrices has extended the range of composite materials even more and this has considerably extended the range of applications. The polyimide resins allow composite components to be subjected to temperatures in the range of 250 °C or more and polyetheretherketone is a thermoplastic matrix with excellent chemical resistance and greater damage tolerance than epoxy based composites.

Composite materials are now used extensively on Class 1 structures for fixed wing aircraft and helicopters, and, as can be seen from Figure 7.1, a modern helicopter has a high proportion of its structure manufactured from composite materials.

7.2 REINFORCEMENTS

Composites can be reinforced with short fibres or with continuous fibres. The short fibres are normally employed for injection mouldings and other similar applications and will not be considered here as it is intended

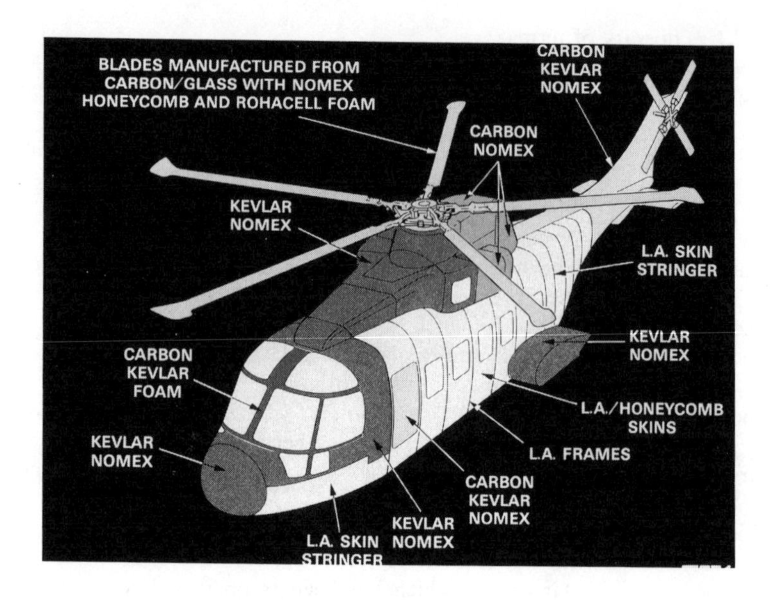

Figure 7.1 Composites in EH101.

to concentrate on continuous fibre reinforced composites for structural applications.

Any fibre which can be formed into long filaments would be potentially suitable for reinforcing composite laminates. There are, however, relatively few which have been developed to a position which makes them suitable for utilization on aircraft components. They are glass, carbon, boron and aramid fibres.

7.2.1 Glass fibre

Continuous filament glass fibres are manufactured by allowing molten glass to pass through minute holes in a metallic block under gravity and drawing the filaments so produced. These are then spun into fibres.

The most commonly used fibre for manufacturing composite components is E glass. This material has a calcium aluminium borosilicate base and is a good general purpose fibre.

Two glass fibres which are used extensively in the aircraft industry are S glass and R glass. S glass was developed in the USA and has a magnesium aluminium silicate base. R glass was developed in Europe and has properties similar to S glass. These materials have a higher strength and modulus than E glass and they also retain their strength to higher temperatures. They are, however, more difficult to produce than E glass and are consequently considerably more expensive.

7.2.2 Carbon fibre

Carbon fibre, which is known as graphite in the USA, is the fibre which is most extensively used in the aircraft industry. This fibre was first developed by researchers at Royal Aircraft Establishment, Farnborough, UK, following work by Shindo in Japan in the early 1960s and it has probably had a greater influence on the use of composites in the aircraft industry than any other factor. Carbon fibres can be produced from a number of precursors to offer different properties but the two which are mainly used are polyacrylonitrile (PAN) and pitch.

(a) PAN-based fibres

A multistage process is used to convert the PAN precursor to carbon fibre. Although the actual details of the process are confidential (because the structures are very sensitive to rates of heating), the basic process is as follows. An initial stage converts the orientated linear fibre to an orientated ladder fibre. It is then necessary to maintain a controlled tension on the fibre during the next two stages. The first is an oxidation process during which the fibres are heated to 350 °C in an atmosphere of oxygen. This is followed by a condensation stage when the fibres are heated to about 1200 °C in an inert atmosphere such as nitrogen. This produces fibres which are strongly orientated but have poor crystal arrangement. The fibres are therefore subjected to a graphitization stage to produce fibres with a well orientated ribbon structure. This is achieved by heating in an argon atmosphere to temperatures above 1500 °C. The temperature of treatment at this stage modifies the fibre properties, therefore it is necessary to tailor the heat treatment to obtain the optimum properties for the fibre.

The fibres produced by this process have the following advantages: they are easily spun into highly orientated very fine fibres, the fibres are stabilized, with the molecular orientation axially parallel to the orientation of the original precursor, and they have a relatively high yield. Improvements in the technology in recent years have resulted in superior strengths and strain to failure for this type of carbon fibre.

(b) Pitch-based fibres

These are also produced by a multistage process. The first stage polymerizes the pitch; this is achieved by heating it to 350 °C when it is melt spun into fibres. During this process, the material is converted into a highly anisotropic product with a high degree of orientation parallel to the fibre axis. The fibres are then oxidized and carbonized by heating to temperatures of the order of 2000 °C. Finally they are graphitized by heating to very high temperatures, 2900 °C. During this process the orientation of the fibres is increased still further. This process produces fibres with very high modulus without the necessity of high temperature stretching. Fibres have been produced which approach the theoretical value for a graphite crystal: 1000 GPa.

7.2.3 Boron fibres

These are produced by vapour deposition of boron onto heated tungsten wire. Tungsten wire, approximately 12 μm in diameter is continuously drawn through a glass reactor. Electrical power is fed into the tungsten wire causing it to heat up and a mixture of boron trifluoride vapour and hydrogen is passed into the tube. The boron trifluoride and hydrogen react and boron is deposited on to the surface of the tungsten. Fibres of between 100 and 150 μm have been produced by this technique. The fibres produced have high strength and modulus with a density similar to that of E glass but are very expensive.

7.2.4 Aramid fibres

These were the first organic fibres to be produced which had properties which made them suitable for reinforcing structural composites and were introduced in the early 1970s. The first fibres produced by Dupont under the trade name Kevlar had the chemical formula polypara phenylene tetraphthalimide.

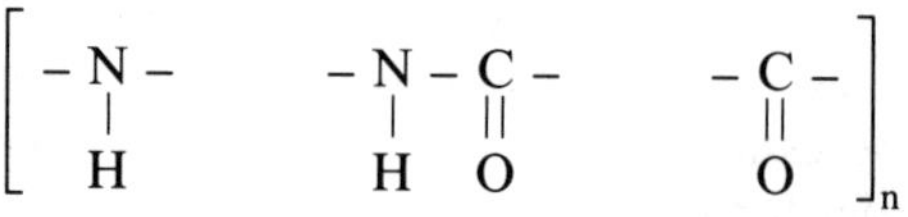

$$\left[\begin{array}{ccc} -\underset{\displaystyle H}{\underset{|}{N}}- & \quad -\underset{\displaystyle H}{\underset{|}{N}}-\underset{\displaystyle O}{\underset{\|}{C}}- & \quad -\underset{\displaystyle O}{\underset{\|}{C}}- \end{array} \right]_n$$

The aromatic ring structure contributes to its high thermal stability and the para configuration leads to stiff rigid molecules resulting in high strength and modulus. It is a liquid crystal polymer, therefore if it is taken into solution then extruded through a spinneret it produces a fibre which has straight polymer chains and is extremely well aligned parallel to the fibre axis.

Aramid fibres have high tensile properties and since they have a low density their specific strength and modulus are very high. They do, however, have rather low compression properties and are affected by moisture.

7.2.5 Comparison between fibre reinforcements

Having established the different types of fibre reinforcements available, it is now necessary to compare their properties. Table 7.2 compares the tensile properties of the different fibres. As stated earlier, specific

Table 7.2 Properties of reinforcing fibres

Property	*Ultra high modulus carbon*	*Intermediate modulus carbon*	*Medium modulus carbon*	*Aramid*	*R glass*	*E glass*	*S glass*	*Boron*
Tensile Modulus (GPa)	724	310	235	124	86	69	88	400
Tensile Strength (GPa)	2.2	5.2	3.8	3.6	4.4	2.4	4.6	3.52
Density (g/cc)	2.15	1.8	1.8	1.45	2.55	2.54	2.49	2.58
Specific Modulus (GPa/g cc)	336	172	131	85	34	27	35	155
Specific Strength (GPa/g cc)	1.02	2.9	2.1	2.5	1.7	0.95	1.85	1.36

strength and modulus are very important factors when choosing materials for aircraft applications. If these were the only factors then intermediate modulus carbon has good all-round properties and for extremely stiff lightweight structures, ultra high modulus carbon would be a strong contender, but obviously there are other things to consider. The following section therefore compares the overall properties of the different fibres.

- E glass fibre. This has the lowest cost; it has high electrical resistivity, good chemical resistance and high durability but its modulus is low and it has a high density.
- R glass fibre. This is considerably more expensive than E glass but it has a higher tensile strength, higher temperature resistance and better chemical resistance. The other properties are very similar to those of E glass.
- Aramid fibres. These fibres are more than twice the price of E glass fibres but they have low density, intermediate modulus and high toughness. They do, however, absorb moisture, they are degraded by strong acids and alkalis, have poor compressive strength and a maximum service temperature of 150 °C.
- PAN-based carbon fibres. These are available in a range of moduli from medium (235) to high (340 GPa). The cost of medium modulus fibres is about five times that of E glass and intermediate modulus is about seven times. They have high specific strength and modulus at both room temperature and elevated temperatures, excellent chemical resistance and do not suffer from stress corrosion. They are, however, difficult to wet out.
- Pitch-based carbon fibres. These are extremely expensive, being more than fifteen times the cost of E glass fibre. They have extremely high specific moduli and excellent specific strength both at room temperature and elevated temperature. They are very brittle in nature, which makes them difficult to process.
- Boron fibres. These are also very expensive, being approximately twenty-five times the cost of E glass fibres. They have high strength and stiffness with a low density, giving good specific properties. Probably because of the cost, composites using boron fibres have not been developed, particularly in Europe, to any great extent.

One of the great advantages of fibre reinforced composites is that the fibres can be mixed to produce a hybrid laminate. In this manner, it is possible to tailor the properties to the requirements of the component.

7.2.6 Forms of reinforcement

All the discussions to date have concentrated on unidirectional properties of composite laminates. This form of reinforcement is used extensively within

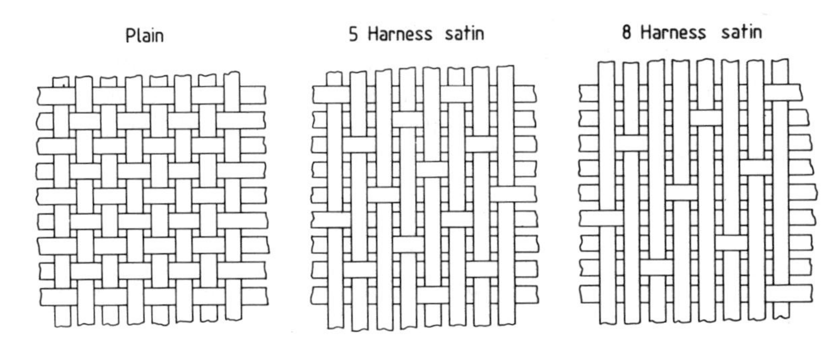

Figure 7.2 Types of weave.

the aircraft industry but other forms are also used and newer ones are being developed. Probably the form most extensively used in the composite industry is in the form of a woven fabric. In fact, when the original fibreglass components were made, this was the only form of continuous fibre reinforcement. Glass, medium modulus carbon and aramid fibres are available in many different weave patterns. A woven fabric consists of warp fibres which are the ones held in a taught position during the weaving process and the weft of fill fibres which are the ones woven through the warp fibres. The two weaves mainly employed in the aircraft industry (Figure 7.2) are plain weave, which has one warp fibre over and under one weft fibre, and satin weave. Two types of satin weave are used extensively: five harness satin which has one warp fibre over four and under one weft. Satin weave fabrics are more drapable than plain weave but are heavier.

The advantages of a woven fabric are as follows:

- The properties of a fabric give similar properties in both directions although those in the weft direction are marginally lower because the fibres are crimped when they traverse the warp fibres.
- They are easier and faster to lay up than unidirectional materials because the equivalent of two layers are laid at a time.
- The thickness of the ply is more controlled.
- It is possible to have a hybrid weave.
- It is easier to form over curved surfaces.
- They have better properties in a notched laminate.
- The cost is lower.

The disadvantages are:

- Low mechanical strength because of the crimped fibres.
- More chance of entrapping air in the laminate.

In addition to unidirectional and woven materials, other forms of reinforcement are being developed for use in composite structures and some of these appear to have some potential for use on aircraft structures. Two which certainly appear to have potential are knitted fabrics and braids. The main advantage of these types of reinforcement is that they provide three-dimensional reinforcement. This could overcome one of the weaknesses of composites – interlaminar.

7.3 MATRICES

There are two basic types of matrix used on composites: thermosetting and thermoplastic.

A thermosetting resin consists of at least two completely separate components – base and a curing agent. When the two components are mixed together, a chemical reaction takes place and a solid resin is formed. With many of these resins, it is necessary to employ elevated temperatures for the reaction to occur. This process is known as polymerization and once a component has been formed it is not normally possible for it to be reshaped.

A thermoplastic matrix is a fully polymerized material and it is solid at room temperature but when heated it will soften and eventually melt. In either of these two conditions it can be formed to shape or it can be laminated. On cooling, the matrix solidifies, keeping the formed shape. This process can, however, be repeated as often as is necessary.

There are also some semi-thermoplastic materials which are polymerized by a chemical reaction similar to the thermosetting matrices but it is possible to soften the polymerized material and reshape it.

Table 7.3 lists typical matrix materials and indicates the category into which they fall.

Table 7.3

Thermosetting matrices	*Thermoplastic matrices*
Epoxy	Poly ether ether ketone
Polyimide (bismaleimide)	Poly ether imide
Polyester	Poly phenylene sulphide
Phenolic	Poly arylene sulphone

7.3.1 Thermosetting matrices

(a) Epoxy resins

These resins must be considered to have been fundamental in the development of structural composite components. All the initial development work, in the 1970s, employed epoxy matrices and they are still used extensively. Epoxy resins are being continually developed and it is probable that they will be in use well into the future.

The base component of an epoxy resin contains an epoxide grouping which can be reacted with compounds containing a reactive hydrogen group, such as an amine, to form the cured resin. A typical reaction is shown below:

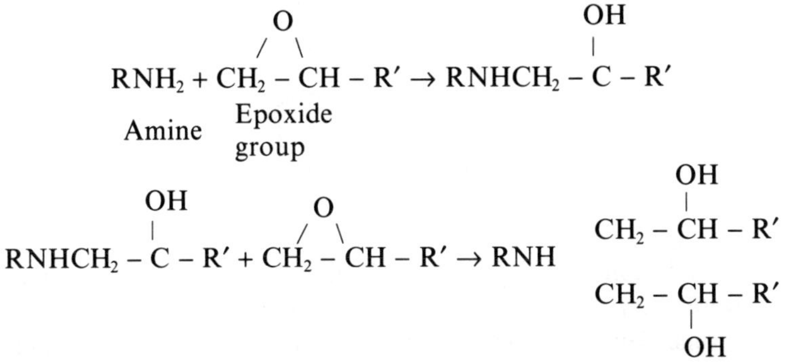

$$\underset{\text{Amine}}{RNH_2} + \underset{\text{Epoxide group}}{CH_2 - \overset{O\ (\text{bridging})}{} CH - R'} \rightarrow RNHCH_2 - \overset{OH}{\overset{|}{C}} - R'$$

$$RNHCH_2 - \overset{OH}{\overset{|}{C}} - R' + CH_2 \overset{O}{-} CH - R' \rightarrow RNH\begin{matrix} CH_2 - \overset{OH}{\overset{|}{C}H} - R' \\ CH_2 - \underset{OH}{\underset{|}{C}H} - R' \end{matrix}$$

Many epoxy bases have been used to produce composite components but currently in the aircraft industry there are two main epoxy bases. The first is the diglycidyl ether of bisphenol/A (DGEBPA) which has the following formula:

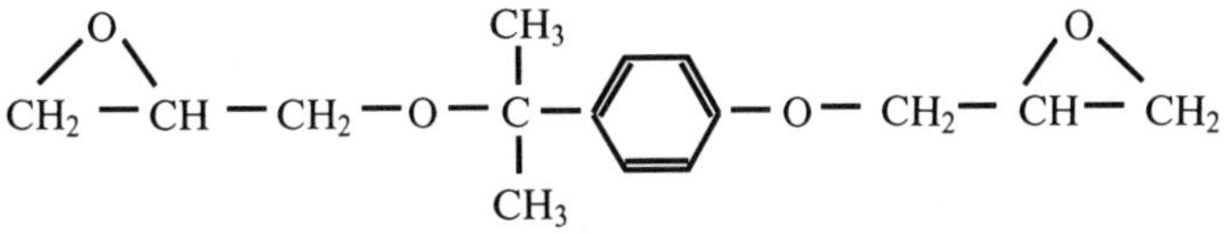

$$\overbrace{CH_2 - CH}^{O} - CH_2 - O - \overset{CH_3}{\underset{CH_3}{\overset{|}{\underset{|}{C}}}} - C_6H_4 - O - CH_2 - \overbrace{CH - CH_2}^{O}$$

The other one is tetraglycidyl 44′ diamino diphenyl methane (TGDDM)

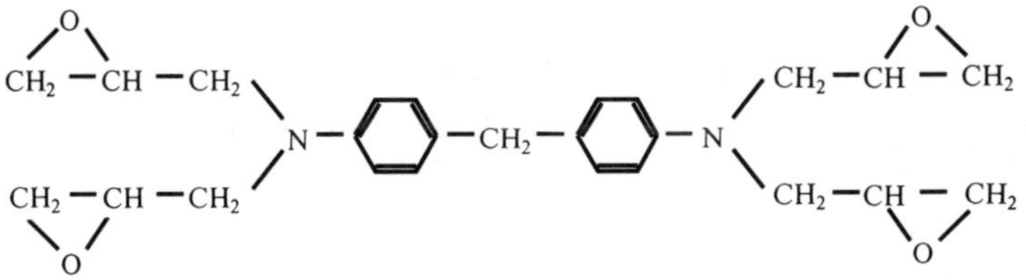

$$\begin{matrix} \overbrace{CH_2 - CH}^{O} - CH_2 \\ \underbrace{CH_2 - CH}_{O} - CH_2 \end{matrix} \gt N - C_6H_4 - CH_2 - C_6H_4 - N \lt \begin{matrix} CH_2 - \overbrace{CH - CH_2}^{O} \\ CH_2 - \underbrace{CH - CH_2}_{O} \end{matrix}$$

TGDDM has become the most widely used resin in the production of aircraft structural components.

Many curing agents have been used within the aircraft industry, particularly boron trifluoride complex and diamino diphenyl sulphone (DDS) but the curing agent most widely employed at the present time is dicyandiamide (dicy) which has the formula $H_2NC(NH)NHCN$.

(b) Polyimide resins

As stated earlier, polyimides can be catagorized as thermosetting resins, which are cross-linked, and semi-thermoplastic resins, which are formed from a condensation reaction between anhydrides and diamines.

If we first consider the thermosetting polyimides, the areas used for the production of aircraft components are the bisimaleimides (BM1) resins. These have the general chemical formula shown below:

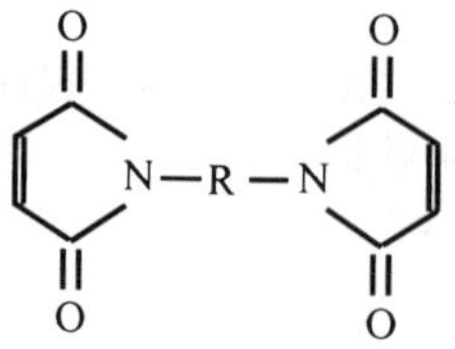

O O

N—R—N

O O

The original resins were based on bismaleimidodiphenyl methane but these were very brittle and were considered unsuitable for aircraft applications. Several companies modified the process to overcome this problem and also to improve the processing. Typical materials used on aircraft structures are Compimide and Kerimid.

The semi-thermoplastic polyimides are fully imidized linear polymers and are produced by a condensation reaction between aromatic dianhydrides and aromatic diamines and have the general formula:

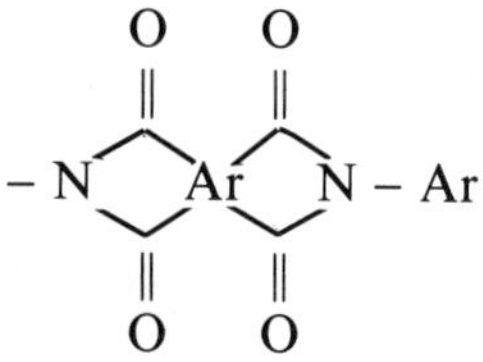

O O

– N Ar N – Ar

O O

Ar = aromatic grouping

The original materials were relatively difficult to process, therefore considerable modifications have been made and materials such as Aramid K from Dupont or Torlon from Amoco have been produced.

7.3.2 Thermoplastic matrices

(a) Poly ether ether ketone (PEEK)

Although thermoplastics composites have been in use in the injection moulding field for many years, these employ very short fibres, of the order of 200 μm long. The first thermoplastic to enter the field of structural composites suitable for the aircraft industry was based on a PEEK matrix. This was introduced by ICI in the early 1980s under the tradename Victrex APC1. This was subsequently replaced by an improved product, APC2 which is still one of the main thermoplastic composites employed in the aerospace industry. The basic chemical structure of PEEK is:

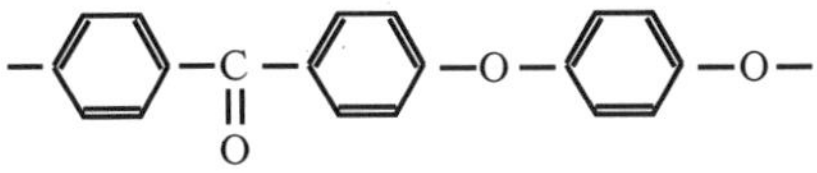

This is a semi-crystalline material which results in excellent chemical resistance. Because of this, it is not practical to solvent impregnate PEEK and it is normally impregnated by a melt process.

(b) Polyetherimide (PEI)

This material has been in use for some considerable time and was first introduced as a woven impregnate. It is normally produced using solvent impregnation. The basic chemical structure is:

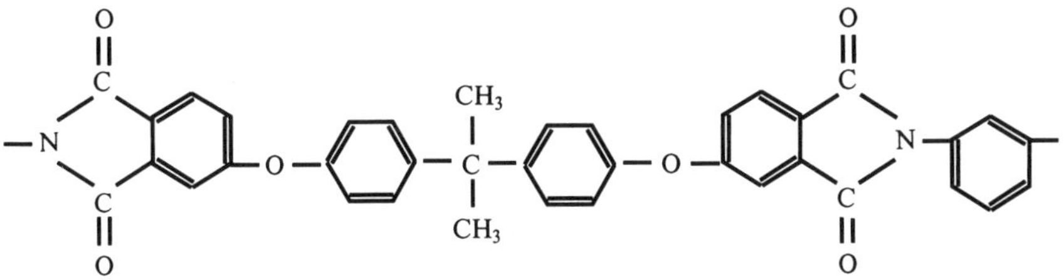

It is an amorphous thermoplastic, therefore its chemical resistance is inferior to PEEK.

(c) Polyphenylene sulphide (PPS)

This was introduced as a structural composite material soon after the introduction of APC1. It is a semi-crystalline polymer but has an inferior chemical resistance to PEEK. The initial usage tended to be restricted to the USA. The basic chemical structure is as follows:

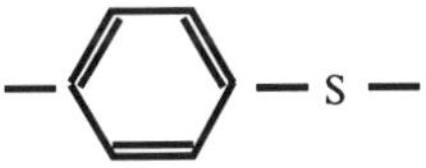

(d) Polyarylenesulfone

This material is a relative newcomer in thermoplastic matrices but it has properties which make it attractive for aircraft use.

Table 7.4 gives a comparison between the properties of the matrices used or with potential for use in the aerospace industry.

7.4 INTERFACE

The fibres in the composite material provide the basic strength of the component but, if the full strength and stiffness is to be fully employed,

Table 7.4 Comparison of matrix properties

Matrix	*Maximum hot/wet working limit*	*Advantages*	*Disadvantages*
Epoxy	120°C	Established technology Excellent fibre adhesion Good shelf stability Low cure temperature Low cure pressure No volatiles Low shrinkage Good tack Good chemical resistance	Long cure times Absorb moisture Requires storage at − 18°C
Thermosetting polyimides	220°C	High service temperature	Long cure cycles Difficult to process and tend to contain voids Brittle, expensive
Semi-thermoplastic polymides	260°C	High service temperature. Can be shaped after curing	Very long cure cycles Expensive
PEEK	120°C	Indefinite shelf life Low moisture absorption Tough Excellent fibre translation of properties Excellent chemical resistance	High processing temperature No tack Very boardy Expensive prepreg
PEI thermoplastic polyimides	120°C	Indefinite shelf life Low moisture absorption Tough Low cost	Very boardy prepreg No tack Poor, solvent resistance

it is essential that they are well bonded to the matrix. The efficiency of the composite is therefore greatly affected by the interface between the matrix and the reinforcing fibres. If there is a weak interface then the laminate will have a low stiffness and strength but it will have a greater resistance to fracture. Conversely, a laminate with a strong interface will have a high strength and stiffness but the composite will tend to be brittle.

In order to enhance the bond between fibre and matrix, it is often necessary to apply a coating of size to the fibres. An efficient interface can also reduce the effect of moisture on the properties of a laminate. The methods of enhancing the interface for the different fibres are given in the following sections.

7.4.1 Glass fibres

These are normally coated with a size which has a two-fold effect: it protects the fibres from damage and it provides a chemical link between the fibre and the matrix. Most of the sizes currently used in epoxy-based composite are silanes.

A typical silane for use as a size is shown:

$$\overset{\displaystyle O}{\overset{/\ \backslash}{CH_2 - CH}}\ CH_2O(CH_2)_3\,Si(OCH_3)_3$$

polymer grouping — silane grouping

The silane grouping will give a chemical link to the glass fibre and the polymer molecules to the epoxy resin.

Although it is more difficult to bond in the same manner to a thermoplastic matrix, it is still normal to size the fibre with a silane.

7.4.2 Carbon fibres

Carbon has a highly active surface and a range of functional groups can be produced at this surface by different oxidation treatment such as heating in oxygen or treatment in nitric acid or sodium hypochlorite. These functional groups can form chemical bonds directly with unsaturated resins and unsaturated groups in thermoplastic resins. Adhesion promotion has also been obtained by using polymer or silane coatings on the fibres.

7.4.3 Aramid fibres

In order to avoid surface damage during processing, aramid fibres can be coated with polyvinyl alcohol size. Conventional coupling agents have not proved very effective but fibres can be given a light coating of an epoxy resin to promote adhesion between the fibres and the matrix.

7.5 PROCESSING

7.5.1 Thermosetting composites

(a) Wet lay-up

This was the original method for producing composite components and is often referred to as the 'bucket and brush' technique. By the nature of the technique, it is really restricted to use with fabrics in conjunction with resins which are liquids at room temperature.

As the bucket and brush technique implies, the method is very simple. The original moulds were made from wood coated with a wax parting agent, and this is still used, but a variety of tooling materials are employed.

A layer of resin is brushed on to the tool and a layer of fabric placed on top of the resin. The resin is absorbed into the fabric by stippling or rolling until the fabric is completely wetted out. More resin is placed on to the wetted fabric, another layer placed on to the resin and the process repeated until the laminate has been built to the required thickness.

The resin can then be cured at room temperature or at elevated temperatures. Non-structural laminates can be cured without pressure but structural laminates will be cured under a vacuum or pressure. Modifications to the process have been made but basically it is still the same process. This method of producing laminates has the following disadvantages:

- The process entraps air during the laminating and this is very difficult to remove in subsequent processing.
- It is extremely labour intensive.
- It is very difficult to accurately control the resin to fibre ratio.
- The resins have to be handled by the operator and most of them do have some health hazards.
- The resins have to be mixed in relatively small quantities because large volumes can result in an exotherm.
- It is a very messy process.
- The laminates produced are to a relatively low standard.

The wet lay-up technique is therefore not employed to manufacture structural components for aircraft use. It is, however, still employed to carry out repairs to both composite and metallic structures.

(b) Preimpregnates (also known as prepregs)

The use of composites on aircraft structures was accelerated by the introduction of preimpregnates. As the name infers, a preimpregnate is a material in which the matrix resin has been impregnated into the fibres and is supplied in a form ready for immediate use. The process includes passing the impregnated fibres through pinch rollers; this helps to wet out the fibres with the resin, removes excess air and also controls the ratio of resin to fibre.

Since it is necessary to have a preimpregnate which can be stored for a considerable length of time, it is essential that they remain in a state suitable for laminating for long periods, i.e. they must have long shelf-lives. In order to achieve this, resins and curing systems are employed which require to be heated to allow the chemical reaction to take place. Also, prepregs are normally stored at depressed temperatures (normally – 18 °C) which will retard the chemical reaction even more. In this manner a typical prepreg currently being used for aircraft components will have a shelf-life of 12 months at – 18 °C with an out-time of 30 days.

One of the big advantages of prepregs was that they allowed the introduction of unidirectional materials. The resin in the preimpregnate acts as a binder which holds the fibres in place. Having unidirectional preimpregnates allows the designer to place the fibres in the direction where the main load carrying path is required.

Another advantage of prepregs is that it is possible to use matrices which are not liquid at room temperature. With this type of matrix, the resin can be taken up in a solvent for the impregnation process. After impregnation, the prepreg is subjected to a drying process to remove the solvent. An alternative method is to form a thin film of the resin and apply this film to the fibres. By passing the combination through rollers the resin is impregnated into the fibres.

There are several methods in which these preimpregnates can be processed to form laminates. The basic requirements are heat and pressure. The most commonly used methods are autoclave, press clave, matched tool using a press, vacuum alone, air bag, consolidation using a foam. The next section will discuss some of these methods in more detail.

Autoclave moulding is probably the most popular method of forming components from preimpregnates. The component is laid up into either a male or female tool. This is then enclosed in a flexible membrane such as a silicone rubber sheet and the assembly sealed so that the component can be subjected to a vacuum. The whole assembly is then placed into an autoclave which is basically a pressurized oven. The gas pressure (this can be air or an inert gas) within the autoclave is raised to the requisite pressure (typical for an epoxy preimpregnate is 550 kPa (80 psi). The temperature within the autoclave is slowly raised to that necessary to cure the resin (typical aircraft epoxy systems cure at 175 °C or 125 °C) and the temperature maintained until the resin is cured. The temperature within the autoclave is then cooled but the pressure is maintained until the resin matrix has reached a safe temperature. This temperature will be dependent upon the resin being cured. It is sometimes necessary to 'post-cure' the component. In this process the component is allowed to free stand in an oven and heated to a temperature above that required to cure the resin. This process will improve the properties of the laminate.

Press-claving is very similar to autoclaving. Figure 7.3 shows a typical layout for this method of processing laminates. Again the laminate is subjected to a vacuum, then pressure is applied by passing gas into the

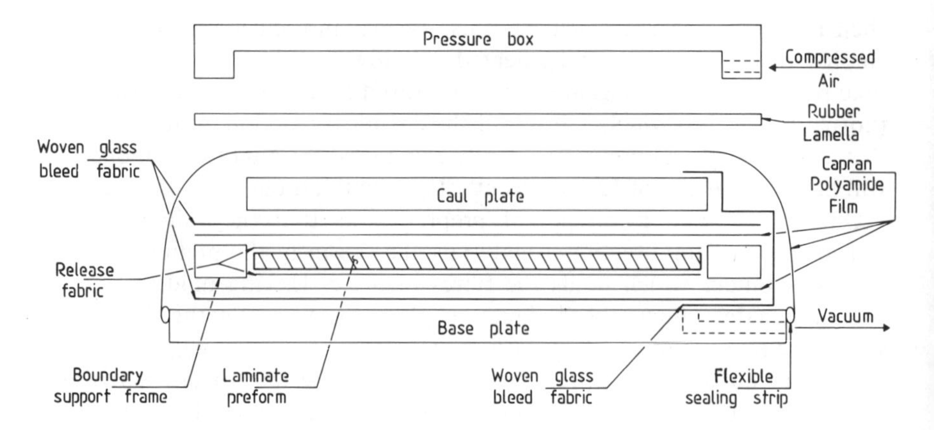

Figure 7.3 Press clave techniques.

pressure box, the whole assembly being placed in a press. This technique is particularly useful for producing flat laminates and is often applied in laboratories for making test laminates.

The matched tool method is only used for large quantity manufacture or for components which have to be produced to very close tolerances regarding profile. As the name suggests, a tool is manufactured to produce the finished shape of the component and the laminate is formed within the tool. The tool is placed into a heated platen press for the curing operation. The tool is normally manufactured from metal or fibre reinforced composite material. This is normally an expensive method of manufacture because of the high cost of manufacturing the tool.

Vacuum moulding is a low pressure technique for producing laminates. This method of manufacture does tend to produce an inferior laminate to one which has been cured under positive pressure, therefore it is not normally employed for producing aircraft components. It is, however, used for carrying out repairs on components where it is not practical to apply positive pressure.

The air bag technique is one which can be employed for manufacturing hollow comonents and has been used to great advantage to produce the main loadbearing component (the spar) of a helicopter main rotor blade. A female tool is manufactured to the outside profile of the component. The composite components are assembled around an elastomeric tube and the whole assembly is placed inside the female tool. The elastomeric tube is sealed to form an airtight tube so that gas pressure can be applied to the inside of the tube. The assembled tool is then placed in a platen press and the temperature slowly raised. At a temperature where the viscosity of the resin is at its most suitable, pressure is applied within the elastomeric tube and this consolidates the laminate to the shape of

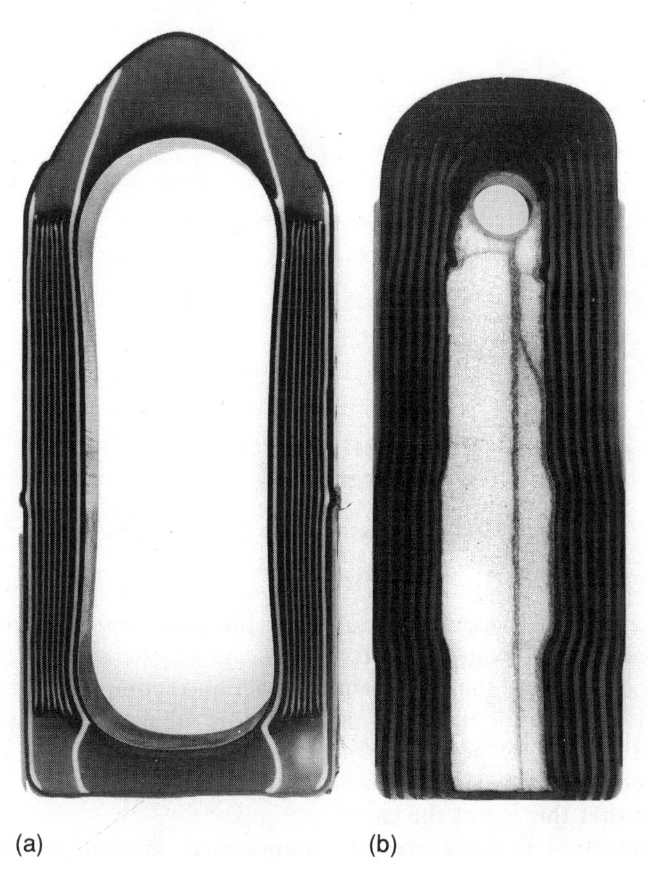

Figure 7.4 A cross section of the spar of a helicopter main rotor blade manufactured using the air bag technique.

the tool. A cross section of spar manufactured to this technique is shown in Figure 7.4 (a).

The final method to be discussed is foam consolidation. This is another technique which has been used to manufacture helicopter main rotor blade spars and an example is shown in Figure 7.4 (b). The same type of female tool is used, but instead of applying pressure by means of an elastomeric tube, the laminate is built around a central foam core. This core has to be manufactured from a material which has good compression properties at the cure temperature of the resin since the consolidating pressure is provided by the foam. This method has one main advantage over the rubber tube technique in that it can be automated and is therefore less labour intensive.

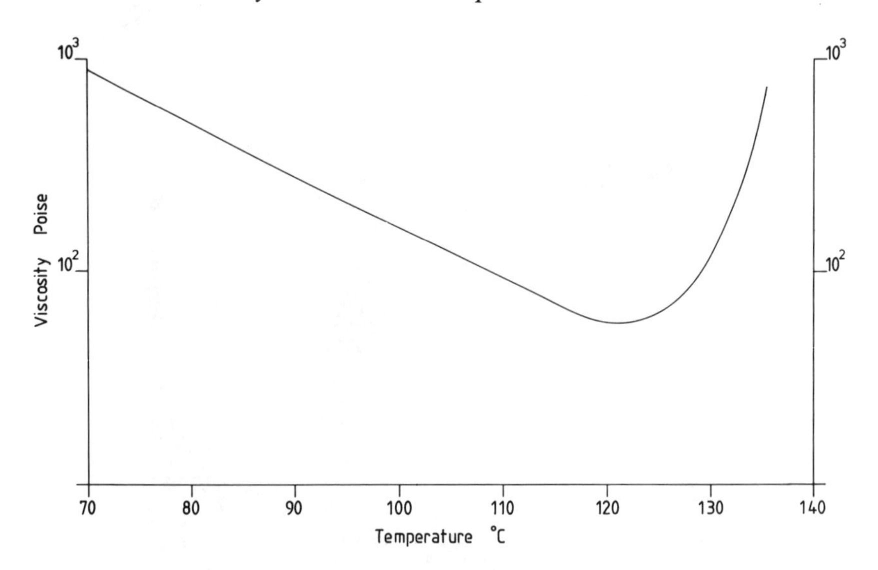

Figure 7.5 Rheometric curve for epoxy resin.

To complete this section on processing thermosetting preimpregnates, it is necessary to consider cure cycles. As stated earlier, the main objective is to ensure that the chemical reaction is complete and the resin is completely cross-linked. To achieve this objective, pressure plus elevated temperature need to be maintained for a certain time period. This sounds relatively simple but a closer scrutiny of a typical cure cycle will show that this is not the case.

It is important that the pressure is applied to the laminate when the resin is mobile and will flow. If we examine a typical rheometric curve for a 125 °C curing epoxy resin (Figure 7.5), it will be seen that the viscosity of the resin is lowered with temperature, but when the chemical reaction commences there is a sudden rise in viscosity.

Another important point to be considered is that the chemical reaction causes an exotherm to occur. This is particularly crucial with thick laminates because composites are generally poor thermal conductors and if the chemical reaction is not controlled there will be a sudden rise in temperature. The poor thermal conductivity means that there can be a considerable time lag between the parts nearest the heat source and those furthest away.

Because of all these complications, it is necessary to employ complicated cure cycles when making large components. This is illustrated in Figure 7.6, which shows a cure profile for a thick laminate curing at 125 °C. It can be seen that, although it is only necessary to maintain 125 °C for 1 hour, the full process is 12 hours long.

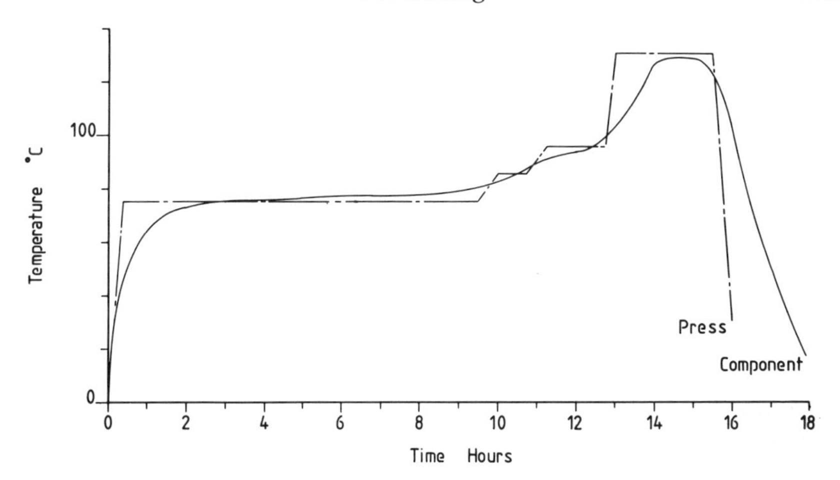

Figure 7.6 Cure profile for MRB.

7.5.2 Thermoplastic matrices

There are considerable differences in the techniques used to process thermoplastic materials because of the difference in their physical nature. When preparing a laminate, it is necessary to tack the layers together by melting the resin locally – this is normally done using a soldering iron or something similar. Component can be made directly from these laminates tacked together, or alternatively the laminates can be consolidated into a flat laminate and then shaped to the component in a separate process. There are many different ways of forming components, some of which are very similar to the thermosets; this section will only describe those techniques which are different: press forming, diaphragm moulding and deep drawing. One of the important factors when processing semi-crystalline thermoplastics is that the morphology of the laminate will be dependent upon the cooling rate and this must be taken into account.

There are two basic methods of press forming: hot compression and cold compression. For the former, the preimpregnate or laminate is placed directly into a matched metal tool. This is placed in the press and the whole assembly heated to the forming temperature. After forming, the whole assembly is cooled, with the pressure being maintained until a temperature is reached below which the laminate will retain its new shape (for APC2 this is 220 °C). For cold compression moulding, the laminated prepreg is heated to the forming temperature and then quickly transferred to a cold tool which has been mounted in a press. The tool is closed rapidly. The tool can be opened within minutes of forming.

Deep drawing is illustrated in Figure 7.7. The laminate is lightly held between two plates and the whole assembly heated to the forming

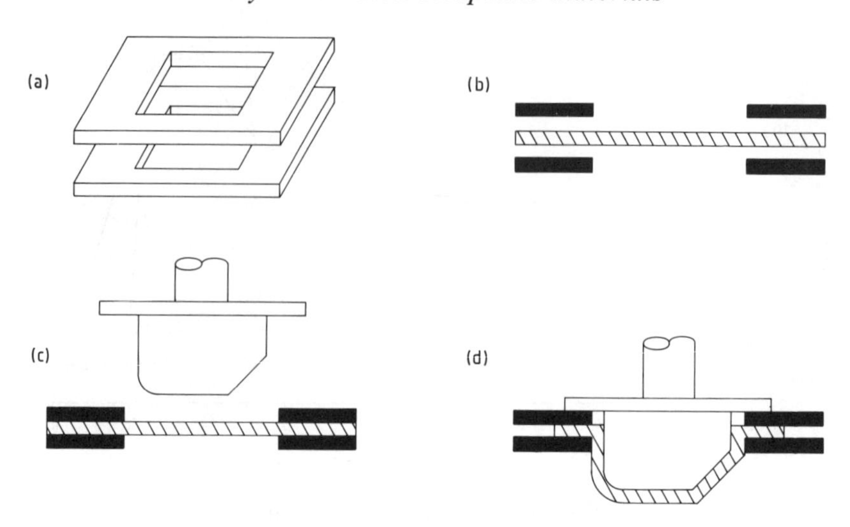

Figure 7.7 Deep drawing technique.

temperature. A cold male tool is then used to form the shape of the component.

Diaphragm forming uses the same principal as superplastic forming for metals. The thermoplastic laminate is sandwiched between two superplastic aluminium sheets. After heating to the forming temperature, the

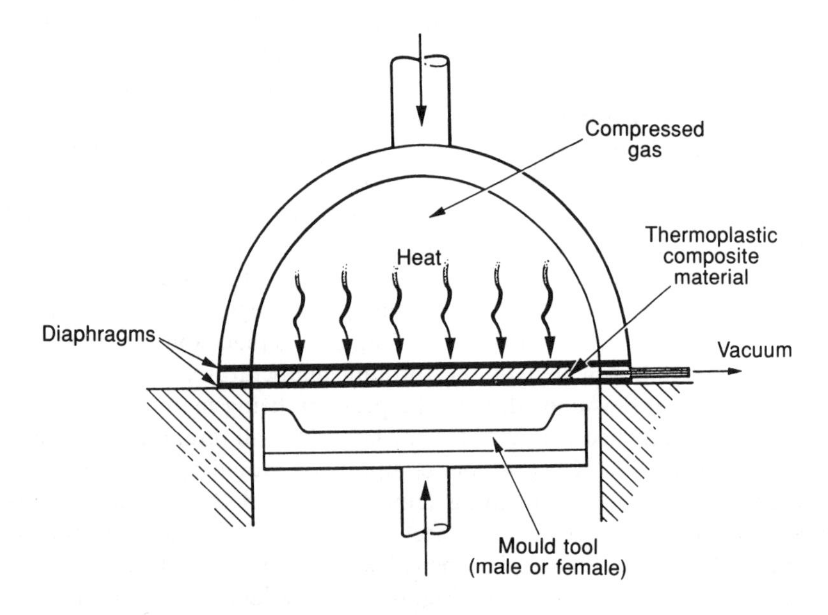

Figure 7.8 Diaphragm forming of components from prepreg.

laminate plus aluminium sheets are deformed to shape by applying differential pressures – Figure 7.8 illustrates. This technique does allow shapes to be formed which would not be possible by other means but, since the aluminium sheets are discarded, it is an expensive method of manufacture.

One of the main disadvantages of composites in the initial stages was that they were very labour intensive, therefore the processing was expensive. Processing has been greatly improved in recent years by the introduction of automated methods of manufacture. The two methods used to the greatest advantage are tape laying and filament winding. Both these techniques are available for thermosetting and thermoplastic laminates. Automated techniques of cutting the prepreg and using robotics for placement of prepreg have also been developed. The use of these automated methods of manufacture have assisted in making composite materials more competitive for producing aerospace components.

7.6 PROPERTIES

The main objective is to obtain a mix of reinforcement and matrix which will produce the optimum properties for the component being manufactured. It is therefore essential that contaminating particles and air are excluded from the laminate. When it comes to examining the mechanical properties of a laminate these can be divided into those which are fibre dominated and those which are matrix dominated. Properties which are measured in the former category are tensile, compression and flexural strength and modulus of unidirectional laminates in the direction of the fibres. Those which are matrix dominated are interlaminar shear strengths, in-plane shear strength and modulus and transverse strength of unidirectional laminates. Other properties which are important to designers are bearing strength, notched tensile strength and residual compression strength after impact.

At the present time there are no internationally accepted methods of measuring these properties and each aerospace programme has its own set of test methods. There are, however, movements to produce national, European and internationally agreed methods. Currently test methods are defined by ASTM, AECMA and CRAG.

Although composite materials are not normally subject to corrosion, some matrices are affected by moisture – this is particularly the case with epoxy resins. There has therefore been extensive research into the effect of hot wet conditions on the properties of composite laminates. In the case of epoxy composites it has been found that the laminate will absorb moisture in accordance with Fick's laws of diffusion and that it is possible to accurately predict the amount of moisture which a laminate will absorb in hot wet conditions. Some of the new thermoplastic matrices such as PEEK are much less affected by moisture and this is one of their main advantages.

7.7 JOINING COMPOSITES

It is possible to use most of the methods associated with joining metals. This includes mechanical fastening using bolts, rivets, etc. but it is very important that the joints are well designed. The method used most extensively for joining composites is bonding. Again it is important that the joints are well designed and individual methods of preparing surfaces are necessary for each matrix. Composite laminates are often employed in conjunction with honeycomb cores to provide lightweight rigid structures. It is also possible with thermoplastic composites to weld joints. This technique, however, still requires some development before joints which are acceptable for aerospace structures are produced.

7.8 NON-DESTRUCTIVE TESTING (NDT)

The use of composite materials on structural components has been greatly assisted by the development of NDT methods for examining the integrity of the component. The main techniques employed are radiography, acoustics, ultrasonics and thermography.

- Radiography highlights cracks, voids, resin rich areas, foreign matter and fibre misalignment.
- Ultrasonics, often in conjunction with a c-scan, will highlight areas containing voids and can also examine bonded areas for integrity of bonding.
- Acoustics and thermography are used on bonded joints to highlight debonded areas or bonded joints containing voids.

7.9 ADVANTAGES OF COMPOSITE MATERIALS

Aerospace manufacturers are always trying to improve the performance of their product which must take into account weight, stress, strain and cost. How do polymeric composite materials compare with other materials with regard to this policy?

As has already been stated, probably the greatest advantage of long fibre reinforced composites is the ability to place the strength and/or stiffness in the direction most suited to an individual component. The fibres being used for reinforcement are extremely light in comparison with metallic components – carbon fibre has a density of 1.8 compared with aluminium with a density of 2.71. Also the matrix materials are very light – epoxy resin has density of the order of 1.2. The fibres also have very high strength and stiffness, therefore the specific strength and modulus of the composite is much greater than a metallic one, as the following comparison shows:

	Specific tensile strength	*Specific tensile modulus*
Aluminium alloy	0.23	26
Carbon reinforced composite	1.1	205

This therefore illustrates one of the major advantages of composite materials.

As far as cost is concerned, this has been a problem in the past, but the industry is always examining the cost of ownership of the composite components. There are many different considerations which have to be taken into account and each component has to be examined independently. It is, however, generally considered that aircraft structural composite components are competitive regarding cost with other materials.

What other advantages are there? A main advantage, which my own company has used to advantage on helicopter main rotor blades, is the ability to produce complicated shapes. The main rotor blade for an EH101 helicopter has a continuous change of profile along its length and has a twist from one end to the other – this would have been impossible to manufacture as a metallic structure.

Composites have a higher fatigue endurance than metals and the damage tolerance is normally better, plus, under normal conditions, when they are impacted by a sharp object they do not suffer catastrophic failure. Even after being struck by a sharp object, a composite component will often have sufficient residual strength to complete a task. These properties are extremely important in the aerospace industry.

It is much easier to tailor properties of the composite component, not only by fibre orientation, but also by hybridizing the fibres. Polymeric composites do not corrode but some matrices do absorb water, which can affect their properties. It has been proved, however, that the extent of the effect on the properties of epoxy laminates can be estimated fairly accurately. Care must be taken when carbon fibre composites are joined to light alloy structures since galvanic cells can be formed and corrosion of the alloy can take place.

A property of carbon fibre which has been used to advantage in the space industry is the low coefficient of expansion of the composite structure. This is very important on structures subjected to large extremes of temperatures such as can be the case on space components. Glass and aramid fibre reinforced composites are often used for radomes because they are electronically transparent.

In summary, these types of composite materials can normally offer a great advantage in respect of weight of the finished component and other properties, but it is very important that the cost of manufacture is controlled in order that they can compete in regard to cost of ownership.

8

Metal-based composite materials

H.M. Flower

8.1 INTRODUCTION

The combinations of strength, ductility and toughness displayed by metal alloys, and discussed in other chapters, are sufficiently wide, and available at sufficiently low cost, to make them very attractive to the aerospace industry and explain why they have been the dominant structural materials for aerospace applications for the last 50 years. The mechanical properties are due to the non-directional nature of the metallic bond which permits plasticity by the slip of planes of atoms over one another via the motion of dislocations. Strengthening is achieved by increasing the resistance to such dislocation motion by alloying. However, the bonding also results in relatively low elastic moduli which, measured as specific modulus (Table 8.1), is very nearly constant for the most widely used metal alloys. The mechanisms of plastic deformation and the use of alloying to increase resistance to slip, and thereby increase strength, are thermally activated processes. As a consequence the strength of alloys decreases as temperature rises and creep processes occur during exposure to stress even at moderate temperatures. These are significant limiting parameters to the use of metal alloys in aerospace applications, both in airframes and propulsion systems.

Routes to overcoming these limitations by alloy (e.g. aluminium–lithium alloys) and process (e.g. rapid solidification) development are dealt with in other chapters. An alternative approach is to devise a composite material which retains the advantages of metal alloys but combines them with desirable properties of other materials. In this chapter the combination of metals with ceramics and with polymer composites will be discussed.

Table 8.1 Properties of selected metal alloys and ceramics

Material	*Density* ($Mg\,m^{-3}$)	*Elastic modulus (GPa)*	*Specific modulus*	*Tensile strength (MPa)*	*Specific strength*
Al alloy	2.75	72	25.8	140–550	50–200
Al–Li alloy	2.53	80	32	350–550	140–220
Ti alloy	4.5	115	24.5	700–1400	155–267
B mono-filament (140 μm)	2.4	390	162	4000	1660
SiC mono-filament (140 μm)	3.0	420	140	4000	1330
SiC yarn (10–13 μm)	2.6	185	71	2700	1040
C yarn (7 μm)	1.75–1.96	230–520	130–265	3500–1860	2030–950
Al_2O_3/15% SiO_2 (20 μm)	3.9	380	97	1400 +	360
Al_2O_3/5% SiO_2 (3 μm)	3.3	250	76	2000	606
Al_2O_3/5% SiO_2 staple (3 μm)	3.3	300	90	2000	600
Particulate TiB_2	4.5	530	117		
Particulate SiC	3.2	450	140		
Particulate Al_2O_3	4	420	105		

The numbers in brackets indicate individual fibre diameters.

8.2 METAL–CERAMIC COMPOSITES

Ceramics are typified by strong directional ionic or covalent chemical bonds. This bonding results in high elastic moduli and a high resistance to plastic deformation which exhibits minimal temperature dependence. Thus they are typically stiff, hard and strong (Table 8.1) but with negligible ductility and low toughness (typically 1–5 MPa $\sqrt{m}$ (Table 6.1) compared to aerospace metals 20–60 MPa $\sqrt{m}$). They are very sensitive to the presence of small flaws which can produce stress concentrations sufficient to cause crack growth at low applied stresses. Ceramics also exhibit low electrical and thermal conductivities and coefficients of thermal expansion compared to metals. A metal–ceramic composite is designed to blend the desirable properties of the ceramic with those of the metal alloy to create a composite material that is stiffer and less temperature dependent in its mechanical properties than the metal matrix while possessing increased strength and adequate ductility and toughness. A secondary advantage, of considerable significance to engineering applications additional to the aerospace sector, is increased wear resistance compared to the unreinforced metal. All of this must be achieved at a cost that industry will be prepared to pay in order to obtain these improved property combinations. In the sections which follow the basic mechanical properties of metal ceramic composites are discussed, the principal production routes are described and the limitations and applications of these materials are considered.

8.2.1 Mechanical properties

(a) Monotonic properties

Theoretical calculation of composite mechanical properties at room temperature, as applied to metal matrix materials, is summarized in Section 8.6. From this treatment it follows that maximum strength requires direct loading of the ceramic and is achieved by use of continuous fibre reinforcement. However, yield and fracture are then virtually simultaneous processes, controlled by fibre breakage, and hence ductility is negligible and toughness is low. The use of shorter fibres permits plastic flow in the metal matrix to precede fibre fracture and hence to restore a differential between yield and fracture stresses. Consequently higher ductility and toughness can be achieved. When fibres are sufficiently short, in relation to their diameter, fibres can pull out from the matrix without breaking. The energy absorbed in this process is large and contributes significantly to the toughness of the composite. However, since the fibres are never fully loaded (i.e. to their breaking stress), the composite strength is reduced. A weak matrix–fibre interface also substantially reduces both tensile strength and ductility transverse to aligned fibres, continuous or discontinuous. Finally, short fibres and particulates will permit more modest increases in composite stiffness and strength but with more metal-like levels of toughness and ductility. Yield in the composite is dependent upon yield in the metal and thus to the strengthening mechanisms that are operative in that system (solution, precipitation, grain and sub-grain and dislocation strengthening) and which may well be modified by the presence of the ceramic (section 8.6) and the detailed nature of the composite manufacturing process.

It must be emphasized that the properties discussed here and the equations given in section 8.6 apply to idealized material in which the ceramic is uniformly distributed, matrix and ceramic are free of flaws and the ceramic–matrix interface is capable of bearing the maximum stress that can theoretically be imposed upon it. Real composites fail to meet these criteria. Bunching of fibres or particles increases the local ceramic volume fraction and hence alters the calculated properties and also provides larger continuous crack paths which can produce sites of failure initiation at much lower stresses than predicted. The production process

Table 8.2 Reported properties of some metal matrix composites

Matrix	*Ceramic*	*Vol% Ceramic*	*Production Route*	*Heat Treatment*	*Modulus (GPa)*	*0.2% Proof Stress (MPa)*	*Tensile Stress (MPa)*	*Strain to Failure (%)*	*Reference Source*
Al-2024 (4.4% Cu, 1.5% Mg, 0.6% Mn)	None	0	Wrought	T4 (solution treated and naturally aged)	~ 72	335	470	20	[10]
Al-2124 (4.4% Cu, 1.5% Mg, 0.6% Mn)	SiC particulate	17	Powder process MMC extruded	T4	100	400	610	7	[11]
Al-2124	SiC particulate	25	Powder process MMC extruded	T4	115	500	700	4	[11]
Al-2014 (4.4% Cu, 0.5% Mg, 0.8% Si, 0.8% Mn)	None	0	Extrusion	T6 (solution treated and artificially aged)	73.1	476	524	13	[12]
Al-2014	Alumina particulate	10	Stir cast MMC extruded	T6	84.1	496	531	3	[12]
Al-2014	Alumina particulate	20	Stir cast MMC extruded	T6	101	503	517	1	[12]
Al-380 (8.5% Si, 3.5% Cu)	None	0	Die cast	None	71	159	317	3.5	[12] (+ [10])
Al ~ 10 Si ~ 3 Cu	SiC particulate	10	Die cast	None	93.8	241	345	1.2	[12]
Al ~ 10 Si ~ 3 Cu	SiC particulate	20	Die cast	None	113.8	303	352	0.4	[12]

	None	0	Wrought	Various	~ 106	700–900	850–940	6–15	Chapter 3
Ti-6Al-4V	SiC 100 μm dia monofilament aligned	32	Diffusion bonded fibre/metal foil	–	215 (0°)	–	1550 (0°)	0.89 (0°)	[13]
Ti-6Al-4V	SiC 100 μm dia monofilament aligned	32	Diffusion bonded fibre/metal foil	–	148 (90°)	–	550 (90°)	2.7 (90°)	[13]
Al-7075 (1.6% Cu, 2.5% Mg, 5.6% Zn)	None	0	Wrought	T6	71	~ 475	~ 540	8	–
Al-7075	Laminate (ARALL 1)	Aramid fibre prepreg	0.4% post stretched 1.3 mm sheet	T6	67.6 (0°)	641 (0°)	800 (0°)	1.9 (0°)	[8] (+ [10])
Al-7075	Laminate (ARALL 1)	Aramid fibre prepreg	0.4% post stretched 1.3 mm sheet	T6	48.3 (90°)	331 (90°)	386 (90°)	7.7 (90°)	
Al-7075	Laminate (GLARE 1)	UD glass fibre prepreg	0.5% post stretched 1.4 mm sheet	T6	64.7 (0°) 49.2 (90°)	550 (0°) 340 (90°)	1300 (0°) 760 (90°)	4.6 (0°) 7.7 (90°)	[8]

employed can introduce flaws into both ceramic and matrix. Imposed stresses, mechanical and thermal, can crack the ceramic, chemical reaction between matrix and ceramic at elevated temperature can degrade interface structure and properties and modify the matrix chemistry and microstructure. Voids can be introduced into the matrix by shrinkage during solidification in production processes involving the use of liquid metal or by incomplete densification in the case of powder products. All of these significantly degrade the properties of current metal matrix composites and are expensive to minimize or eliminate.

Aerospace interest in metal-based composites is centred around the use of monofilaments for continuous reinforcement and particulates: some representative property data relating to a range of such metal based composites, produced commercially, is presented in Table 8.2.

(b) Creep and fatigue

In the above discussion it is evident that the constraint placed upon the plastic flow of the metal by the ceramic is greatest for continuous fibres and diminishes as fibre length shortens and aspect ratio moves towards unity (= particulate). This must remain true at elevated temperatures where the ceramic continues to deform essentially elastically while the metal displays a decreasing yield stress (which can be substituted into the equations in section 8.6 to determine composite behaviour). Additionally, at elevated temperature the metal will undergo time dependent (creep) deformation under an applied stress. There is as yet little experimental data available upon the creep behaviour of metal matrix composites. The process has been modelled theoretically by Goto and McLean [1] for fibrous composites and the greatest benefit is demonstrated for continuous fibre reinforcements in which the creep rate continuously declines with time (primary creep). The model predicts little effect of matrix–fibre interface strength on the creep behaviour of continuous fibre reinforced materials whereas weak interfaces are deleterious for discontinuous reinforcements in which a secondary creep regime of constant strain rate with time is observed.

In particulate composites the particles induce early initiation of voidage and tertiary creep in which strain rate increases with time without an intervening secondary creep regime [2]. However, under conditions of an imposed strain rate some fine grained particulate reinforced composites has been shown to undergo superplastic deformation involving diffusionally accommodated grain and matrix–particle interfacial sliding [3,4]. This presents a potential advantage for forming the composite into a component (Chapter 9).

Creep in composites can also arise due to the disparity between the coefficients of thermal expansion of the matrix and the ceramic. The metal expansion on heating is constrained by the ceramic. Indeed this can be exploited to deliver a composite coefficient of thermal expansion tailored to match that of another material. However, the stress created in the metal can induce creep to take place. Thermal cycling of a fibre reinforced composite can then give rise to thermal fatigue and creep deformation of the metal during hold at the high temperature and hence to property degradation and dimensional instability in the absence of any externally applied load.

Fatigue resistance is generally improved, reflecting the higher stiffness of the composite compared to the unreinforced metal. In fibre reinforced materials the greatest gain is in the direction of the fibre axis. Transverse properties can be severely degraded if the matrix–fibre interface is weak or if interfacial reaction between matrix and fibre produces interfacial cracking. This can also degrade properties in short fibre and particulate reinforced materials, as can other processing defects such as inclusions and porosity. If the composite is free from such defects, the *S/N* characteristics of the composite are superior to those of the matrix metal. However, while the threshold stress intensity for crack growth is increased the lower toughness of the composite is generally reflected in increased rates of crack growth at higher values of stress intensity, and thus to a low tolerance of defects, as noted in Chapter 1.

8.2.2 Production Routes

There is an extremely wide range of potential routes by which a ceramic can be combined with a metal matrix. The route employed is, to some extent, determined by the nature of the reinforcement and component design. Continuous fibres of large ($\sim 10^2\,\mu m$) diameter (monofilaments) can be manipulated and placed in desired patterns at required locations, whereas very fine fibres ($\sim$ 1–10 μm), which cannot, must be handled as fibre tows. Large diameter fibres can also be coated with the matrix material, for example by plasma spraying, and the composite is then made by compressing, at elevated temperature, bundles of coated fibres so that the matrix material deforms to fill the voids between the fibres and bonds across the interfaces by diffusional mechanisms (Figure 10.19). If the coating can be made uniform in thickness, extremely uniform, high fibre volume fraction composites can be achieved in this way, even with matrix materials that are difficult to thermo-mechanically process by more conventional metal forming techniques. This is graphically demonstrated in Figure 8.1, which shows a very high volume fraction of SiC fibre in a titanium alloy matrix which was deposited by electron beam evaporation [5].

Finer fibres can only be handled as tows and are typically made into a shaped preform, held together with a binder, prior to composite production

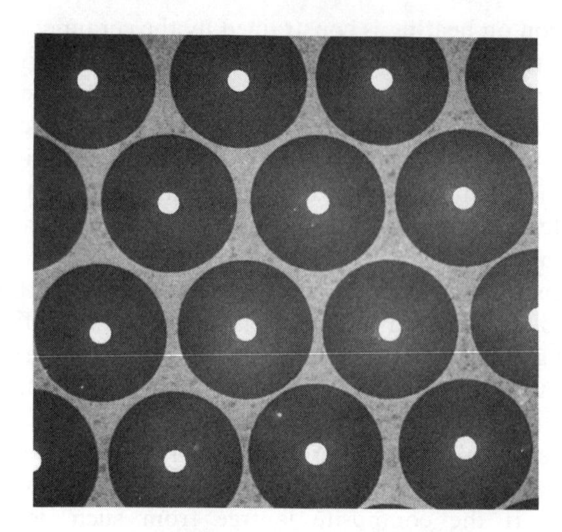

Figure 8.1 A scanning electron micrograph of a polished section of a Ti 5% Al 5% V 80% volume fraction SiC monofilament reinforced composite. It was prepared by solid state consolidation of 100 μm SiC fibres, coated with electron beam evaporated matrix alloy. The technique has permitted an unusually high volume fraction of fibres to be incorporated while retaining a very uniform distribution and avoiding deleterious fibre–fibre contact. The bright central region of each fibre is a tungsten core upon which the SiC was grown by chemical vapour deposition (courtesy of G.M. Ward-Close and P.G. Partridge).

via infiltration with liquid or powder matrix metal. It is much more difficult to achieve uniformity of fibre distribution in this case and the risk of voidage is also increased. However, using liquid metal infiltration, a near net shape product is obtained and the production cost is substantially reduced. Pressure is applied during infiltration to ensure complete filling of the space between the fibres and to minimise the risk of voidage on solidification: the pressure required increases with increasing ceramic volume fraction [6] and must be optimized to achieve full infiltration without compressing or damaging the preform. The high temperatures involved increase the risk of adverse chemical reactions between matrix and ceramic and in the case of titanium-based materials their extreme reactivity precludes the use of liquid metal infiltration entirely.

In the case of short fibres, preforms can still be made and liquid metal infiltration remains a viable route for component production. For short fibres and particulates cost can be reduced further, avoiding the need for a preform, by stir casting the ceramic with a matrix melt held between its liquidus and solidus temperature so that it is semi-solid. Only relatively unreactive metal–ceramic combinations can be employed and uniformity and volume fraction of ceramic are limited: fibre damage must also be expected. The other principal particulate composite production methods are cospray deposition and powder blending. In the former the matrix alloy is melted and sprayed, in a manner similar to that employed for producing powder (chapter 12) and a stream of ceramic particles is injected into the spray. The mixture of metal droplets and ceramic particles is deposited on a substrate where a billet of composite of more than 90% theoretical density is built up. Full densification requires additional thermo-mechanical processing which is also employed to produce the product shape. This method has several advantages. Firstly, the ceramic is only in contact with molten metal for a very short time, limiting the risk of interfacial reaction. Secondly, there is no requirement for the handling of fine metal powders which requires care both with regard to the risk of human ingestion and from the fire or explosion hazard which finely divided reactive metals such as aluminium represent. However, the highest particle volume fractions and most uniform dispersions are achieved by blending metal and ceramic powders and the best combinations of mechanical strength and toughness are obtained via this route, which also involves high cost. It is generally true for all metal matrix composites that the best properties are associated with the highest product costs (see below).

A very different approach to composite production to all those listed above is typified by the XD [TM] process patented by Martin Marietta [7]. In principle this involved the mixing of powders of reactants together and at an elevated temperature carrying out a controlled chemical reaction to produce a product consisting of an alloy matrix with a dispersion of ceramic particles within it. The ceramic is, necessarily, in equilibrium with the matrix so no adverse interfacial reactions are to be expected. Close control of the process conditions can permit the production of composites with volume fractions of up to 0.4 of the ceramic uniformly distributed. The technology has been demonstrated for aluminium, titanium and titanium aluminide matrices with titanium diboride together with a number of other matrices and ceramics although the volume fraction of ceramic employed in most cases is less than 10%. Other *in situ* composite production methods under development include reactive gas–melt reaction to produce the ceramic, mixed salt reactions within a melt to create the ceramic (with removal of by-products as a slag) and self-sustaining combustion synthesis in which compacted reactive powders are 'ignited' to set up a reaction front which propagates through the mass to create a metal–ceramic reaction product.

8.3 LAMINATES

This group of materials represents a distinctive category of composite, distinct from both metal matrix and polymer matrix composites. They consist of sandwich structures incorporating fibre reinforced polymer layers between metal alloy sheets. A typical sheet material would incorporate three (or four) aluminium (2000 or 7000 series) alloy sheets with two (or three) epoxy/glass or aramid fibre composite layers separating them. The original development work was carried out in Holland and the materials are marketed jointly by Azko and Alcoa as ARRALL (incorporating aramid fibres) and GLARE (incorporating glass fibres) [8].

They are characterized by higher tensile strength than unreinforced aluminium (Table 8.2) although with slightly reduced stiffness and, in common with metal matrix composites, greatly reduced tensile ductility. As with all fibre reinforced materials the mechanical properties also show anisotropy although the use of cross plies is being developed to overcome this. However, a major attribute is their exceptional resistance to fatigue in tension, in the case of ARRALL, and also in compression in the case of GLARE (ARRALL is weaker in compression because of the poor compressive strength of the aramid fibres). During curing of the laminates a stretch is introduced which puts the fibres into tension and the metal into compression in the finished product: this stress distribution contributes to the fatigue crack growth resistance. Both materials exhibit a very high damage tolerance compared to conventional polymer matrix composites, cost only 25–30% as much as carbon fibre reinforced polymers and can be machined and riveted using essentially conventional aerospace industry methods. These properties make them very attractive for fatigue critical airframe applications, and commercial applications have begun with the specification of ARRALL for the 5.6 × 9.3 m cargo door of the Douglas C-17 cargo aircraft [9]. A weight saving of about 25% compared with a conventional aluminium alloy structure is thereby achieved. Further development, via the use of other metal–fibre combinations, for example aluminium–lithium alloys and carbon fibres, appears likely in the future (e.g. Chapter 1, section 1.3.1).

8.4 COST

Many of the production routes for composites are intrinsically highly expensive. Unfortunately these are also associated with the best mechanical properties. For example the production of particulate reinforced aluminium alloys by powder technology involves the following steps:

- production of a metal billet
- gas atomization to fine powder
- blending the metal powder with fine ceramic powder
- canning the powder mixture
- degassing and sealing of the can (usually by electron beam welding)
- hot isostatic pressing
- decanning (by machining) to produce a billet of composite

The billet must then be subjected to themomechanical treatment to obtain the desired product form.

Although a very high quality product is obtained the production cost is high. Cost is substantially reduced by employing production routes involving liquid metal, either to infiltrate a fibre preform or to mix with ceramic particles. A near net shape product can also be achieved by such casting methods. However, problems of metal–ceramic interaction, non-uniformity of ceramic distribution and defects in the metal matrix due to shrinkage and incomplete infiltration can result in degradation of the final mechanical properties. However, if the process is used to create a billet rather than the final product containing short fibres or particles, subsequent hot working, e.g. by forging or extrusion, can close up defects, increase uniformity of ceramic distribution and improve mechanical properties.

Because of the very many variables involved, and the relatively low production volumes to date, it is very difficult to give an accurate guide to price, but taking the cost of wrought aluminium alloys as unity a very rough order of magnitude estimate of cost per unit weight can be made as shown in Table 8.3.

Table 8.3 Relative Costs (per unit weight) of Alloys and Composites

Material	*Price Index*
Monofilament fibre-reinforced titanium	10^3–10^4
Powder-based particulate aluminium-based MMC	10^2
Titanium alloys	10^1
Laminates	
Liquid metal route continuous fibre aluminium-based MMC	
Liquid metal route discontinuous fibre and particulate aluminium-based MMC	↑
Aluminium alloys	10^0

The increased cost per unit weight is offset to some degree by the lower component weight that can be achieved by exploiting the MMCs mechanical properties and, to a much greater degree, by the savings in the running costs of the aircraft accruing from reduced weight.

8.5 APPLICATIONS

At present applications are almost etirely potential rather than actual. This situation is the result of a number of factors, as follows.

Cost and the generally low levels of toughness obtained compared to traditional wrought aluminium and titanium alloys present barriers to applications. The relative novelty of the materials and the need to build up a substantial property data base, as with any new engineering material, also limit the initial rate of uptake of MMCs by the aerospace industry. The great diversity of potential matrix and reinforcement combinations and production routes necessarily inhibits this process and it is inevitable that many candidate material–process combinations will fail to achieve the breakthrough to commercial applications, leaving a small number of MMC materials to become established in aerospace applications.

A very early use of continuous monofilamentary fibre reinforced MMC was in the form of boron stiffened aluminium tube structure in the American space shuttle to provide the required rigidity in the massive cargo bay. Demonstrator beam, strut and tube structures employing both monofilament silicon carbide and graphite fibres have also been produced with space applications in mind. Continuous fibre reinforcement of titanium and titanium aluminide alloys is being actively pursued, the latter as a means of enhancing the high temperature properties of gas turbine blade and disc assemblies and achieving very substantial weight reductions and corresponding increases in engine fuel efficiency.

Compressor designs incorporating bladed discs and rings made of SiC reinforced TiAl indicate weight savings in excess of 70%. Such a dramatic reduction would result in a commanding competitive advantage over more conventional engines in the marketplace. However, to obtain maximum weight saving the engine must be designed to incorporate the MMC. Should its development fail it would not be possible simply to revert to the use of conventional alternatives. Engine development costs are so high that this would pose severe financial difficulties for the company undertaking the development. Less high risk applications, in which both fibre and particulate MMCs can be exploited, are very high stiffness aerofoil surfaces, including missile fins (Figure 8.2), which can withstand higher turn rates than those made from conventional materials.

Particulate MMC applications include the use of particulate aluminium alloy MMC in an extruded form to manufacture aircraft instrument racks which are more resistant to distortion under flight loadings than equivalent unreinforced aluminium alloy racks. This advantage may also be exploited in more major structures such as floor beams. The high stiffness and low thermal expansion coefficients of particulate MMCs offer opportunities for dimensionally very stable electronic packaging and in optical platforms and missile guidance systems, replacing toxic, brittle and highly costly beryllium components. The greater wear resistance of such MMCs may also find applications in situations where sliding wear can be a problem, such as slat tracks. Laminates have already achieved commercial application in the C-17 transport aircraft, as noted above.

Figure 8.2 A missile fin manufactured from BP 2124 aluminium alloy containing 17% volume fraction of SiC particulate. The powder processed MMC was rolled to plate and hot forged to shape. The fin is approximately 12.5 cm × 12.5 cm along the orthogonal edges and 12.5 mm in thickness (courtesy of Dr A. Begg, BP Composites Ltd)

8.6 APPENDIX

The following text is based on the standard treatment of fibre reinforcement as presented, for example, by Ashby and Jones [14]. It is an idealized treatment which assumes fibres are uniform in size, shape and strength and lie in parallel arrays. A more sophisticated treatment in which variation of some of these parameters is considered is given by Kelly and Macmillan [15]. Consideration of low aspect ratio fibres and particulates is based on additional references which are indicated at relevant positions in the text.

When considering mechanical properties it is not simply the absolute values that matter. The specific properties, normalized by dividing the absolute value by the material density, are important comparators between competing materials, particularly in aerospace where weight reduction in structure is of prime importance. In the case of metal–ceramic composites density may be increased or decreased compared to the unreinforced metal depending upon the combination of metal and ceramic chosen.

The composite density is readily calculated for a given volume fraction of ceramic, V_c, using the following equation:

$$\rho = V_c\rho_c + (1 - V_c)\,\rho_m$$

where ρ is the density of the composite, ρ_c is the density of the ceramic and ρ_m is the density of the metal.

It is notable that for the ceramics most commonly employed with aluminium alloys (alumina and silicon carbide), density is increased, whereas in the case of titanium the favoured reinforcement (silicon carbide) reduces the density.

8.6.1 Modulus

If the ceramic is present as an array of parallel, continuous fibres the metal and fibres experience equal elastic strains in the fibre axis direction. In this case the elastic modulus ($E_{comp}(0\,°)$) along the fibre axis may also be calculated from a 'rule of mixtures' equation equivalent to that employed for the calculation of density:

$$E_{comp}(0\,°) = V_cE_c + (1 - V_c)\,E_m \qquad (8.1)$$

If the fibres are discontinuous and embedded in the matrix metal, equation (8.1) is modified to:

$$E_{comp}(0\,°) = V_cE_c(1 - \alpha) + (1 - V_c)\,E_m \qquad (8.1a)$$

where α (< 1) is a function of the geometry of the fibre distribution, fibre radius and fibre and metal moduli.

It is evident that considerable increases in modulus can be achieved by incorporation of ceramic fibres into metals. However, the increase in modulus is smaller transverse to the fibre axis as the fibres and matrix experience essentially equal stresses rather than the equal strains (a situation also applying to particulate ceramic composites), the relevant equation for calculation of the modulus being:

$$E_{comp}(90\,°) = [V_c/E_c + (1 - V_c)/E_m]^{-1} \qquad (8.2)$$

In practice the predictions of equation (8.1) are found to be accurate, while equation (8.2) provides only approximate values of modulus due to the oversimplification involved in the derivation of the equation.

8.6.2 Strength and toughness

The factors of importance in determining strength and toughness differ depending on whether the ceramic is directly load bearing (continuous fibres loaded along the fibre axis) or indirectly loaded by transfer of stress via the metal matrix (continuous fibres loaded transverse to the fibre axis, discontinuous fibres and particulates).

The easiest case to consider is that of directly loaded (i.e. continuous) fibres. In this case the metal matrix will undergo plastic yield while the fibres remain elastically loaded. The matrix yields when the strain in the composite (ε_{comp}) equals that at which the metal yields. This strain (ε_{my}) is given by:

$$\varepsilon_{my} = \sigma_{my}/E_m \tag{8.3}$$

The stress in the composite at this point (σ_{compy}) is thus:

$$\sigma_{compy} = E_{comp}\varepsilon_{comp} = \{1 + [V_c E_c/(1 - V_c) E_m]\} (1 - V_c) \sigma_{my} \tag{8.4}$$

Further increases in load produce work hardening in the metal and eventually brittle fracture of the fibres. In the metal work hardening range the effective modulus of the composite can be obtained by substituting the metal work hardening rate ($d\sigma_m/d\varepsilon_m$) for E_m in equation (8.1). Since the work hardening rate in the metal is much less than the elastic modulus of the ceramic the effective composite modulus approximates to $V_c E_c$. When the first fibre fractures, the stress on the remaining load bearing cross section necessarily increases and, since all the fibres are fully loaded, they will also fracture. At fibre volume fractions in excess of a critical value (see below) the metal cannot bear the increased stress imposed on it and it undergoes local plastic deformation between the broken fibres and also fractures. The tensile strength (σ_{compTS}) of the composite, therefore, can be taken as the stress required to fracture the fibres. This can be calculated as follows:

$$\sigma_{compTS} = V_c \sigma_{cf} + (1 - V_c) \sigma_m \tag{8.5}$$

where σ_{cf} is the fracture stress of the fibres and σ_m is the tensile flow stress of the metal.

The toughness will be low since the fracture energy of the fibres is limited and the uniform plastic deformation of the metal is very small and the final plastic deformation, post fibre fracture, is very localized.

σ_m is generally significantly lower than the metal fracture stress σ_{mf}. Consequently, at low fibre volume fractions it is possible that the stress predicted by equation (8.5) is lower than the stress required to break the metal. This results in the composite strength being given by:

$$\sigma_{compTS} = [\sigma_{mf}(1 - V_c)] \tag{8.6}$$

In such cases the strength of the composite initially decreases from that of unreinforced metal as fibres are added and only rises once the stress obtained from (8.5) exceeds that from equation (8.6). The point where the stresses calculated from equations (8.5) and (8.6) are equal defines the volume fraction of fibres (V_{cmin}) at which strength is a minimum. To obtain strength above that of the unreinforced metal the fibre volume fraction must be such that the stress obtained from equation (8.5) exceeds σ_{mf}. Fibre volume fractions greater than about 0.1 are generally required.

In the case of discontinuous fibres the stress is transferred to the fibre through the matrix metal. Clearly, good adhesion between matrix and fibre is required for load transfer to be possible. This has important implications for the processing of both fibre and particulate reinforced composites (as discussed in this chapter). The stress is transferred in shear across the metal–fibre interface. The maximum value of the shear stress at the interface is equal to the metal shear flow stress – roughly $\sigma_m/2$. The tensile force generated in the fibre via this shear stress ($\int_0^1 \pi d(\sigma_m/2)\,\delta 1$) transfer is zero at the fibre ends and rises linearly with distance, l, from the ends of the fibre of diameter d. The average stress in the fibre is given by:

$$\sigma_{c.av} = \sigma_{cf}(1 - l_{crit}/21) \tag{8.7}$$

where l_{crit} is the fibre length at which the tensile stress at its midpoint just reaches σ_{cf}. The stress at the midpoint is simply $\sigma_m l/d$, hence l_{crit} is defined by:

$$\sigma_{cf} = \sigma_m l_{crit}/d \tag{8.8}$$

The average stress, obtained from equation (8.7), must be substituted into equation (8.5) to obtain the composite strength:

$$\sigma_{compTS} = \sigma_{cf}(1 - l_{crit}/2l)\,V_c + \sigma_m(1 - V_c) \tag{8.9}$$

In practical composites the fibre length need only be $5l_{crit}$ for the contribution of the fibres to the overall composite strength to reach about 90% of that obtained with continuous fibres.

When the first fibre fractures, only those whose ends are more than $l_{crit}/2$ away from the plane cross section of the first fracture will experience a stress equal to σ_{cf} in that same plane. If the volume fraction of fibres is low enough the metal plus these fibres can bear the applied stress. Equation (8.5) must, therefore, be modified by adding the contribution of the unbroken fibres to the composite strength:

$$\sigma_{compTS} = \sigma_{mf}(1 - V_c) + \tfrac{1}{2} l_{crit} V_c\, \sigma_{cf}/l \tag{8.10}$$

It is evident, therefore, that if $\tfrac{1}{2} l_{crit}\sigma_{cf}/1 > \sigma_{mf}$ the composite will exhibit greater strength than the metal at all volume fractions. However, in all other cases (as with continuous fibre reinforcement) there will still be a critical volume fraction of fibre required to raise the strength above that of the metal.

If the fibre length is decreased below l_{crit} then the tensile stress generated in the fibres by shear transfer from the metal can never reach

σ_{cf} (equation (8.8)). The fibres do not fracture and will pull out of the matrix metal when the metal undergoes plastic failure. At this point the maximum tensile stress in the fibre is simply $\sigma_{mTS} l/d$ and the average value is half of this. Substituting this stress into equation (8.5) gives:

$$\sigma_{compTS} = \tfrac{1}{2}(l/d)\, V_c \sigma_{mTS} + (1 - V_c)\, \sigma_{mTS} \qquad (8.11)$$

The yield stress (defined as the point at which the metal yields) is given by:

$$\sigma_{cy} = \tfrac{1}{2}(l/d)\, V_c \sigma_{my} + (1 - V_c)\, \sigma_{my} \qquad (8.12)$$

For alloys with a low yield stress σ_{my} is generally much less than σ_{mTS} and there will, consequently, be a similar difference between the yield and fracture stress of the composite. The plastic work in the metal between yield and fracture is substantial and will increase toughness. However, the work done in pulling the fibres out of the metal represents a significant toughening mechanism. The work done per unit area of fracture surface (WD) is given by:

$$WD = \tfrac{1}{2}\, V_c (l^2/d)\, \sigma_m \qquad (8.13)$$

To maximize this term the fibres should be increased to l_{crit} in length. Substituting for l_{crit} from equation (8.8) yields:

$$WD_{max} = \tfrac{1}{2}(V_c d \sigma_{fc}{}^2/\sigma_m) \qquad (8.14)$$

This equation also indicates that to maximize toughness a low strength metal matrix is required. However, the composite strength will (from equations (8.11) and (8.12)) be least in this case.

As fibres become shorter it is no longer possible to ignore the tensile component of stress transfer from matrix to fibre which takes place across the fibre ends which becomes an increasing component of the total stress transfer process. Various modifications to the fibre stress term in equation (8.12) have been proposed to account for the tensile component of stress transfer. For example, Nardone and Prewo [16] assumed a simple cylindrical geometry for the fibre and incorporated the tensile component of the stress on the fibre ends ($\pi d^2 \sigma_m/4$) into the calculation of composite yield strength and obtained:

$$\sigma_{compy} = \tfrac{1}{2}[(l/d) + 2]\, V_c \sigma_{my} + (1 - V_c)\, \sigma_{my} \qquad (8.15)$$

However, when fibres become very short, with aspect ratios (l/r) of 5 or less the approximations of geometry and stress state involved in modifying equation (8.12) become so large that its use becomes inappropriate and it becomes necessary to consider the micromechanisms by which strengthening takes place.

Yield takes place in the composite when the metal yields, i.e. when dislocation motion takes place in the metal. The stress required for dislocation motion depends on the nature and density of barriers to slip present in the material. These have been considered by Miller and Humphreys [17] and are summarized below.

The ceramic particles will have negligible direct influence on yield strength since the typical interparticle spacing is large in relation to the scale upon which strengthening can be obtained by the Orowan mechanism (in which dislocations loop around the particles under the applied stress) for which:

$$\sigma_{\text{Orowan}} \approx 2Gb/s \tag{8.16}$$

where G is the metal shear modulus, b is the Burger's vector of the dislocations ($\approx 3 \times 10^{-10}$ m) and s is the particle separation (of the order of microns).

However, the presence of the ceramic exerts an indirect influence on the yield strength by modification of the microstructure produced by processing. The grain and sub-grain size of the metal can be very much smaller than in unreinforced metal. Strengthening is inversely proportional to the square root of grain and subgrain size. If grain size is reduced to the scale of $10^1\,\mu$m the contribution to strength can be very substantial (e.g. of the order of 10^2 MPa for an aluminium matrix).

Processing inevitably involves the use of elevated temperatures, and differential thermal contraction of the metal and ceramic, due to differing thermal expansion coefficients, leads to dislocation generation in the metal and residual elastic stresses in the composite. The dislocations act as barriers to slip while, for aligned fibres and platelets the residual elastic stresses lead to the yield stress in compression differing from that in tension. Strengthening via dislocations is given by

$$\sigma_{\text{dislocations}} = \alpha b\sqrt{\rho} \tag{8.17}$$

where α is a constant of value 0.5–1 and ρ is the dislocation density. For values typical of particulate aluminium alloy composites the excess dislocation density is estimated to be between 10^{12} and 10^{13} m^{-2} and the contribution of the dislocations to the strength of the composite is of the order of 20 MPa.

In addition to the dislocations present at the start of deformation additional dislocation generation will take place at low strains (below the macroscopic yield stress) due to the presence of the particles. If this leads to very high rates of work hardening it will contribute to the observed yield stress.

In alloys which are conventionally heat treatable to produce age hardening by the precipitation of a fine dispersion of second phase precipitate particles (for example, Chapter 2) the modification to the microstructure due to the presence of the ceramic particles can alter the precipitation via inducing particle nucleation on the dislocations or the ceramic interfaces present. Precipitation within the matrix and precipitation kinetics generally can also be modified if the excess vacancy concentration retained on quenching post solution treatment is lost to the dislocations and ceramic interfaces present. Chemical reaction between the matrix alloy and the ceramic can also modify the alloy chemistry and further alter the nature of precipitation reactions. Thus substantial changes may be observed in the precipitation hardening achievable in the composite from that associated with the parent alloy alone. Such modifications have been widely reported in the research literature for many alloy/ceramic combinations and process histories.

All the above factors must be accounted for in assessing the yield strength of particulate composites. Assuming a high matrix–ceramic bond strength fracture will not involve particle decohesion. Fracture involves the build up of high local stress concentrations at the particles causing voidage or particle fracture. As matrix strength or particle volume fraction increases the plastic relief of the stress concentrations becomes more difficult and fracture will occur at lower macroscopic strains. Inhomogeneity in the particle distribution with some locally high volume fractions will necessarily reduce the ductility further.

REFERENCES

1. Goto, S. and McLean, M. (1991) *Acta Met. Mater.*, **39**, 153–77.
2. Lilholt, H. (1991) *Mater. Sci. Eng.*, **A135**, 161–71.
3. Mishra, R. S. and Mukherjee, A. K. (1991) *Scripta Met.*, **25**, 271–6.
4. Yu, M. Y. and Sherby, O. D. (1984) *Scripta Met.*, **18**, 773–6.
5. Ward-Close, G. M. and Partridge, P. G. (1990) *J. Mat. Sci.*, **21**, 4315–23.
6. Mortensen, A., Masur, L. J., Cornie, J. A. and Flemings, M. C. (1989) *Met. Trans*, **20A**, 2535–57.
7. Christodoulou, L. and Bruchpbacher, J. M. (1990) *Materials Edge*, December, 29–33.
8. *Advanced Composites Engineering*, June 1991, pp. 23–5.
9. *Advanced Composites Engineering*, September 1991, p. 4.
10. American Society for Metals (1985) *Metal Handbook* (Desk edn), ASM., Ohio.
11. *Particulate Metal Matrix Composites: Data Sheet*, BP Metal Composites Ltd, Farnborough March 1991.
12. *Duralcan Composites: Data Sheets*, Alcan, Geneva, 1991.
13. *Sigma Monofilament Metal Matrix Composites: Data Sheet*, BP Metal Matrix Composites Ltd, November 1991.
14. Ashby, M. J. and Jones, D. R. H. (1986) *Engineering Materials*, **2**, 241–4, Pergamon Press, Oxford.
15. Kelly, A. and Macmillan, N. H. (1986) *Strong Solids* (3rd edn), Clarendon Press, Oxford, pp. 241–377.
16. Nardone, V. C. and Prewo, K. M. (1986) *Scripta Met.*, **20**, 43–8.
17. Miller, W. S. and Humphreys, F. J. (1991) *Scripta Met.*, **25**, 33–8.

9

Superplastic forming

D. Stephen

9.1 INTRODUCTION

Since the mid-1970s, the superplastic characteristics associated with a limited range of structurally significant aluminium, titanium and nickel based alloys have been exploited in the manufacture of aerospace components. The principal manufacturing route used in the production of these components has been the method of superplastic forming (SPF). Latterly, however, for titanium alloys, SPF has been combined with diffusion bonding (DB), which, because of the compatibility of the process conditions, enables a concurrent manufacturing process (SPF/DB) to take place. This chapter deals with the application of SPF and SPF/DB to aerospace component manufacture and as such relates to the use of superplastic alloys in thin sheet form. It should be noted, however, that net shape forming of bulk material by the isothermal forging route exploits the superplastic characteristics of the alloys to which it is applied and that this process, which is not discussed in this chapter, is also being used in successful aerospace component manufacture [1].

The principal reasons for selecting the SPF and SPF/DB processes are the savings in cost and weight that can be achieved relative to alternative conventional manufacturing routes. These savings arise in the main from the ability of SPF and SPF/DB to reduce 'parts count' and eliminate joints. However, the simple transfer of processes does not of itself achieve these benefits but is subject to overriding considerations of:

- the careful selection of candidate components
- the evolution of *a specific* SPF or SPF/DB design which satisfies the component's structural and functional objectives

A fundamental feature of SPF component manufacture is the variability in the wall thickness of the finished component. In addition, for

SPF/DB components the internal structural forms achieve their shape and position from the stability of and interaction between adjacent 'free blowing bubbles'. In consequence most of the components which are currently in production have been established using empirical data, supported by pre-production trial and error forming. This, in most cases, and for SPF/DB in particular, has resulted in costly and time consuming development activities, which has to a degree limited the number of components which have matured through to production. This situation is changing, however, as effective and accurate finite element modelling (FEM) techniques have now been developed which have eliminated or significantly reduced the pre-production development requirements [2]. The advent of these CAD/CAM systems promises to assist greatly the wider application of SPF and SPF/DB in aerospace component manufacture. Designers in particular will find these FEM techniques of major importance in their acceptance of those processes.

As previously indicated, the nature of these processes is such that the physical restraints that can be applied to the material during SPF and SPF/DB manufacture are limited, and therefore an essential prerequisite for successful production is the consistency in the superplastic characteristics of the material being formed. The fact that many thousands of components have now been produced and have accumulated thousands of hours of satisfactory in-service experience on both military and civil aircraft, is a clear demonstration of the ability of the material producers to supply a consistent quality product in support of this method of manufacture.

Further recognition of the maturity of SPF and SPF/DB for aerospace component manufacture is exemplified by the fact that it has now become standard practice for the superplastic and diffusion bonding characteristics of new aerospace materials to be evaluated. Such characteristics have and are now being actively examined for:

- aluminium lithium
- metal matrix composites
- titanium aluminides
- ceramics

In all cases the superplastic and diffusion bonding characteristics already exhibited could result in SPF and SPF/DB becoming attractive routes for future component manufacture in these materials, some of which are known to be incompatible with conventional manufacturing processes.

9.2 SUPERPLASTICITY AND ITS CHARACTERISTICS

Superplasticity is a term used to describe a characteristic of certain metallic alloys which exhibit exceptional ductility. Such ductility is only achieved at high temperatures and the phenomenon shows a sensitivity to strain rate. A highly significant practical feature of superplasticity is the fact that deformations are achieved by the application of relatively low levels of stress either in tension or compression with stress levels of 10–20 MPa being typical. By contrast, conventional methods of forming are associated with stresses in the material which are at least one order greater.

9.2.1 Types of superplasticity

There are several classifications of superplasticity which are defined in terms of the deformation characteristics and/or the microstructural mechanisms that apply, including the following:

- fine grain superplasticity
- transformation superplasticity
- internal stress superplasticity

The specific type of superplasticity discussed in this chapter relates to fine grain superplasticity as this is the type currently exploited in aerospace component manufacture. However, reference is made to the other types of superplasticity as these may grow in significance in future manufacture. Evidence in support of this has already been provided experimentally where metal matrix composites and ceramics have been deformed in a superplastic mode. One example reported in the literature [3] refers to a silicon carbide (SiC) whisker reinforced aluminium alloy (6061) exhibiting 1400% elongation when subjected to thermal cycling between 100 and 430 °C at a rate of one cycle in 100 s. By contrast this compares with only 12% elongation when this material is strained at a constant 430 °C. The superplasticity exhibited in this case is attributed to internal stress cycling induced by the different coefficients of expansion associated with the aluminium alloy matrix and the SiC whisker reinforcement.

9.2.2 Elongations

In general, the deformation limits associated with the current range of superplastic aerospace alloys range from 600 to 1000% elongation, i.e. an ability to increase their original length by a factor of 10. By contrast, these represent elongations which are one to two orders greater than those associated with non-superplastic alloys and hence suggests the possibility of a manufacturing route which far exceeds the 'shape making' capability of conventional processes. It is this basic characteristic which is now exploited in the process of SPF.

The preservation of constant volume obviously demands that elogations of the material are matched by reductions in the material thickness.

The factors affecting this reduction in thickness are discussed later in this chapter.

The phenomenon of superplasticity is now the subject of considerable metallurgical and scientific interest world-wide, and has achieved the distinction of being accorded a world record featured in the *Guinness Book of Records*. Although not of specific interest in aerospace manufacture, the highest recorded elongations achieved to date are almost 8000% for commercial aluminium bronze [3].

9.2.3 Effect of temperature on ductility

Because of its influence on the accommodation mechanisms involved in the exceptional plastic straining of superplastic metallic alloys, temperature is a first order variable in the achievement of superplastic flow. For a typical superplastic alloy, two temperature regimes can be defined, each of which exhibits fundamentally different ductility levels. The transition between these two regimes occurs at a temperature of $\simeq$ 0.4 TM, where TM is melting temperature in kelvins.

(a) < 0.4 TM

In this regime, the ductility is dominated by strain hardening and only limited improvements in ductility are achieved with rising temperatures. In this respect, there is little distinction between the behaviour of superplastic and non-superplastic materials.

(b) > 0.4 TM

In this regime and specifically for superplastic materials, temperature sensitive mechanisms which accommodate grain movement start to play a dominant role and result in dramatic increases in ductility with increasing temperature. The continuity of this improvement is usually interrupted at phase transformation temperatures, because of the associated change of the alloys microstructure to a form which does not accommodate superplastic flow.

Figure 9.1 illustrates clearly the features discussed above for titanium 6% aluminium, 4% vanadium (Ti6/4) alloy, which, because of its significance in aerospace manufacture, is probably the best characterized superplastic material to date. Reference is made to this alloy throughout this chapter as its main characteristics can be considered as being typical of fine grain superplasticity. Where appropriate, specific reference will be made to characteristics exhibited by other alloys which are not found in Ti6/4. The characteristic illustrated in Figure 9.1 shows a significant rise in ductility commencing between 600 and 700 °C (0.4 TM = 690 °C) and

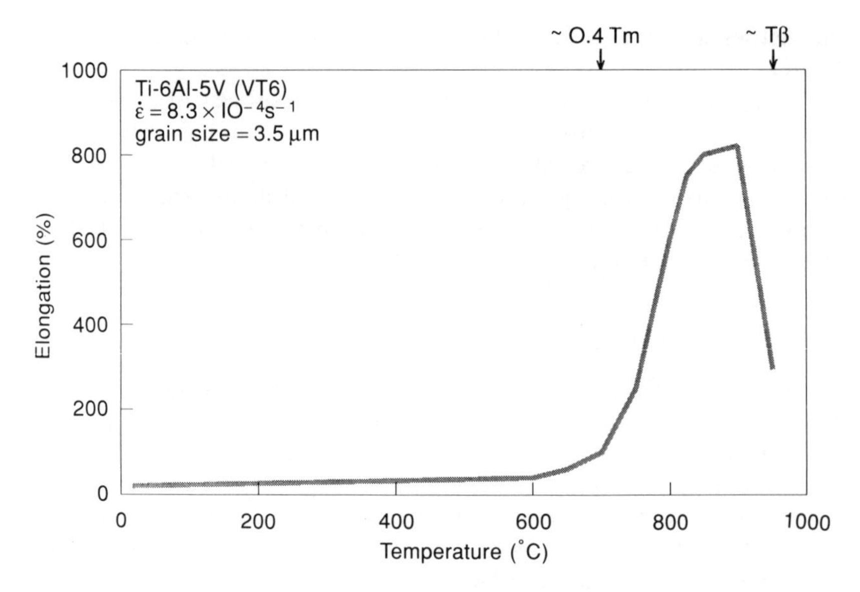

Figure 9.1 Ductility of Ti6/4 as a function of temperature.

reaches a peak at 927 °C (0.5TM = 931 °C). The decay in ductility beyond 927 °C is associated with the β transus for this alloy of 985 °C. These data relate to a constant strain rate and clearly illustrate the importance of temperature in the attainment of optimum fine grain superplasticity. In addition, the form of the curve highlights the importance of temperature control in SPF processing.

9.2.4 Strain and strain rate sensitivity effects on elongation

We will now consider the effect of mechanical loading on superplastic performance, in particular in relation to the ductility levels achieved. Also of significance is the level of mechanical loading that is necessary to induce superplastic flow and the rates of deformation that arise from the action of these loads; these factors have a significant bearing on the process equipment specification.

The characterization of fine grain superplastic materials is usually conducted at constant temperature and involves the application of tensile loads to standard test pieces manufactured from bar or sheet. Because of the nature of superplastic deformation, the equipment used for such characterization needs to accommodate large elongations, and, in addition, the rates of movement of the loading cross head need to be controllable.

The methods of analysis of the data derived from these tests and the accuracy and interpretation of the resultant predictions is an extensive subject which is beyond the scope of this chapter, but they can be studied in some detail in the references [4,5]. From a practical standpoint, the essential characteristics which need to be established because of their relevance to the SPF processing of the material are:

- the stress levels applied to deform the metal – flow stress σ
- the rate at which the material will deform under the action of the flow stress – strain rate $\dot{\varepsilon}$
- the maximum ductility that can be achieved – strain to failure ε_{max}
- the uniformity of the stretching/thinning of the material

These parameters are discussed in the sections that follow.

9.2.4 Flow stress–strain

The relationship of true stress to true strain for Ti6/4 is shown in Figure 9.2. These data relate to a fixed strain rate applied at the optimum superplastic strain temperature. The form of the relationship indicates an insensitivity of flow stress to strain but there is an indication of some 'strain hardening'. In addition, Figure 9.2 shows a clear dependence of flow stress on the initial grain size of the material with flow stress increasing with grain size. This latter effect provides an explanation for the strain hardening phenomenon as grain growth does occur in Ti6/4 as a function of time at elevated temperature, and, in addition, there is a strain rate dependence. This is clearly demonstrated in Figure 9.3. Thus the apparent strain hardening is in fact time and not strain dependent, a

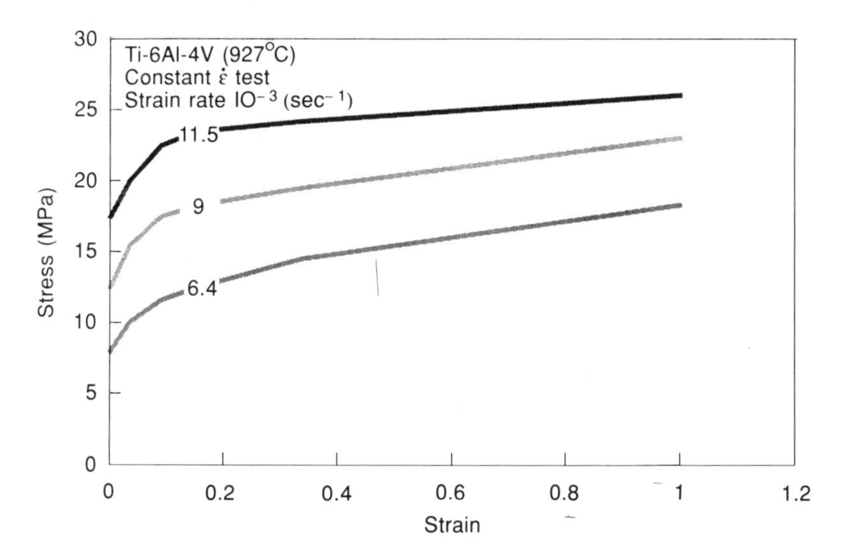

Figure 9.2 Flow stress vs. strain for Ti6/4.

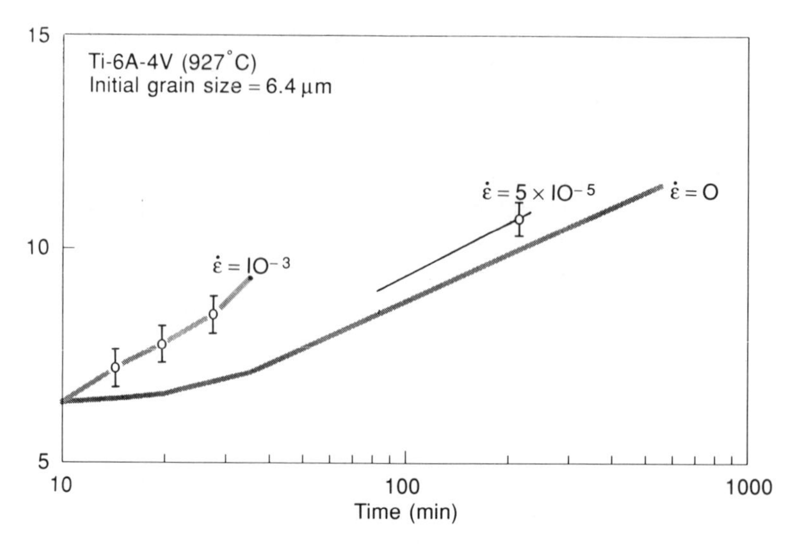

Figure 9.3 Grain growth in Ti6/4.

factor which needs to be taken into account in the manufacturing process, especially in relation to the total time spent at elevated temperature. (In the case of SPF/DB manufacture this can be measured in hours.)

9.2.5 Flow stress–strain rate

Given the exceptional ductilities displayed by superplastic materials, it is not surprising that the microstructural accommodation mechanisms associated with deformation are dependent upon time and therefore a strain rate sensitivity is to be expected. This fact is clearly demonstrated in Figure 9.4 which provides data for Ti6/4 when strained at 927 °C. The form of this relationship is sigmoidal with effectively three distinct regimes. Much debate and metallurgical experimentation has been conducted in an effort to explain the fundamental metallurgical reasons for the existence of these three regimes [6,7], but for the purpose of this chapter all comments will be limited to regime II, which is the regime of practical significance for SPF. The flow stress/strain rate relationship depicted in Figure 9.4 can be expressed in the form:

$$\sigma = k\,\dot{\varepsilon}^m$$

where m is the slope of the flow stress strain rate characteristic and is defined as the strain rate sensitivity factor. It is of interest to note that for $m = 1.0$ the relationship given above represents ideal Newtonion viscous flow.

The relatively low flow stresses exhibited in Figure 9.4 represent the fundamental difference between SPF and conventional forming, the

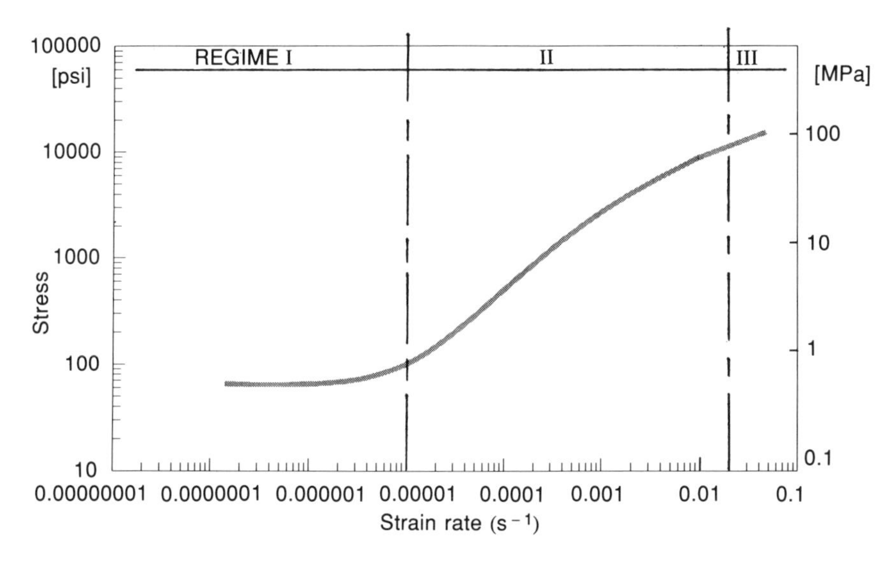

Figure 9.4 Flow stress vs. strain rate for Ti6/4.

latter requiring stresses at least one order higher. This fact is of major significance in the specification of the press equipment associated with the respective processes.

9.2.6 Strain rate sensitivity factor *m*

Figure 9.5 shows the variation of *m* with strain rate for Ti6/4 when strained at the optimum temperature of 927 °C. The form of this relationship reflects the sigmoidal character of Figure 9.4 and exhibits values of *m* in the superplastic regime in excess of 0.5, with a peak value of 0.8 occurring at a strain value of $1.5 \times 10^{-4} s^{-1}$. This strain rate is equivalent to approximately 100% elongation in one hour, which highlights a fundamental feature of SPF, that although high levels of deformation are achievable at low levels of stress, the process is relatively slow in comparison with conventional forming processes. Figure 9.5 also demonstrates the significance of grain size in superplastic performance, showing a trend of decreasing *m* and associated strain rate with increasing grain size. As will be discussed later, grain movement plays a significant part in superplastic flow and larger grains tend to inhibit the flow.

9.2.7 Significance of *m* in SPF performance

There are two important superplastic performance parameters which have been shown to be related to a first order to *m*. These are:

- the ability of superplastic materials to resist non-uniform thinning

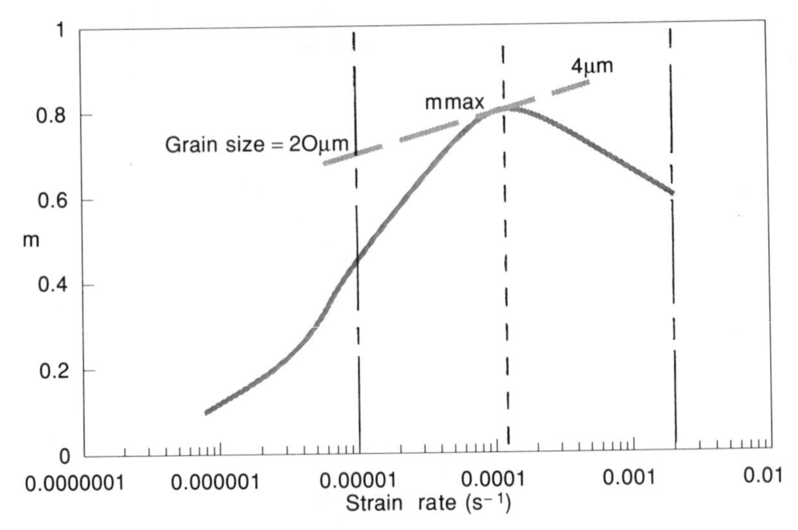

Figure 9.5 Strain rate sensitivity factor for Ti6/4.

- the total elongation to failure achievable in a superplastic material

The influence of m in relation to the first parameter has been established analytically by modelling the strain conditions in the region of a defect in a standard tensile test piece [8]. The results of this analysis are presented in Figure 9.6 and show a trend of improving accommodation of defects with increasing m and a total insensitivity to defects at a value of $m = 1.0$. Thus thinning of the material during SPF is best accommodated at the highest m value conditions.

It obviously follows that the ability of a material to resist non-uniform thinning as displayed in Figure 9.6 must also be related to the total strain that can be achieved before failure (ε_{max}). The relationship of ε_{max} to m can be established analytically [9] but has also been supported by analysis of test results from a wide range of metallic alloys [10]. The relationship established from this test data is presented in Figure 9.7, and while this exhibits a wide scatter band there is an obvious trend of increasing ductility with increasing m. One of the factors which is thought to contribute to the scatter band is a phenomenon termed cavitation, which can arise with the superplastic straining of some materials. The origins of this phenomenon are discussed later in this chapter.

9.2.8 Material requirements for superplasticity

As the name implies, the essential material requirement for fine grain superplasticity is a fine grain microstructure. (Grain sizes of 10 μm or less are desirable.) This basic requirement does, however, need to be expanded to include the following additional qualifications:

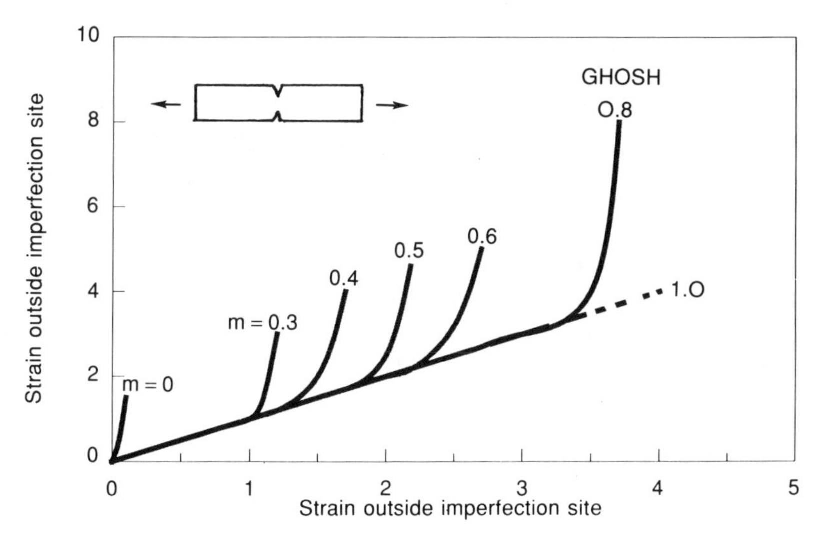

Figure 9.6 Influence of strain rate sensitivity factor on the accommodation of defects

- The fine grain structure is required only during superplastic processing.
- The microstructure needs to remain stable during superplastic processing.
- The grains need to be uniform and equi-axed to accommodate the mechanisms of superplasticity.

These conditions can be achieved in two major groups of alloys which are classified as:

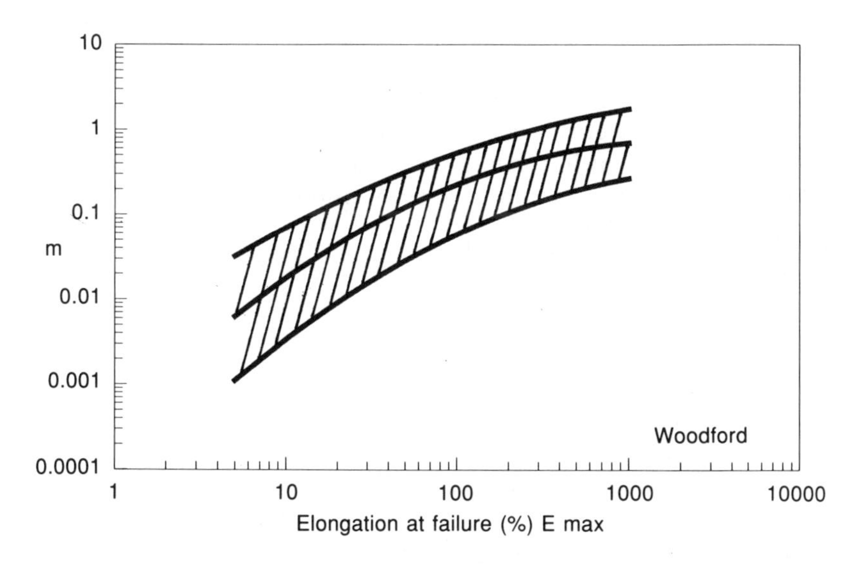

Figure 9.7 Influence of strain rate sensitivity factor on the elongation to failure.

- microduplex
- pseudo-single phase

Titanium and stainless steel superplastic alloys are in the microduplex category and their microstructural stability and superplastic performance is dependent upon the presence of at least two roughly equal phases in the microstructure at the processing temperature.

By contrast the pseudo-single phase alloys achieve stability from the presence of dispersed precipitates which define and 'pin' the structure. This category encompasses the current aerospace aluminium alloys which are heavily dependent upon thermo-mechanical processing to achieve the desired microstructural conditions [11].

9.2.9 Mechanisms associated with superplasticity

Much debate continues in the metallurgical field on the subject of the mechanisms which accommodate the extremely high levels of ductility experienced in superplastic metals. There appears to be broad agreement, however, that the dominant accommodation mechanism is associated with grain boundary sliding (GBS). This mechanism has been studied extensively in microduplex alloys and is supported by experimental evidence [12].

The dependence of superplasticity on GBS is clearly consistent with:

- the essential material characteristics described in the previous section, in particular in relation to grain size and its uniformity
- the grain size dependence exhibited in the straining characteristics shown in Figures 9.2 and 9.5

Although GBS is recognized as a first order accommodation mechanism it cannot be effectively accomplished without further mechanisms, in particular those which relate to the grain boundaries. The further metallurgical processes which have been proposed [10] are:

- grain boundary diffusion
- stress induced diffusion
- solute diffusion
- dynamic recovery
- dynamic recrystallisation

These processes vary in significance, dependending upon the particular alloy being strained. Dynamic recrystallization is of particular significance, for example in the superplastic behaviour of pseudo-single phase alloys (aluminium). In these materials recrystallization occurs at the superplastic temperature with grain size and stability being provided by a uniform dispersion of fine particles which pin the grain boundaries.

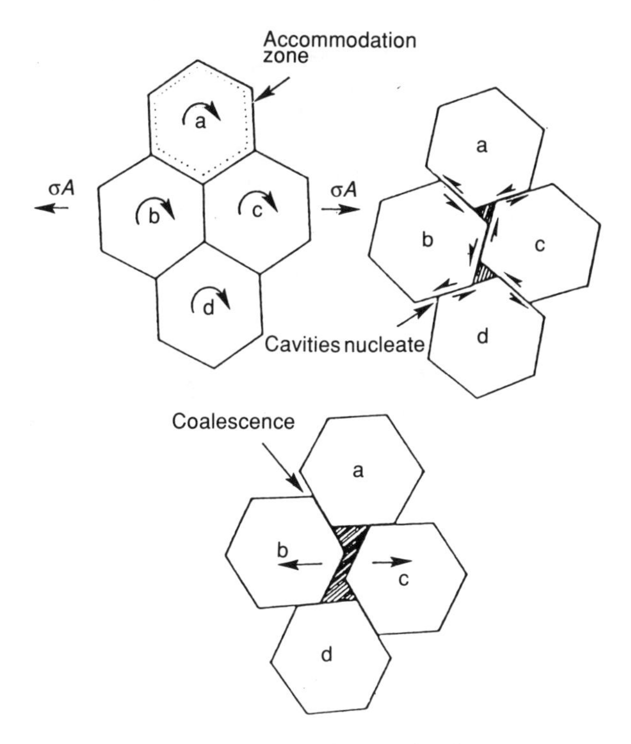

Figure 9.8 Cavitation mechanisms.

9.2.10 Cavitation

As the main accommodation mechanism for fine grain superplasticity is GBS, then inherent in this process is the possibility of grain boundary dislocations occurring in conditions which inhibit sliding (hard particles at grain boundaries) or limit the diffusion mechanisms (low temperature, fast strain rates). During straining, these dislocations can coalesce to form voids at the boundaries in the manner illustrated in Figure 9.8 and it is this phenomenon which is defined as cavitation. On the basis of experience to date this is found to be a fundamental feature of superplastic aluminium alloys, in particular when strained at standard atmospheric pressure.

The generation of such voids is obviously undesirable as this can have a detrimental effect on the mechanical properties of the material after straining and in addition limits the maximum superplastic ductility. Fortunately, complete suppression of cavitation can be achieved by straining the material in a high pressure atmosphere which corresponds to 0.5–0.75 of the uniaxial stress [13]. Cavitation suppression by this method is currently being used in the production of aluminium aerospace

Figure 9.9 Superplastically formed particulate reinforced component.

components and is defined as 'back pressure' forming. In practice the level of back pressure applied can be adjusted to provide a cavitation level which is acceptable to the requirements of the final product, in particular in relation to the mechanical properties required. It is important to realize, however, that many secondary applications of SPF aluminium have been successfully adopted for production without back pressure forming.

It should be noted that back pressure forming has been shown to have benefits in forming non-superplastic materials. A particular example which is of major significance for future aerospace applications is the use of back pressure forming in the manufacture of components in MMC particle reinforced materials. Figure 9.9 shows an experimental component in aluminium lithium MMC which has been formed successfully by this method by British Aerospace Military Aircraft Division.

9.3 AEROSPACE SUPERPLASTIC ALLOYS

Currently there are three categories of superplastic metallic alloys used in aerospace component manufacture. These are:

- titanium based alloys
- aluminium based alloys
- nickel based alloys

Table 9.1 provides a listing of the alloys in these three categories and provides information relating to those alloys which are currently used in volume component production together with those which have been shown experimentally to have superplastic properties. In addition to the alloy composition, Table 9.1 lists key parameters relating to superplastic

Table 9.1 Aerospace superplastic alloys: typical superplastic and mechanical properties

Base	*Alloys*		*T* (°C)	*m*	ε (Sec^{-1})	ε *max* (%)	*f* tu (*MPa*)
Titanium	6Al–4V	(IMI318)	927	0.8	5×10^{-4}	1000	950
Titanium	4Al–4Mo–25Zn–0.5Si	(IMI550)	–	–	–	–	1000
Titanium	6Al–2Mo–4Zr–2Sn	(6242)	900	0.7	2×10^{-4}	538	–
Aluminium	6Cu–0.25Mg–0.4Zr	(Supral 100/150)	460	0.6	1×10^{-3}	1600	350
Aluminium	6Cu–0.35Mg–0.4Zr–0.1Ge	(Supral 220)	450	0.5	1×10^{-3}	1100	470
Aluminium	5.8Zn–1.6Cu–2.3Mg–0.22Cr	(7475)	516	0.8	2×10^{-4}	1000	540
Aluminium	2.4Li–1.2Cu–0.6Mg–0.1Zr	(Al Lith 8090)	510	0.6	2×10^{-4}	1000	450
Nickel and	26Cr–6.5Ni	INC0744	1000	0.5	1.5×10^{-3}	350	–
Duplex	21Cr–55Ni–5Nb–3Mo	INCO718SPF	950	0.4	1.3×10^{-4}	400	1114
Stainless	23Cr–4Ni–1.2Mo	AVESTA 2304	900/1000	0.5	1×10^{-4}	450	640
	22Cr–5Ni–3Mo	AVESTA 2205					
		(SANDVIK 2304)					
		(SANDVIK 2205)					
	22Cr–55Ni–9Mo–5Fe	INCO 625LCF	–	–	–	–	980

performance together with typical mechanical properties for the most significant alloys.

In general, it has been necessary to specifically develop the alloys given in Table 9.1 to achieve an acceptable superplastic performance. This is particularly true of the aluminium and nickel based alloys. In the case of the titanium based alloys, Ti6/4 in particular, a fine grain microstructure is an inherent feature but this alloy was originally developed for high strength before being identified and characterized as superplastic. Ti6/4 was first characterized by Backhofen in 1965 and the results of this work provided a major stimulus to the development of SPF and SPF/DB in aerospace.

Superplastic aluminium alloys with acceptable in service properties were first developed by British Aluminium Company (BACO) in the early 1970s. These are now widely recognized as Supral alloys and are characterized by relatively high levels of zirconium. The microstructure of these alloys corresponds to the pseudo-single phase category for which superplastic properties are achieved by dynamic recrystallization [14] with the grain size and stability being conferred by a dispersion of very fine particles of Al_3Zr.

As the initial (100 series) Suprals were of medium strength, subsequent developments in superplastic aluminium alloys have been largely directed towards the achievement of superplasticity in higher strength alloys. This has led to the development of the 200 series Suprals, which are basically

Figure 9.10 Sink unit.

alloy developments of the 100 series and achieve their superplasticity in a similar manner.

In addition to the Supral developments, methods of thermomechanical processing of high strength conventional aluminium alloys have been developed, notably by Rockwell [11], and successfully applied to the existing high strength 7475 alloys.

The latest developments in aluminium superplastic alloys involve the aluminium lithium alloys which can be made to be superplastic by either the Supral or thermomechanical routes. These alloys are currently in a state of development but are scheduled to find applications on the next generation of civil and military aircraft.

A number of superplastic stainless steels have been developed which exhibit microduplex structures. The densities of these materials limit their application in aerospace but it should be noted that these materials are currently being used for volume production of toilet and galley units for civil aircraft which fully exploit the extreme forming capability of these superplastic alloys (Figure 9.10).

9.4 POST-SUPERPLASTIC STRAINING MECHANICAL PROPERTIES

The post-straining mechanical properties of the superplastic alloys of primary importance in aerospace manufacture have been extensively studied and reported in the literature [15]. A detailed analysis of these results would be too extensive for this chapter and therefore only the broad conclusions are reported here.

(a) Titanium (Ti6/4)

Relative to the as-received properties some limited degradation in tensile and fatigue strength is experienced post-straining. Analysis shows that this degradation is largely due to the thermal cycling experienced during the SPF process and only minor effects appear to be attributable to straining. Some recovery of these properties can be achieved by heat treatment after forming but it is worth noting that the present practice associated with the components produced to date does not involve heat treatment even when these are primary structural components.

(b) Aluminium

Significant reductions in properties are experienced post straining due to:

- thermal cycling
- cavitation

These properties can be largely recovered by reheat treatment and the use of back pressure forming to suppress cavitation. Developments in aluminium lithium alloys (8090) show this material to have a low quench rate sensitivity and therefore it is likely that reheat treatment of this alloy may not be necessary to recover mechanical properties.

9.5 SUPERPLASTIC FORMING (SPF)

SPF is a manufacturing process which exploits the superplastic behaviour of metallic alloys. However, the principles of the process relate to and have been largely adopted from the manufacturing methods associated with **thermoplastic** materials. The basic principle involved is the stretching of a membrane of superplastic material from its initial flat condition when placed in the tool and clamped around its edge, to achieve a final shape which is defined by the tooling cavity. The force used to stretch the membrane is provided by a differential pressure applied by the introduction of a gas on one or both sides of the membrane – the latter case applies specifically for back pressure forming. It follows from the earlier discussion of the superplastic characteristics of a metallic alloy that ideally the optimum processing conditions need to be appropriate to maximum m if minimum thinning variations and maximum ductility are to be achieved.

On the basis of the conclusions of these earlier discussions there are essentially three critical factors associated with the achievement of maximum m. These are:

- the maintenance of an appropriate processing temperature
- the achievement and maintenance of an appropriate flow stress in the membrane throughout its forming cycle
- consistency in the material's metallurgical quality in terms of grain size and distribution, in particular between different batches of material

In satisfying the third factor the fact that many thousands of SPF components have now been produced successfully, in particular using Supral, Ti6/4 and duplex stainless steels, provides full support to the ability of the material supplier to provide a consistent product. This is, however, sometimes supported by microstructural examination of received material, carried out by the component manufacturer. The achievement of the appropriate temperature conditions is usually associated with the use of a heated platen press of a type similar to that shown in Figure 9.11. These presses are designed to have a high thermal inertia to ensure temperature stability. The preferred method of operation is one which involves maintaining the platens at the SPF operating temperature and cycling the components, and in some cases the tools, during manufacture.

Figure 9.11 SPF press.

Although the inducement of a flow stress in a forming membrane is readily accomplished by the application of a pressure differential, the maintenance of the optimum flow stress throughout the forming operation demands the application of a variable pressure, which relates to the changes in shape and thickness of the membrane. The pressure/time profile (PTP) which results is therefore unique to every component

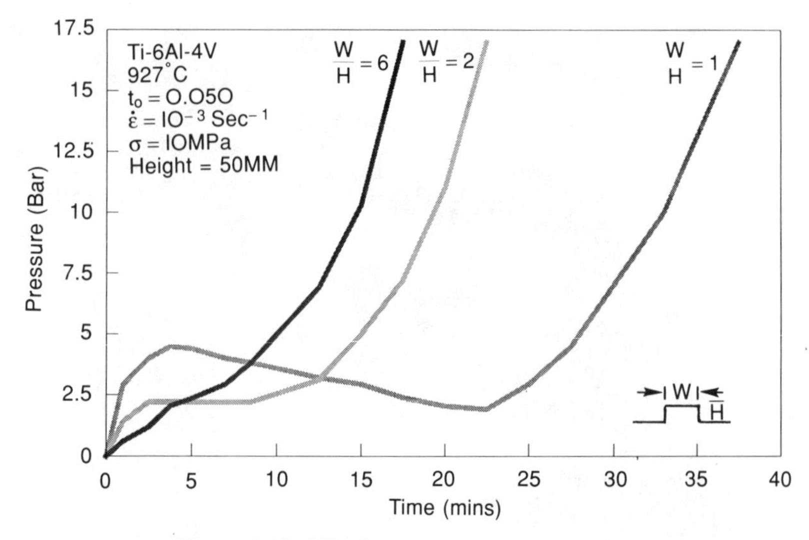

Figure 9.12 PTP for rectangular box sections.

design. This is illustrated in Figure 9.12 which shows the PTP that would be required to form rectangular box sections of different aspect ratios using a female tool. It should be noted that the PTP would be different for a given aspect ratio if a male tool was used, and in addition the thickness distribution in the finished component would also differ (Figure 9.13). The calculation of the PTP is complicated by the need to establish the shape of the forming component and its associated thickness distribution throughout the forming operation, and while the PTP can be established with acceptable accuracy for simple shapes using empirical methods the advent of FEM promises to be a considerable boon to the SPF designer/manufacturer in aiding the establishment of the appropriate PTP for complex components.

Even using the best superplastic material and applying optimum flow stress, non-uniform thickness distributions are an inevitable consequence

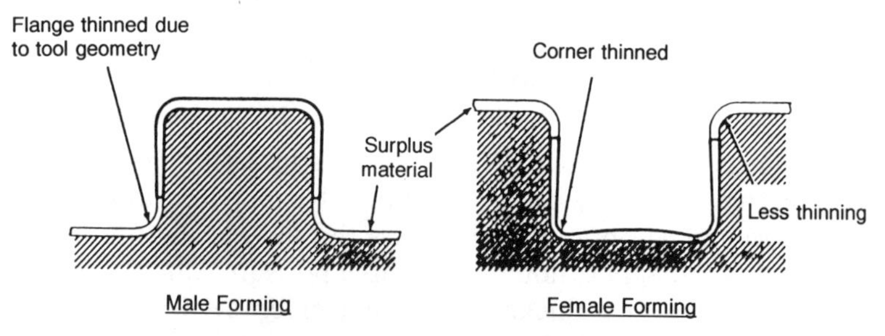

Figure 9.13 Male/female forming effect on thinning.

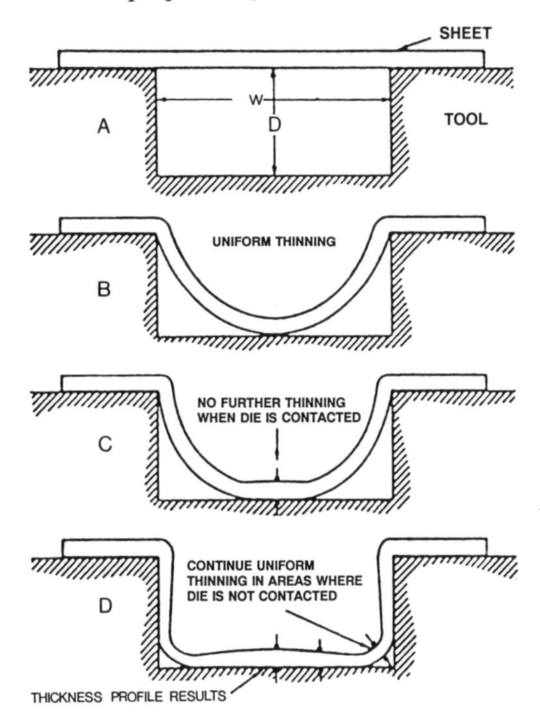

Figure 9.14 Rect box forming thinning distribution and evolution during forming.

of SPF. This is further complicated by the interaction of the forming membrane and the tool surface, which introduces contact friction to the membrane stress field which can result in forming ceasing at the point of contact. This is illustrated clearly in Figure 9.14 which shows the thickness variation and the influences which affect the thickness variation during the forming of a simple rectangular channel section.

While thickness variations cannot be completely eliminated, certain techniques have been developed which can be employed to improve the thickness variations or alternatively ensure that the material in the finished component is dispersed in the optimum manner. These methods involve:

- use of lubricant coatings on the tool and forming blank; this permits slippage between the tool and the forming material
- use of preforms which are pre-shaped by conventional means with minimal thinning and then formed into the fine detail by SPF
- use of variable thickness starting blanks
- male or female forming (Figure 9.13)

Table 9.2

Characteristics	*Cost benefits*	*Structural benefits*
(A) *Advantages of superplastic forming (SPF)*		
• Forming capability in excess of conventional forming	• One piece forming rather than multipart + joining	• Reduction in joints
	•	• For compression structure low cost stabilizing features
• 'One-shot' process	• Reduction in tooling and or intermediate heat treatment	• ====
• Accuracy/repeatability	• Reduction in assembly/fitting costs	• Low variability in structural performance
• Simple cavity tooling	• No requirement for matched tooling	• ====
• Negligible tool wear	educed tool refurbishment/replacement	• ====
• Process time independent of size or number	• Increasing benefit with size or use of multicavity tooling	• ===
(B) *Advantages of diffusion bonding (DB)*		
• Process time independent of bond area or numbers	• Increasing benefits with sizes or multiples off in one press cycle	• ====
• Joint strengths approaching or equal to parent metal	====	• Improved joint efficiency relative to conventional joining
• Low strain/distortion	• No correction after joining	
(C) *Advantages of SPF/DB*		
• Simple starting blanks	• Of particular importance for titanium	====
• Concurrent SPF/DB processes	• oining and forming as combined operation	====
• Complex monolithic structures	• duction in details and fasteners	• Reduction in joint lands
• Combines the benefits of SPF and DB as independent processes	Refer to advantages of the separate processes	• Refer to advantages of the separate processes
• High material utilization	• Of particular importance for titanium	====

9.6 ADVANTAGES OF SPF IN AEROSPACE STRUCTURAL DESIGN/MANUFACTURE

The primary design and manufacturing objectives associated with the selection of any processing route and associated material are the achievement of:

- maximum structural efficiency
- minimum cost
- a compromise based upon maximum structural efficiency and minimum cost

We live in an increasingly cost conscious world and hence the latter two factors are gaining in significance as primary design and manufacturing objectives. The generally accepted benefits provided by SPF in relation to the structural efficiency and cost objectives are summarized in Table 9.2.

9.6.1 Structural efficiency advantage provided by SPF

The main structural efficiency advantages provided by SPF can be categorized under three headings all of which exploit the superior–low cost–accurate–shape making capability of the SPF process. These are:

- shape or form which improves structural stability and therefore load-carrying capacity in compression and shear carrying structure – e.g. use of corrugations
- shape or form which reduces joints, thus saving weight penalties which would otherwise be associated with overlaps and stress raisers
- consistent shape and thickness in the finished component which therefore minimize variability in structural performance, allowing lower 'scatter' factors to be used in design (compression structure in particular)

9.6.2 Cost advantages provided by spf

The features of SPF which can be exploited to achieve cost benefits are:

- Forming capability in excess of conventional forming. This can be exploited to reduce the number of parts required to make a component and is clearly illustrated by the component shown in Figure 9.15.
- One-shot process. This provides an advantage over conventional forming process which because of the severity of the shape to be produced would require intermediate forming and heat treatment operations to achieve the final shape.
- Accuracy and repeatability. The cost implications associated with these particular attributes of the SPF process are a potential to reduce assembly times where a number of sheet metal parts are to be fitted together to create the final product.

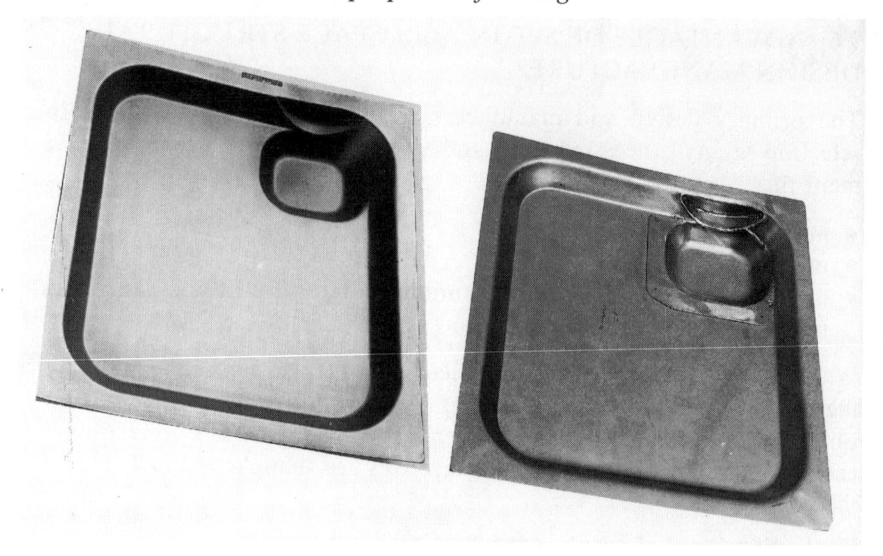

Figure 9.15 SPF and conventional panel.

- Process time independent of size and number of parts produced in one press operation. SPF is a strain-rate sensitive process and therefore the process time is dictated by the strain level to be achieved in the finished component and is not influenced by size or quantity produced in one press cycle.
- Simple cavity tooling. This compares with a requirement for matched tools in most conventional forming operations and therefore results in a considerable cost saving, especially at the initial stages of a project.
- Negligible tool wear. Because the slippage between the forming membrane and the tool is minimal a wear action is therefore absent which in turn results in excellent tool life, in particular where metal tools are used. Frequently ceramic tools are used for SPF but these in general have limited production life due to structural failure. They are, however, still viewed as being viable relative to metal tools because of the ease of re-casting a tool.

9.7 AEROSPACE APPLICATIONS OF SPF

Many applications of SPF are now incorporated into both civil and military aircraft and have now accumulated many thousands of successful flying hours. The material used in these applications are in the main Ti6/4 and the Supral range of aluminium alloys but a limited range of stainless steel applications have also been produced. Applications in the high strength aluminium alloys have yet to achieve full production status, but a number have been identified for early introduction in the future.

From an analysis of the wide range of applications which have now been developed by the world's aerospace industry it is possible to identify a generic range of applications for SPF. These are:

(a) Stability designed applications

- ribs
- frames
- beams
- compression struts

(b) Complex multi-element sheet components

- panels
- cowlings
- mounting brackets and supports

(c) Complex envelopes

- ducting
- tanks
- vessels

(d) Decorative panels and finishings

An example of a production SPF component which incorporates many of the benefits of SPF discussed in section 9.6 is provided in Figure 9.16. This component, which is fitted to the Airbus A310 aircraft, has the distinction of being the first primary structural application of SPF on a civil aircraft and since its introduction in 1981 has accumulated many thousands of hours of successful operations. This component is manufactured in Ti6/4 by BritishAerospace and forms a fuel tank boundary at four locations on the A310 wing. Because of the implosive nature of the fuel loads applied to this component the corrugated shapes are an essential feature of the design to provide structural stability, and because of the SPF ability to achieve these difficult forms, a 35% weight saving results relative to the conventional equivalent, which was also made from titanium. In the case of the conventional component only limited shape could be achieved, thus making it less structurally efficient.

Significant cost savings in transferring this component to SPF manufacture were achieved by:

- the ability of SPF to achieve the required shape in one operation by contrast with eight intermediate heat treatment/forming operations for the conventional equivalent

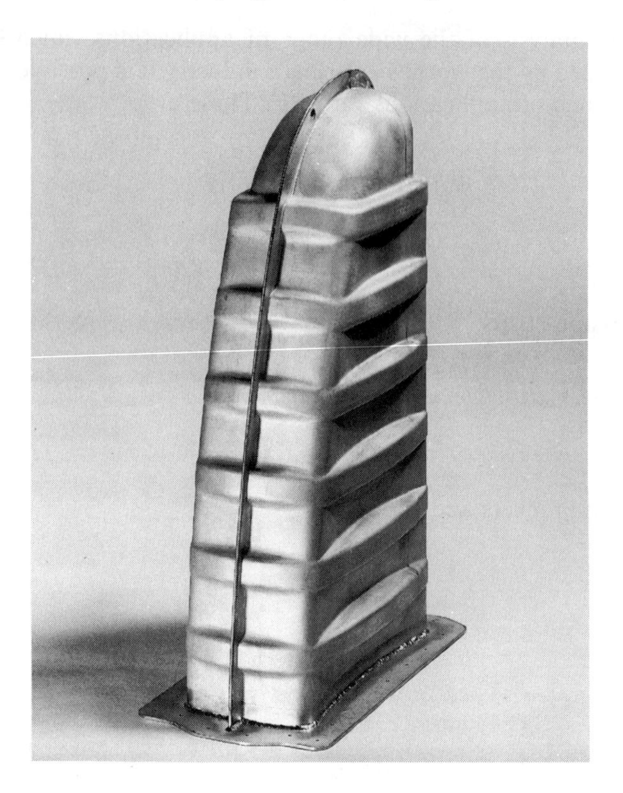

Figure 9.16 A310 jack can.

- reduction in parts count and fitting time by blowing the two halves of the can body simultaneously from an envelope made up of two flat profiled plates (Figure 9.17); this method of manufacture also ensured an accurate profile created by the forming tool which in turn facilitated accurate fitment of the mounting flange, thereby saving cost in assembly

9.8 SPF/DB

The fortunate coincidence that Ti6/4 exhibits SPF and DB characteristics, both of which can be achieved at the same temperature, and can be driven by the application of gas pressure, provided a spur to the process development engineers to explore the possibility of a combined joining and forming process (SPF/DB) which could lend itself to aerospace component manufacture. Diffusion bonding is considered in its own right in the following chapter.

The earliest patents on the combined process [16] appeared in the early 1970s and since then a range of generic structural types has emerged which

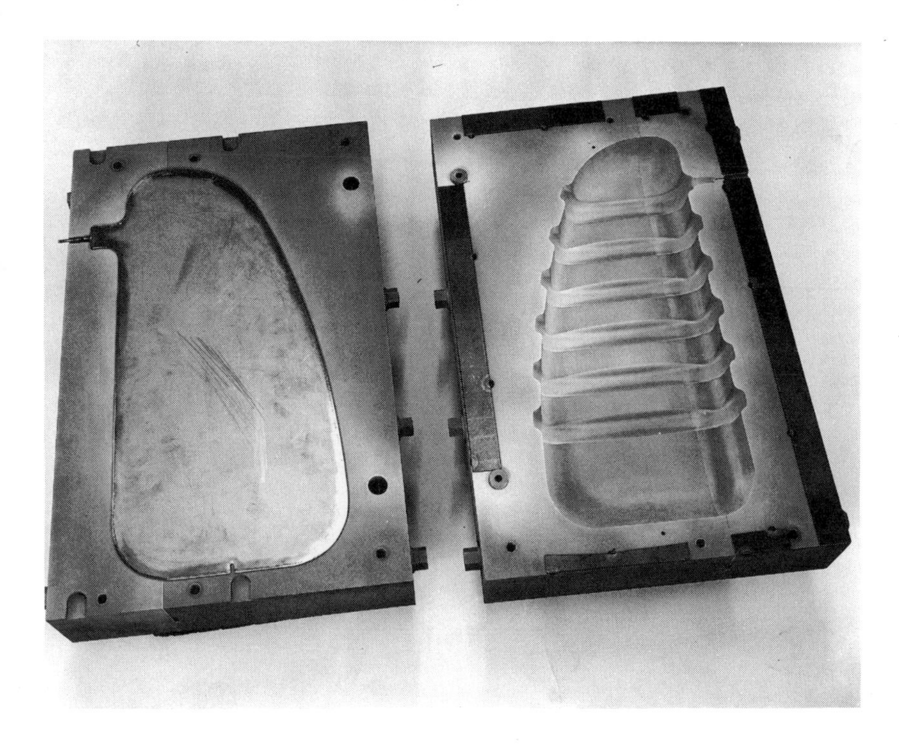

Figure 9.17 A310 jack can blank/forming tool.

have formed the basis for a wide range of aerospace structural applications, many of which have now progressed through to full production.

These generic types of SPF/DB structure have been universally classified in terms of the numbers of sheets that are used in their production. One particular form has, however, proven to be extremely versatile and promises to achieve the widest application in future aerospace component design and manufacture. This form is the four sheet cellular structure, the salient features of which are shown in Figure 9.18, and the steps in its manufacture are described below.

9.8.1 Four sheet cellular structure – method of manufacture

(a) Core preparation

The first stage in manufacture is the preparation of the core element. This consists of two core sheets which are joined in a selected and closely controlled pattern using line DB or alternatively by conventional welding methods such as spot, laser or electron beam. The purpose and function of these joint lines will be explained and become more readily understood

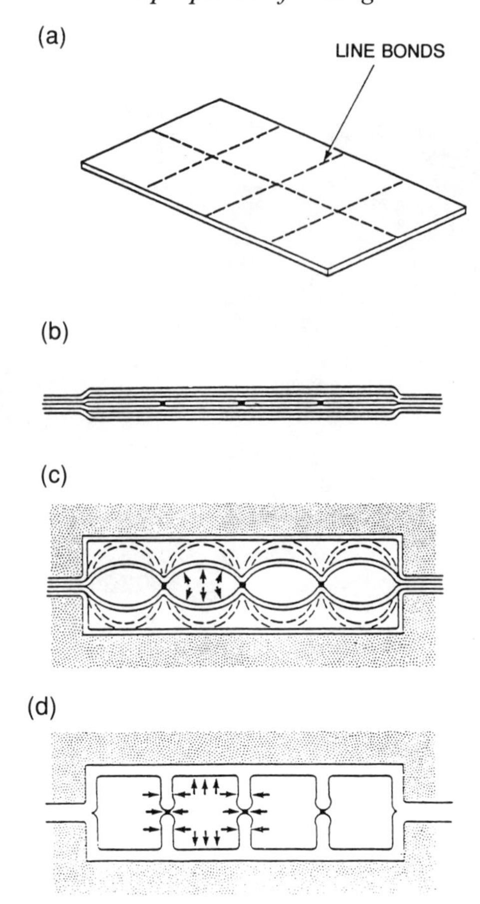

Figure 9.18 Evolution of a SPF/DB four sheet cellular structure: (a) core envelope; (b) blank; (c) forming; (d) bonding.

in the description of the later stages of the forming operation. It should be noted, however, that as illustrated in Figure 9.18, the joint lines associated with this structural form need not be unidirectional. This feature is of major significance in providing versatility in its structural application as it enables orthogonal stiffening to be employed if required.

Completion of the core element manufacture involves welding around its periphery to create a gas-tight envelope but to facilitate SPF of the core, a gas entry connector is fitted to enable the **core envelope** PTP to be applied.

(b) Blank completion

The final preparation of the SPF/DB blank involves enclosing the core envelope described in section 8.1.1 between two further sheets which are again welded around their periphery to form a second envelope. As in the case of the core, a gas entry connector is fitted to the skin envelope to enable the **skin** PTP to be applied.

(c) Forming

To commence the SPF/DB component forming operation, the completed four sheet blank is placed into a tool, the cavity of which provides the external shape of the component to be formed. Gas lines are attached to the skin and core gas entry connector points and the tool and blank are raised to the SPF/DB operating temperature.

The forming sequence starts by expanding the skin envelope until it conforms to the shape of the tool. The expansion of the core envelope then commences, but unlike the skin the core is constrained by the line bonds/welds and takes up a shape which is similar to that associated with an **air bed**. Continued inflation of the core causes the 'bubbles' to expand until they are eventually constrained by the tool. Continued expansion is then controlled by the tool surfaces in the horizontal plane and the equilibrium between adjacent bubble contact surfaces in the vertical plane. The process continues until adjacent bubbles are in full contact, creating vertical members spanning between the two tool faces. These vertical members act as stiffeners in the final structure and are of major significance in component design. Their location and orientation is defined by the original pattern of core joining and as stated earlier orthogonal stiffening patterns can be created in this SPF/DB process.

(d) Bonding

Once full contact has been made between the core and skin surfaces and between adjacent core cells then the final process involves raising the pressure in the core to the bonding pressure, and after an appropriate time all surfaces in contact will have merged by DB to create a monolithic structure.

Currently the process detailed above can only be applied to titanium based SPF alloys. This is due in the main to the ease with which titanium alloys can be bonded. SPF/DB of aluminium and nickel based alloys is currently limited to experimental activities but titanium SPF/DB has reached full maturity as a production process.

In the author's view the four sheet cellular structure and its method of manufacture provides designers with the widest range of freedom from which to establish an optimum design. These freedoms can be defined as:

- the ability to change the skin envelope thickness and reinforcements independently of the core, including the deletion of either or both skins to produce two or three sheet cellular forms

Table 9.3 SPF and SPF/DB development components

Component/aircraft	*Form (titanium)*	*Replacing*	*Cost saving (%)*	*Weight savings (%)*	*Test conditions*	*Comment*	*Reference*
BLATS – wing carry through structure	SPF/DB assembly	Titanium fabrication	41	30	2 lives fatigue 2 lives damage tolerance	Experimental design	[19] [23]
Nacelle frames – Bl	SPF SPF/DB	Titanium fabrication	55 43.5	33 40		Integrally stiffened webs	[19] [23]
APU door – B1	SPF/DB	Titanium machining	50	31	Acoustic test: duration 15 hrs; levels 159–167dB	In production Bl-B but SPF only	[19] [23] [24]
Engine bay – B1 door	SPF/DB	Aluminium honeycomb	34	29		Largest component made to date 105 in × 60 in	[19] [23] [24]
Hot air – Bl windscreen blower	SPF/DB	Steel fabrication	40	50	Fully qualified for B-l	In production for B-1B	[19] [23] [24]
Undercarriage – T38 door	SPF/DB	Aluminium honeycomb	2	\$5m life cycle costs	Fully qualified for flight and flight tested since early 1980's	Likely full scale production as replacements	[18] [19]
Horizontal – F4 stabilizer TE	SPF/DB	Steel honeycomb				Proposed for flight test	[19]
T E Flap – C-17	SPF/DB		15	26	Fully qualified for flight		[18] [19]
Nozzle fairing – F-15	SPF/DB	Titanium honeycomb	40		Acoustic testing levels up to 165 dB. In excess of 3 years flight testing	1st SPF/DB component to be flight tested. Features in redesign of F-15 nacelle area	[18] [19]
Wing glove fairing – F14A	SPF/DB	Aluminium fabrication	30	10		Proposed as flight demonstrator	[18] [19]
Escape hatch – BAe 125/800	SPF/DB	Aluminium fabrication	30	10	Fatigue test to 120 000 pressure cycles	Refer to figure 9.19	[19]

Access panel – harrier AV8B	SPF/DB	Aluminium fabrication plus titanium skin		12	Acoustic testing,, duration 40 h; levels – 165 dB	In production	
Slat surface refale	SPF/DB	Aluminium fabrication			Fully qualified for flight test	Flight tested	
Hot ducting – Tornado	SPF/DB	Titanium fabrication	34			In production	[21]
Missile wings	SPF/DB	Various					[20]

- the ability to change the core thicknesses and associated reinforcements independent of the skin
- the ability to change the number form and position of the stiffeners
- the ability to interrupt stiffeners
- the ability to have orthogonal stiffening patterns

In addition, but not illustrated in this chapter, techniques have been developed to form integral fittings and apertures through the stiffeners, to facilitate the passage of control runs, wires and conduits through the structure. These fittings and apertures can be created as a concurrent operation during SPF/DB.

All of these variations can in most cases be accomplished without changing the forming tool which has the sole function of defining the external shape of the component.

9.9 ADVANTAGES OF SPF/DB IN AEROSPACE STRUCTURAL DESIGN/MANUFACTURE

The principal economic and structural advantages provided by SPF/DB relative to alternative forms of manufacture are presented in Table 9.2(C). The content of this table is reasonably self explanatory. It should be emphasized, however, that the savings both in cost and weight are highly dependent upon the specific component under consideration, its equivalent conventional method of manufacture and associated material. Nevertheless, taking all of these factors into consideration, considerable cost and weight saving is achievable in Ti6/4 SPF/DB when compared with conventional manufacture in titanium, aluminium and CFC. In support of this, Table 9.3 provides details of the results of a wide range of development programmes which were undertaken during the years 1970–90.

9.10 AEROSPACE APPLICATIONS OF SPF/DB

On the basis of an assessment of civil and military aircraft, an estimate of the potential embodiment levels of Ti6/4 SPF/DB is viewed as being between 8 and 10% of the structure weight. Almost half of this application is seen in the replacement of aluminium fabricated structure. Areas of application derived on a generic basis from these components which have been successfully developed to date are:

- control surfaces
- smaller flying surfaces (military fins/tailplanes)
- access panels/doors
- engine bay components
- hot ducting
- engine rotating parts (blades)

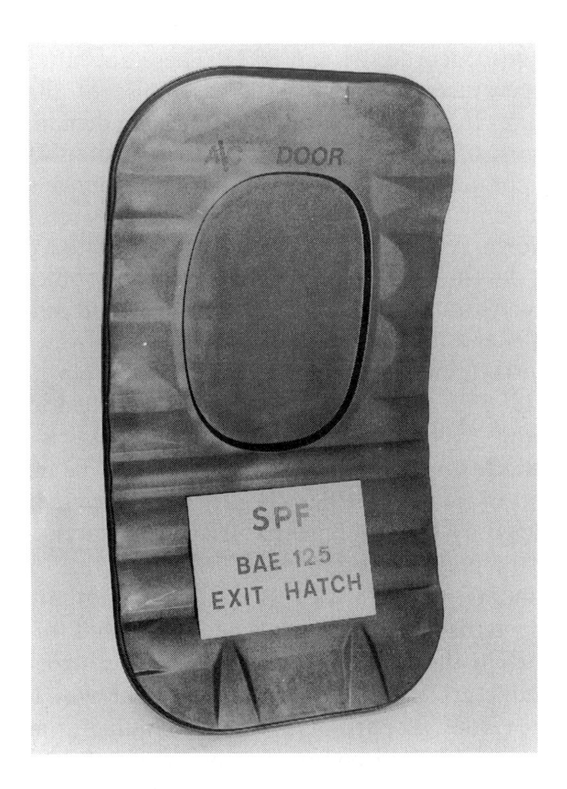

Figure 9.19 Aircraft cabin escape hatch four sheet cellular structure.

As an example of the significant reduction in parts count which is achievable using the SPF/DB process, the pressure shell door shown in Figure 9.19, when conventionally fabricated in aluminium, required 76 detail parts and 1000 fasteners in its manufacture. This compares with 14 details and 90 fasteners for the equivalent SPF/DB design. The significant difference in parts count which results is a major contributor to the 30% cost saving associated with this component in SPF/DB. In addition to the cost saving the titanium SPF/DB door has a weight advantage over the aluminium fabrication of some 10%. This fact has been fully substantiated by structural tests on this component in static strength and fatigue.

9.11 BACKGROUND TO THE APPLICATION OF SPF AND SPF/DB IN AEROSPACE

9.11.1 The beginning

The major stimulus to the development of SPF and SPF/DB in aerospace component design and manufacture was the advent in the 1960s of concepts for sustained supersonic cruise aircraft with mach numbers greater than 2.0 (Concorde, B1 and SST). These aircraft, with their high structural temperatures, demanded increased use of titanium in their primary structure. In recognition of this increasing demand, coupled also with the high cost of **conventional** manufacture of titanium components, the aerospace companies who were primarily involved with the major supersonic cruise projects at that time – British Aircraft Corporation (Concorde), North American Rockwell (B1) – initiated programmes of manufacturing development which had as a primary objective the establishment of new manufacturing methods which would reduce the cost of manufacturing titanium components.

The characterization of titanium alloys as superplastic materials by Backofen in 1965 in the USA was timely and led directly to the development and exploration of titanium superplastic forming as a potentially cost effective manufacturing technique. The main stimulus to the development of titanium SPF in the UK aerospace industry was work initially conducted by Dr R Johnson at CEGB Research Laboratories at Capenhurst in the mid to late 1960s.

Diffusion bonding has been used as a method of joining from ancient times and in the aerospace industry has been exploited for a considerable number of years in the cladding of materials for corrosion protection. DB as a manufacturing process in its own right was therefore being explored in the 1960s as part of the new manufacturing development effort for low cost titanium component manufacture. By the 1970s the combined processes of SPF/DB for titanium structures had been established by British Aircraft Corporation and North American Rockwell [16,17].

Superplasticity had of course been recognized and demonstrated by metallurgists for a considerable number of years prior to the 1960s but these demonstrations were mainly associated with materials and alloys which were of little interest in aerospace structural design. In recognition of the potential improvement in forming capability provided by superplasticity, some aluminium alloy developments were, however, carried out in the 1960s by the then British Aluminium Co (BACO). These alloys were the forerunner of the present day Superal 100 and 150 which are now in widespread use in secondary aerospace components.

9.11.2 Development phase

The decade from the early 1970s to the early 1980s was one in which the SPF and SPF/DB processes were developed from the small laboratory scale basic process demonstrations, to full size component design, manufacture, structural and flight testing. This decade also saw the spread of the technology to all of the world's major aerospace companies.

The largest programmes of development were conducted in the USA, involving substantial government funding which was reported in 1984 [18] as being in excess of $23m. These programmes involved to a greater or lesser extent most of the US major aerospace companies. Parallel developments also took place in the UK commencing in the late 1960s, in particular involving British Aerospace (titanium and aluminium) and Superform (aluminium). Government sponsorship for this activity was also provided in the UK. Work has now been reported from Germany, France and Japan [19–22]. Although not intended to be an exhaustive list, Table 9.3 provides a summary detailing the most significant components associated with these development programmes. This table has been compiled from a review of the literature and attempts to bring together the salient comments with respect to the structural and design aspects associated with each component developed.

Without exception the primary objectives of these programmes were to demonstrate:

- the cost and weight advantage provided by SPF and SPF/DB
- the practicability of SPF/DB in the manufacture and design of full size components

These programmes showed a consistency in their conclusions and positive results with respect to their main objectives. This provided the stimulus to move forward into production.

Apart from the attainment of these objectives, it is of interest to note from a design point of view the wide range of component types which were considered in these programmes and the range of loading and test conditions which were successfully applied. More particularly, however, it is important to note the number of components for which titanium SPF/DB provided cost and weight advantages over the equivalent component conventionally fabricated in aluminium. Latterly, this case especially in relation to cost has also been made for titanium SPF/DB in competition with highly loaded CFC components. These latter facts are perhaps the most significant from the point of view of the level of embodiment of SPF/DB titanium that might be anticipated on future aircraft, and Table 9.3 also provides a useful guide to the selection of suitable SPF/DB components on the basis of generic types. A further significant factor which is in favour of titanium SPF/DB and one which is gaining in prominence in our increasingly cost conscious environment, is the favourable 'cost of ownership' or 'life cycle cost' which results from the corrosion resistance and fatigue qualities of titanium. These claims were made in particular in respect of the replacement of aluminium honeycomb as in the case of the T-38 main landing gear door developed by McDonnell/Douglas [19].

During the period of major development in titanium SPF/DB, work was also undertaken in the development of high strength SPF aluminium alloys. Cavitation and its effects on material properties had been a fundamental problem which inhibited the successful exploitation of high strength aluminium, but the development of back pressure forming has enabled the full exploitation of high strength SPF aluminium to be made.

While DB of aluminium has now been demonstrated to be technically possible, the feasibility of a cost effective combined SPF/DB process with the range of structural possibilities provided by titanium has yet to be fully demonstrated.

9.11.3 Production phase

The period from 1980 onwards has seen the increasing exploitation of SPF in the manufacture of production components in both the USA and UK. It was reported [18] that some 230 titanium parts were in production by 1984 for the F-15, F-18, AV8B and B1. In the UK in 1982, 4 primary structural SPF parts had been introduced for the first time on to a civil aircraft Airbus A310, and 12 parts had been introduced on to Tornado by 1983. SPF/DB production parts were successfully introduced on to B1, Airbus A310, A320, A330/340 Tornado and Harrier/AV8B. The number of SPF/DB parts in production greatly increased as a result of the redesign of the F15 rear fuselage/nacelle area to take account of the advantages offered by these processes.

Perhaps one of the most significant applications from the point of view of its importance as a guide to the future exploitation of titanium SPF/DB is the adoption of a titanium SPF/DB control surface (fore plane) for the EFA aircraft. This component has been developed by British Aerospace and was chosen for its cost and weight advantage relative to a CFC alternative.

Production exploitation of the low strength Supral 100, 150 alloys has taken place since the mid-1970s and a wide range of secondary structural applications has been established on both civil and military aircraft. Full production of components in high strength 7475 and 8090 aluminium alloys has yet to mature, but applications on civil and military aircraft are anticipated in the near future.

Although not employing SPF, it is nevertheless highly significant from the point of view of confidence in the ability of these processes to produce production components to meet the widest range of demanding conditions, that Rolls-Royce have introduced DB into the manufacture of their wide chord fan blade for their RB211-535E4 engine. Further developments in the use of titanium SPF/DB in engine components (blades in particular) is being actively pursued and represents an area of application with huge production potential.

SPF and SPF/DB of structural alloys, in particular Ti6/4, have come of age inasmuch as these processes have now been introduced into the volume production of a significant number of primary components for airframe and aero engines. The processes are well understood and produce components which are superior in strength, cost and quality to their conventionally manufactured equivalents. To exploit these processes, an in-depth understanding of their flexibilities and limitations is required by the designer if the maximum benefits are to be realized. The designer capability has now been greatly enhanced by the ability to model the process using FEM methods.

REFERENCES

1. Bridges, P. J., Smith, D .J., Brookes, J.W. and Pale, P.S. (1989) Isothermal forging of turbine disc alloys, *2nd Parsons Int. Turbine Conference 1989*, Institute of Metals, London.
2. J. Bonet and R. D. Wood (1989) Solution procedures for the finite element analysis of superplastic forming of thin sheet, *Proc. Int. Conference on Computational Plastic Models, Software and Applications, Swansea, 1987*, Pine Ridge Press.
3. O. D. Sherby and J. Wadsworth (1989) Advances and future directions in superplastic materials, *AGARD Lecture Series No 168*, pp. 3.1, 3.12.
4. A. K. Ghosh (1982) Characterisation of superplastic behaviour of metals, *Proc. AIME Met. Soc. Conference on Superplastic Forming of Structural Alloys, San Diego, 1982.*
5. Suery, M. and Baudelet, B. Problems relating to structure, rheology and mechanisms of superplasticity, *Proc. AIME Met. Soc. Conference on Superplastic Forming of Structural Alloys, San Diego, 1982.*
6. Langdon, T. G. Experimental observations in superplasticity, *Proc. AIME Met. Soc. Conference on Superplastic Forming of Structural Alloys, San Diego, 1982.*
7. Gifkins, R. C. Mechanisms of superplasticity, *Proc. AIME Met. Soc. Conference on Superplastic Forming of Structural Alloys, San Diego, 1982.*
8. Gosh, A. K. (1977) *Acta METAL*, **25**, 1413.
9. Pearce, R. (1989) Superplasticity – an overview, *AGARD Lecture Series No 168.*
10. Woodford, D. A. (1969) *Trans. ASM*, **62**, 291; (1976) *Metal Trans.*, **7A**, 1244.
11. Hamilton, C. H., Bampton, C.C. and Paton, N. E. Superplasticity in high strength aluminium alloys, *Proceedings of AIME Met. Soc. Conference on Superplastic Forming of Structural Alloys, San Diego, 1982.*
12. Naziri, H. Pearce, R. and Henderson-Brown, M. (1975) *Acta Metal*, 489.
13. Ridley, N. (1989) Cavitation and superplasticity, *AGARD Lecture Series No. 168.*
14. Grimes, R. (1989) The manufacture of superplastic alloys *AGARD Lecture Series No. 168.*
15. Partridge, P. G., McDarmaid, D. S., Bottomley, I. and Common, D. (1989) The mechanical properties of superplastically formed titanium and aluminium alloys, *AGARD Lecture Series No. 168.*
16. British patent No 1 398 929.
17. US patent No 3 927 817.

18. Pellerin, C. J. (1984) *Proc. ASM Symposium on Superplastic Forming, Los Angeles, March 1984.*
19. Williamson, J. R. *Proc. AIME Conference on Superplastic Forming of Structural Alloys, San Diego, 1982.*
20. Boire, M. (1985) Advanced joining of aerospace metallic materials, *AGARD Conference Proc. No. 398.*
21. Beck, W. and Winkler, P. J. (1985) *Proc. 5th Int. Conference on Titanium, Munich, 1984*, pp. 1229–1236.
22. Ohsumi, M., Shimizu, M., Takahashi, A. and Tsuzuku, T. (1984) *Mitsubishi Heavy Industries Technical Review*, **2** (1).
23. Bampton, G., McQuilkan, F. and Stacher, G. (1984) *Proc. ASM Symposium on Superplastic Forming, Los Angeles, March 1984.*
24. Weisert, E. D. and Stacher, G. W. (1982) *Proc. AIME Conference on Superplastic Forming of Structural Alloys*, San Diego, 1982.

10

Joining advanced materials by diffusion bonding

P.G. Partridge and A. Wisbey

10.1 INTRODUCTION

The selection of suitable joining techniques is vital for the successful utilization of advanced materials in engineering. Sophisticated mechanical fasteners have been developed and these may be combined with adhesive bonding, but they usually incur a weight penalty and the associated fastener holes can limit the fatigue strength of the joint. The principal alternative joining techniques are summarized in Figure 10.1. In all fusion welding techniques the base metal is melted and the fusion zone will have a cast microstructure. The high temperature required can lead to a wide heat affected zone either side of the fusion zone associated with a steep temperature gradient. With high cooling rates, residual stresses and component distortion can occur. These are reduced in the much narrower fusion and heat affected zones produced with laser and electron beam welds. Fusion welding is usually unsuitable for joining dissimilar materials or for metastable alloys produced by rapid quenching if the quenched microstructure must be preserved.

In conventional brazing a flux is used to disrupt surface oxide films and facilitate wetting of the surfaces by an alloy with a melting point below that of the base material. To avoid the risk of remelting the braze alloy, all joints must be made in one operation. There is little diffusion between the braze metal and the metals being joined, and joint strength relies on the braze metal.

Solid state joining techniques can be divided into those using high pressure and high deformation and those associated with very little deformation. Of those depending on high deformation, roll bonding and explosive bonding involve bulk deformation of the parts being joined,

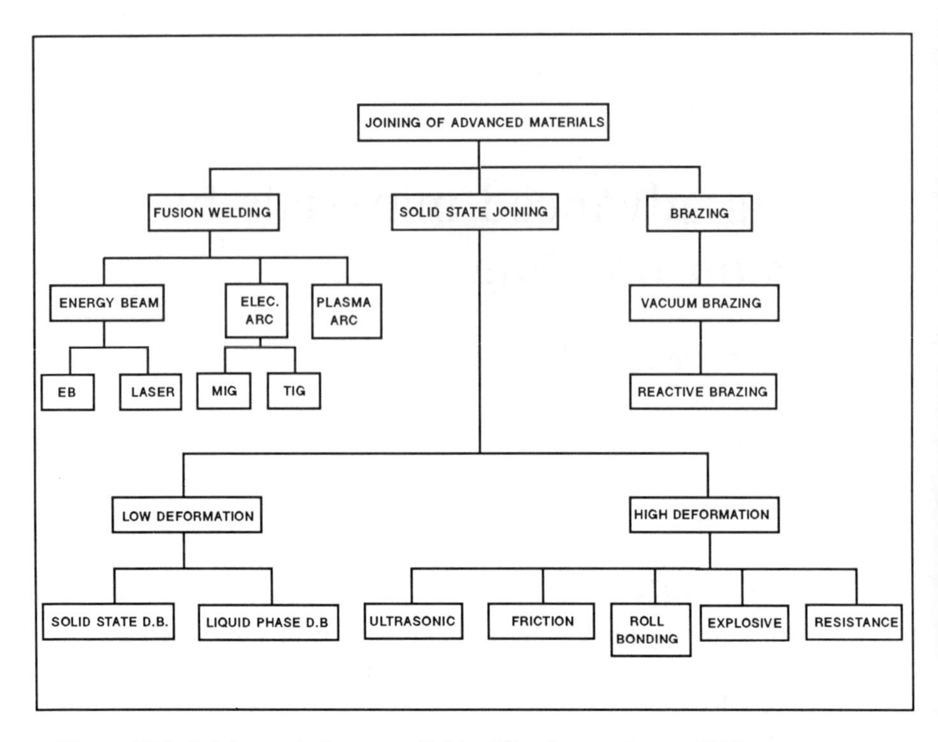

Figure 10.1 Joining techniques available: EB, electron beam; TIG, tungsten inert gas; MIG, metal inert gas.

and are used to produce semi-finished products such as clad sheet or plate. In friction and ultrasonic welding, intense deformation confined to the bond interface region raises the temperature and disrupts and disperses the surface oxide films prior to welding. The above techniques are described in reference [1].

In low deformation diffusion bonding (DB) the surfaces are carefully cleaned prior to pressing the surfaces together under a low pressure at high temperatures for relatively long times. This allows close control over the joining process, with overall deformation of < 1–10% and minimum distortion. Joints with 100% efficiency, i.e. joints having mechanical properties identical to those of the materials being joined, are possible with some materials. Hot isostatic pressing (HIP) under inert gas pressure has the unique advantages of high pressure (up to 300 MPa) and temperatures (up to 2000 °C) and more rapid bonding than under unidirectional pressing with minimum deformation. It is widely used for consolidating, in the solid state, powders and chopped melt spun ribbons.

Some of the many factors that affect the selection of a particular joining process in production are plant acquisition costs and utilization, unit production time, joint quality and reproducibility. For example, in EB welding carried out in vacuum, plant costs are high and production rates are relatively low and consequently production costs tend to be high. DB involves long bonding times and whole components generally have to be heated to the DB temperature. Equipment costs therefore tends to be high because of the time at high temperatures and pressure, usually in vacuum or inert gas environment. This often limits the size of components that can be accommodated. However, for critical aerospace components requiring accuracy, reproducibility, and freedom from defects these costs may be justified. Other processes lend themselves to high production rates, such as brazing, friction, resistance and laser welding.

Diffusion bonding, in both the solid and liquid states, can be used for joining a very wide range of similar or dissimilar materials. This flexibility together with other advantages (Table 10.1) is expected to make DB increasingly attractive for advanced materials. The theory and practice of diffusion bonding is therefore reviewed in this chapter.

Table 10.1 Characteristics of diffusion bonded joints

1	Cast, wrought and sintered powder products and dissimilar metals can be joined. May be only viable method for metal matrix composites and rapidly quenched materials with metastable microstructures.
2	Joints have corrosion resistance of parent metal or of selected interlayer combinations. No fluxes are required.
3	Small plastic deformation, low residual stresses and close dimensional control is possible.
4	Joint strengths approaching or equal to the parent metal.
5	Large area bonds can be made to give improved joint efficiency.
6	Absence of rivet holes and residual stresses increases fatigue strength of joint.
7	Bonded sheet structures can have high acoustic fatigue resistance
8	Thick and thin sections can be joined to each other.
9	Process time independent of bond area or numbers of components.
10	Machining costs may be reduced.
11	More efficient design and smaller buy to fly ratio may be possible.
12	May be combined with superplastic forming of thin sheet under gas pressure (SPF/DB). Sheet surface can have any curvature.

10.2 DIFFUSION BONDING MECHANISMS

Advances in process control and microanalysis have allowed a systematic investigation of the effects of the major bonding parameters (temperature, time, pressure, surface roughness, surface deformation) and much improved bond quality and joint strengths can now be achieved with confidence for many materials. Solid and liquid phase DB techniques are

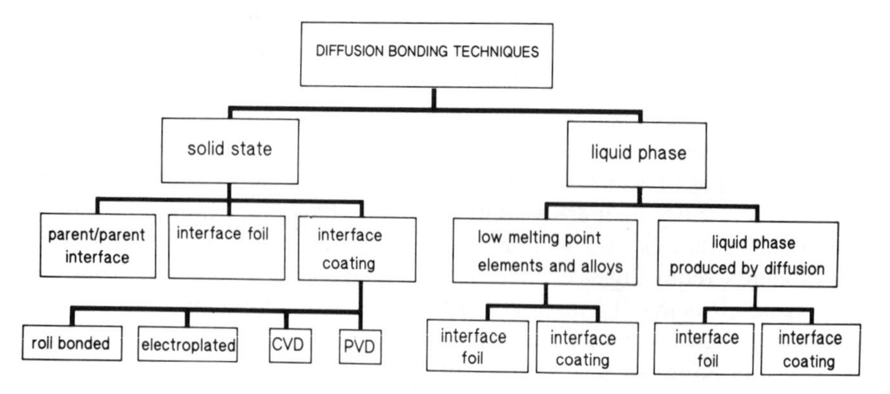

Figure 10.2 Summary of diffusion bonding techniques.

summarized in Figure 10.2. The presence of stable surface films (present on aluminium alloys) or brittle intermetallic compounds (formed between dissimilar materials) can lead to low joint strengths. To permit DB of such materials, interface coatings or foils may be employed. With dissimilar materials suitable interlayers may also reduce stresses caused by thermal expansion mismatch.

Diffusion bonding in the solid state involves various mechanisms operating in the following sequence:

- local plastic deformation of surface asperities on loading) Stage 1
- bulk creep deformation)
- diffusion across the bond interface) Stage 2
- grain boundary migration across the interface)

At the low pressures (much less than the macroscopic yield stress) and high temperatures (> 0.5 T_M), where T_M is the absolute melting point) used in bonding the deformation is confined primarily to the surface asperities as shown in Figure 10.3 [2]. Stage 1 involves instantaneous deformation on loading followed by power law creep according to a relationship of the type:

$$\dot{\varepsilon} = A\sigma^n \exp\left(-\frac{Q_c}{RT}\right) \tag{10.1}$$

where $\dot{\varepsilon}$ = creep rate, σ = stress, Q_C activation energy for diffusion and A and n are constants with $n \sim 3$–4 for a typical titanium alloy (Ti–6Al–4V). For fine grained material superplastic deformation with little strain, hardening ($n \sim 1.5$–2) can accelerate the deformation of the asperities in stage 1. In ultra high carbon (UHC) steel a decrease in the grain size from 2 to 0.4 μm was predicted to increase the strain rate by a factor of 1000.

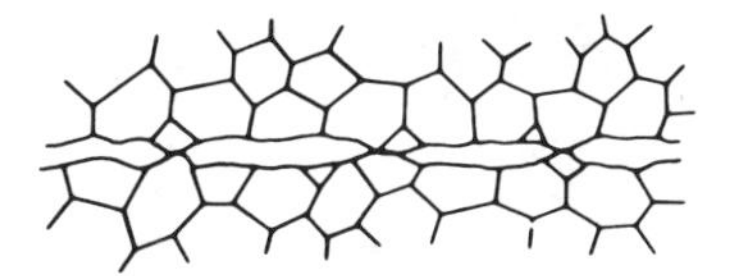

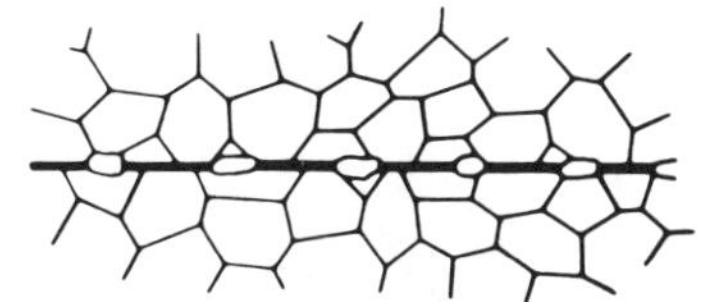

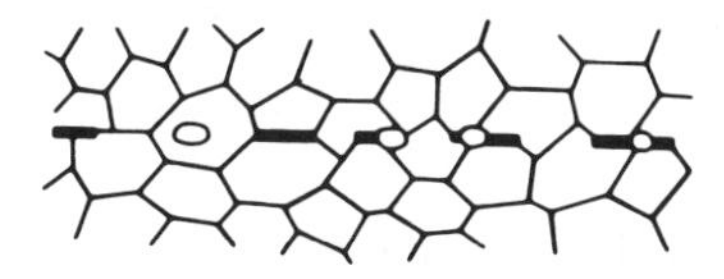

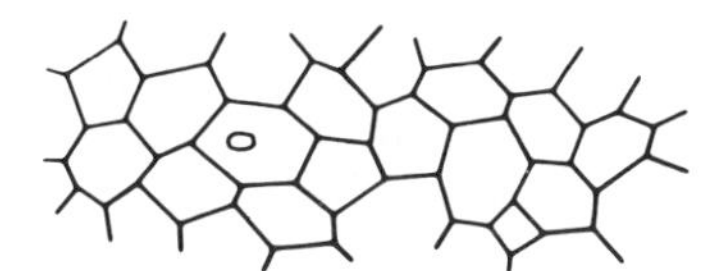

Figure 10.3 Two stage mechanistic model of diffusion bonding [2].

At the end of stage 1 the bond interface consists of bonded areas separated by areas containing small voids (Figure 10.4). Hydrostatic pressure can accelerate void closure by diffusion and plastic deformation for voids $> 20\,\mu m$, but below this size diffusion alone controls the elimination of voids by surface, grain boundary and volume diffusion mechanisms. Fast grain boundary diffusion is favoured by a small grain size but in practice only time and temperature are important variables for many materials. Small empty voids may be removed by isothermal anneals in the absence of pressure. Bonding in stage 1 is sensitive to stress and occurs much faster than in stage 2. Recrystallization and grain boundary migration which operate in the final stages of bonding may be essential for high strength joints since it leads to the elimination of the

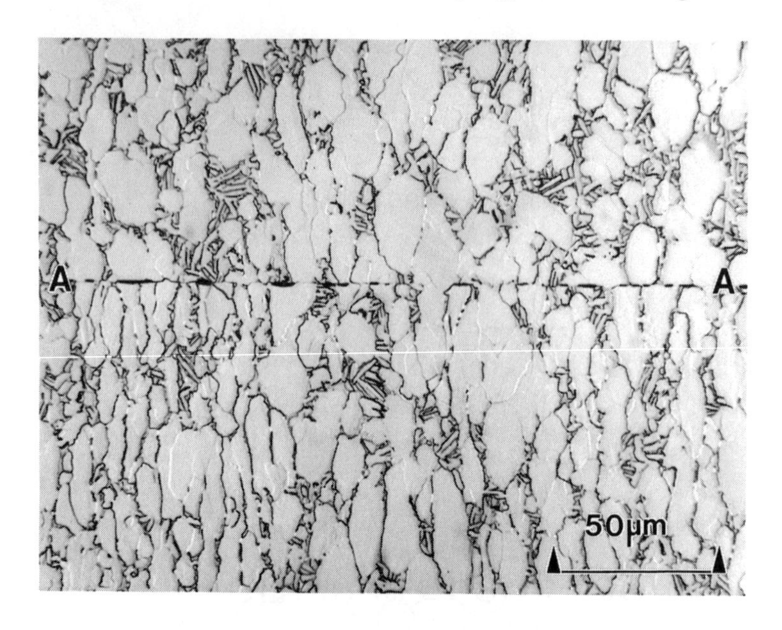

Figure 10.4 Porosity in a diffusion bonded joint in Ti–6Al–4V after 0.5 h at 800°C and 10 MPa. Bond interface at A–A.

planar grain boundary interface. Diffusion bonding models based upon an idealized surface geometry and the above mechanisms have been reasonably successful in predicting the effect of pressure on time to produce a 95% parent metal bond strength in titanium and copper. More complex models may be required to explain bonding in the presence of oxide films and for dissimilar materials.

When the temperature or pressure required to obtain good surface contact prohibits diffusion welding, diffusion brazing offers an alternative joining method. This process has some of the advantages of conventional brazing (low pressure, short times) and of diffusion bonding (base metal microstructure and strength). Diffusion brazing has been described using the phase diagram in Figure 10.5 [3]. An interlayer or coating with composition 1 is heated to the bonding temperature and melted. During isothermal annealing, diffusion changes the composition from 1 to 3 in Figure 10.5 and the joint solidifies. Further diffusion results in a large dilution of the interlayer elements. This technique has been applied to both titanium and aluminium alloys using copper interlayers to form low melting point eutectic phases. Residual interlayer after bonding should be avoided since it may lead to electrochemical corrosion.

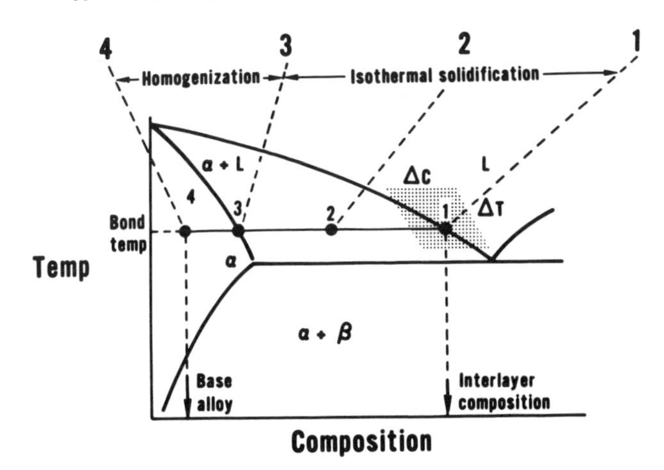

Figure 10.5 Phase diagram showing the microstructural changes that occur during isothermal LPDB [3].

10.3 EFFECT OF SURFACE ROUGHNESS AND CONTAMINATION ON BOND INTERFACE DEFECTS

Surface roughness is one of the important variables affecting the quality of diffusion bonded joints. The roughness affects the time to achieve complete contact between the surfaces being bonded. In practice a metal surface has asperities which are small in terms of height and wavelength (surface roughness) superimposed on longer wavelength asperities (surface waviness) (Figure 10.6) [4]. The surface waviness is particularly important in the bonding of thick section machined parts, i.e. massive DB.

Thin sheet has the advantage that in the as-rolled state the surface finish is usually good. The strength of aluminium alloy joints has been found to decrease with increasing surface roughness. The finish obtained for titanium alloy in the as-received state or after careful grinding is typically better than $R_a = 0.5\,\mu m$.

The defects associated with DB differ from those found in conventional joining processes and particular problems may arise when sheet is bonded under argon gas pressure as in SPF/DB processing (section 10.11).

The defects found may be divided into:

- large voids
- microvoids
- intimate contact disbonds

Large voids or disbonds may be associated with argon gas entrapment and can be avoided by progressive venting. In massive DB similar voids are produced by poor surface finish, tool misalignment or machining

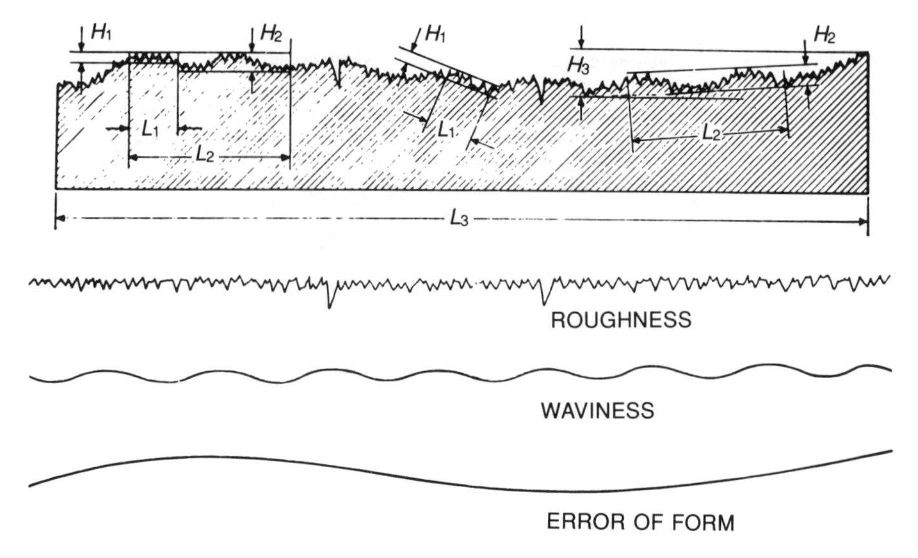

Figure 10.6 A surface texture representing the combined effects of several causes [4].

debris, and for these reasons massive DB tends to be more difficult and costly than gas pressure bonding of thin sheet. Microvoids or intimate contact disbonds, i.e. interfaces in contact but not diffusion bonded, can be caused by surface contamination, oxide films or incorrect bonding conditions and are the most difficult to detect by NDE. These defects can, however, lead to very low bond strengths.

Surface contamination is an important practical consideration for DB. The most common surface contaminants encountered are inorganic or organic films; these may be reduced by solvent cleaning and pickling. Organic films (oil or fingerprints) and water vapour may be removed by heating to ~ 300 °C but at higher temperatures surface reactions can occur. These can be important for aluminium alloys since stable oxide films may be produced which seriously reduce bond strength. At high temperatures many metal oxides, e.g. Ag_2O, dissociate or dissolve in the base metal (oxides of Cu Zr, Ti, Nb, Ta, and Ni). The rate of dissolution of 10 nm thick oxide films is virtually instantaneous for all these metals at $0.5\,T_M$ except for nickel, which requires ~ $0.92\,T_M$; the time to dissolve an oxide film increases rapidly with increasing film thickness. Stable oxides, carbides or nitrides may be difficult to remove except by sputter cleaning at low temperatures.

Although thin oxide films on Ti alloys are readily removed by dissolution at the bonding temperature (~ 925 °C) further oxygen pick-up from the environment can seriously effect the quality of the bonds. Increasing oxygen in solid solution significantly increases the hardness, and excessive pick-up produces a hard oxygen-rich stabilised α-phase surface layer (α-case) which prevents bonding and leads to surface cracking. The argon gas used to flush or pressurize the die chamber can be contaminated by residual O_2 and H_2O or by outgassing of stop-off compounds (used to mask areas from DB). The stop-off compounds (BN, Y_2O_3) may also become less adherent at high temperatures as the binder is burnt off, and they may then become displaced on to the surfaces to be bonded.

10.4 TESTING OF DIFFUSION BONDED JOINTS

DB joints are needed that can be easily made and tested but which may also need to be representative of production components. Testing is also a vital part of joint quality control in a production environment, since conventional non-destructive tests may not be sufficiently sensitive to detect the very small defects that effect bond strengths. The strengths reported for DB joints show a wide variation depending on bonding parameters, bonding technique and test piece design. The latter is particularly important since a test piece must provide meaningful strength data to enable the effects of metallurgical and processing variables on bond strength to be assessed.

A simple low cost DB butt joint produced with bar stock allows standard tensile, impact and fatigue test pieces to be manufactured with the bond interface at the centre of the gauge length. In the bonding of unconstrained cylinders the maximum bonding pressure may be limited by bulging, which may be quite different for each material in a dissimilar material bond. In tensile tests on these joints the bond is loaded uniformly and poor bonds may be reflected in low reduction of area values. The bond quality can vary across the bonded area however, especially if the faces to be bonded are not normal to the axes of the cylinders or if axial alignment is poor. For DB joints in titanium alloys tensile strength is a poor measure of bond quality whereas impact properties are particularly sensitive to bond quality. Fatigue is particularly sensitive to porosity and defects at the test piece surface. To obtain fracture toughness data of DB joints in sheet a stack of sheets can be bonded to provide thicker test pieces. The fracture appearance of DB joints can provide information on the quality of the initial bond; a good bond will reveal uniform ductile shear or tensile fracture cusps.

For thin sheet overlap shear test pieces are often used; in both tensile (Figure 10.7 (a) and (b)) and compressive shear tests (Figure 10.7 (c)) the bond strength is given by the failure load/bonded area [5,6]. For gas pressure bonded thin sheet an overlap test piece can be produced by machining two flat bottomed grooves in the sheet surfaces, as shown in Figure 10.7 (b); the bottom of the grooves should coincide with the

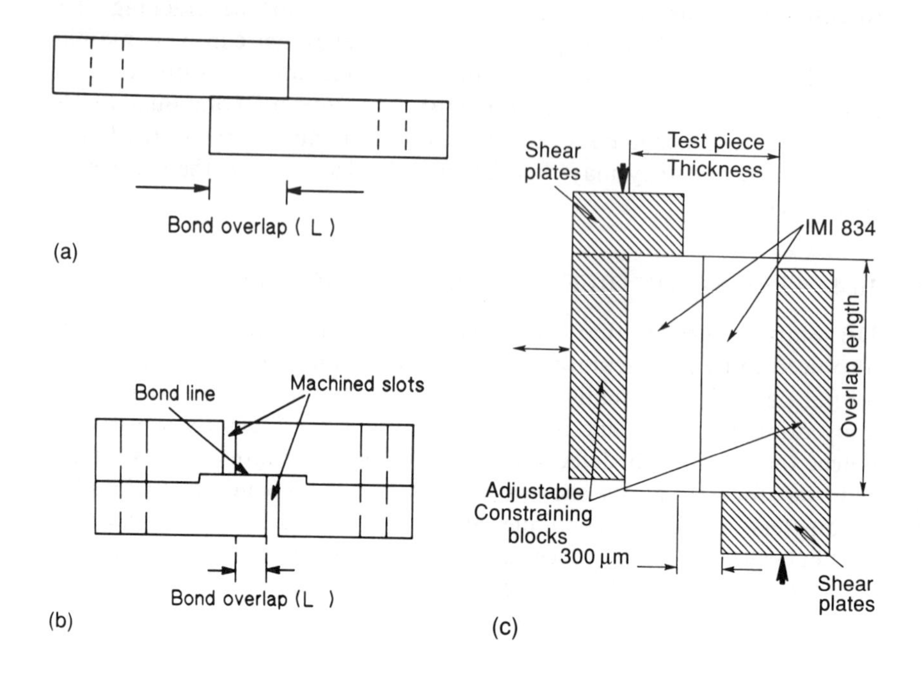

Figure 10.7 Single overlap shear test pieces: (a) tensile, simple overlap; (b) tensile, stepped and slotted overlap; (c) compression simple overlap [5,6].

centre of the bond. Under load the stresses are greatest at the ends of the bond region and unconstrained joints in tension tend to fail under peel stresses (Figure 10.8) [7].

Joint tensile shear strengths have been compared in terms of the joint efficiency (JE), where

$$\mathrm{JE} = \mathrm{P} \times \frac{100}{\sigma A}, \tag{10.2}$$

P = maximum load, and σ and A are the tensile strength and cross-sectional area of the sheet respectively. Failure in the parent sheet gives JE = 100%. The parameter JE does not require bond shear fracture and therefore does not actually measure the bond interface strength. This is best obtained by measuring the bond shear ratio R_B where

$$R_B = \frac{\tau_B}{\tau_p} = \frac{P_B}{P_p} \tag{10.3}$$

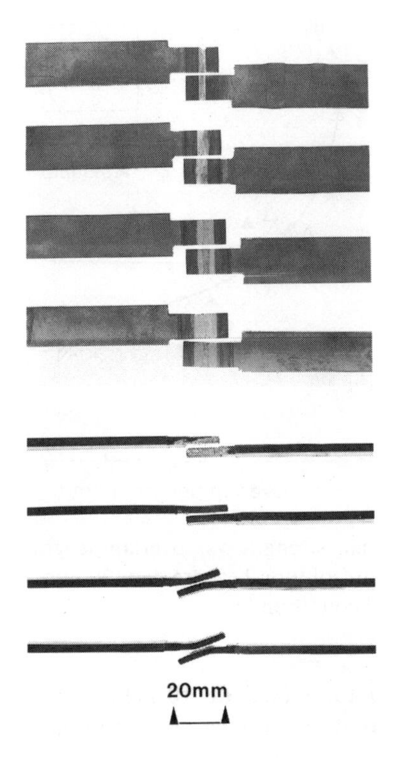

Figure 10.8 Effect of increasing overlap length on bending of DB simple overlap tensile shear test pieces [8].

τ_B and τ_p are the shear strengths of the bond and parent metal respectively, and P_B and P_p are the corresponding failure loads under shear conditions. However, for diffusion bonds between 3.2 mm thick aluminium alloy sheet (and also for liquid phase DB joints between thin sheet) with decreasing overlap L, shear strength increased (Figure 10.9) [5]. This is caused by the increasing effect of the peel stresses for $L > 2$ mm. It is often difficult to control the overlap to within 1 mm and this contributes to the scatter in shear strength data.

Compression overlap shear tests have also been used with thin sheet (Figure 10.7 (c)) and with good constraint on the test piece very little peel can occur. This test piece can be small, permitting many more test pieces for a given bond area, and may be used for bonds between thick and thin sheet. Since peel is avoided the shear strength is largely independent of overlap length.

A peel test measures the resistance to crack growth at the bond line and peel strength (load/bond width) data is vital for the application of DB joints to thin sheet structures. Hot peel strength is required for

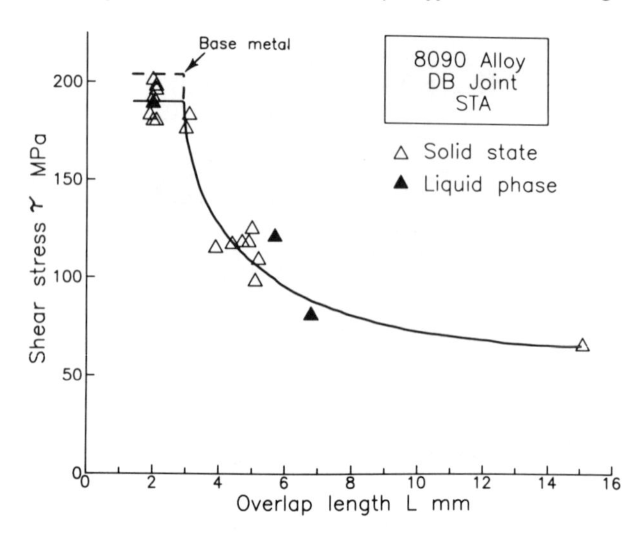

Figure 10.9 Tensile shear strength vs. overlap length for Al–Li 8090 alloy diffusion bonded joints in solution heat treated (20 min at 530°C, water quench) and aged (5 h at 185°C) condition [5].

DB/SPF of sheet structures (section 10.11) and peel strength at ambient temperature is required in service. For comparison of peel strength identical test pieces and test conditions must be used to avoid excessive scatter in the test data. Peel strength can be a sensitive measure of bond quality, particularly in aluminium alloys.

Examples of peel test pieces are shown in Fig. 10.10 [5]. A typical peel load–time curve is shown in Figure 10.11. The shape of the curve has been shown to depend on plastic deformation in the arms and on the bend radius, sheet thickness and loading rate. The peak load on the curve is associated with the crack initiation, whilst the load plateau corresponds to stable crack growth, and this should be taken as the peel strength of the joint.

10.5 DIFFUSION BONDING TECHNIQUES OF METALS

10.5.1 Titanium alloys

Titanium alloys are readily bonded in vacuum or inert gas atmospheres; the conditions are given in Table 10.3. As the β transus is approached rapid grain growth occurs, creep rate decreases and the bonding time may increase. The effect of grain size on the time and pressure required to produce a pore free bond in Ti–6Al–4V alloy is shown in Figure 10.12 [8]; at a typical bonding pressure for this alloy (300 psi, 2 MPa) the

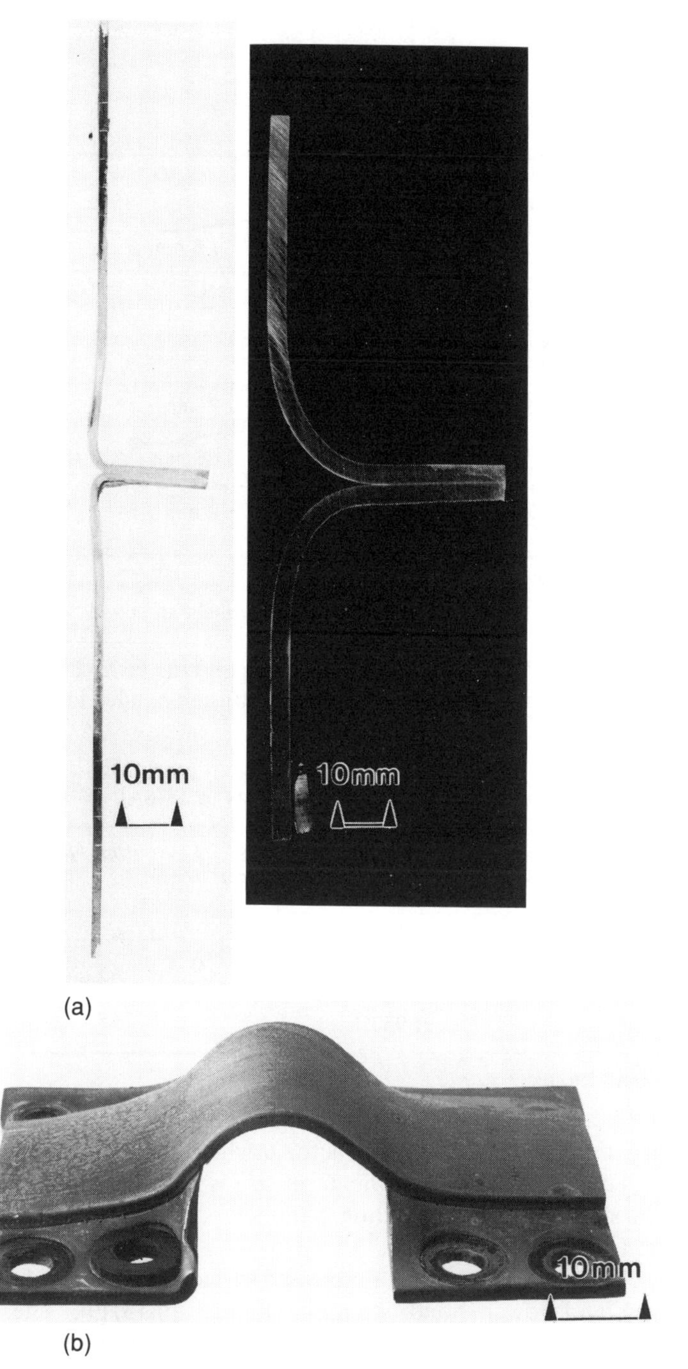

Figure 10.10 Peel test pieces: (a) 90°T peel test piece, 1.6 mm thickness sheet; (b) 90°T peel test piece, 4 mm thickness sheet; (c) variable angle peel test piece [5].

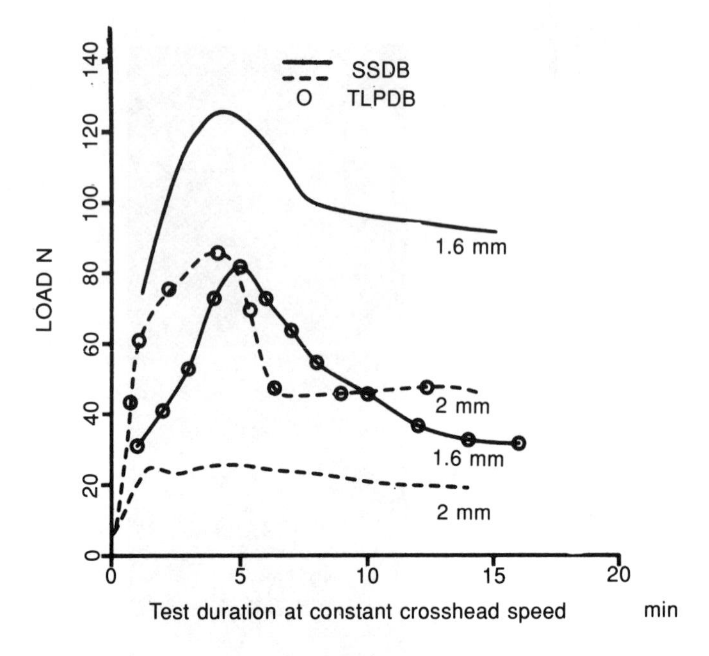

Figure 10.11 Load vs. test duration for 90°T peel tests in Al–Li 8090 alloy solid state (SS) and transient liquid phase (TLP) diffusion bonded joints [5].

Table 10.2 Mechanical properties of DB joints in Ti–6Al–4V

Bonding conditions	*Bulk deformation (%)*	*Condition*	*UTS (MPa)*	*Elongation (%)*	*Impact strength (J)*	*Void density (%)*
850°C, 5 MPa, 0.25 h	0.43	As bonded	791	0	0	7–17
		After 24 hrs at 950°C	950	19	17	ND

(ND – none detected)

bonding time is increased by a factor 6 when the grain size is increased from 6.4 μm to 20 μm. Since the grain size is coarser in plate or forgings than in sheet, different bonding parameters will be required for these different product forms.

An acicular microstructure or increased surface roughness led to increased porosity and lower bond strength (Figure 10.13) [9]. The enhanced bonding associated with fine grained material may be exploited with a fine grained interlayer foil for bonding thick sections of coarser microstructure or by heavy machining to induce fine recrystallised grains at the

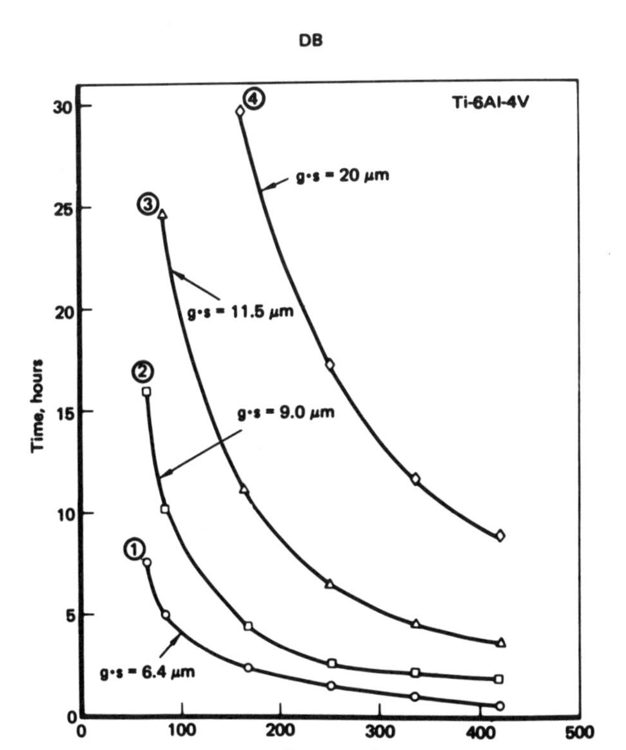

Figure 10.12 Effect of grain growth on time versus pressure curves for the bonding of Ti–6Al–4V [8].

bond interface at the DB temperature. Accelerated bonding can also be obtained under conditions that cause the alloy to deform superplastically; this has been reported to reduce the bonding pressure by a factor 4, the welding time by a factor 6–30 and the temperature required by 50–150 °C. Crystallographic texture may also affect the rate of bond formation.

Small bond defects may have little effect on the tensile or shear strength of the bond but may have a significant effect on other mechanical properties such as fatigue or impact strength. The joint shown in Figure 10.4 exhibited parent metal tensile strength but markedly reduced fatigue and impact strength. Impact properties are more sensitive to bond quality than other mechanical properties. Fine pores, as small as 1 μm, can reduce the impact strength. The distribution of pores is known to affect both the impact and fatigue properties; for example, clusters of pores reduce fatigue properties.

To obtain low bulk deformation during DB, joints may be made at low temperatures and short times (800 °C, 15 minutes and 5 MPa) to produce

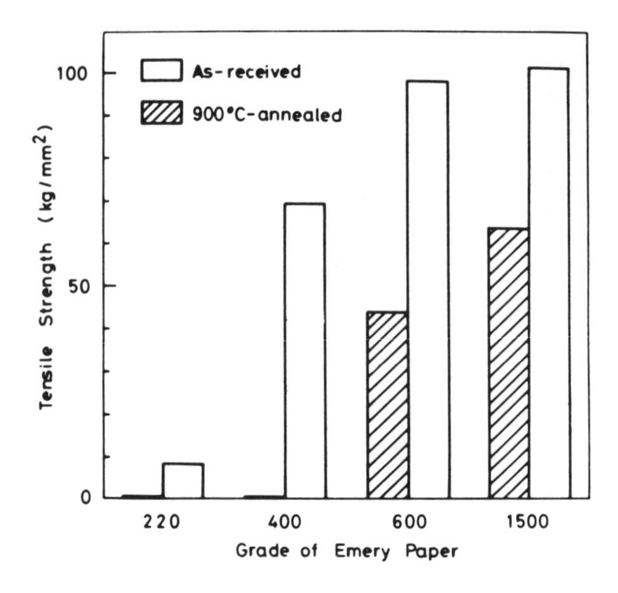

Figure 10.13 Effect of initial surface roughness on the tensile strength of DB joints in Ti–6Al–4V, in as-received condition (fine grain size) and after a 900°C anneal (coarse grain size). It was bonded at 850°C for 10 min and 2 MPa [9].

significant porosity, and then annealed at temperatures just below the β transus for extended times. Voids may then be eliminated and parent metal tensile strengths and increased impact values obtained (Table 10.2).

Liquid phase DB of titanium alloys is possible with Cu, Cu-base and other alloy interlayers to produce almost parent metal mechanical properties. A β alloy, Ti–21V–4Al, bonded with a Ti–20Zr–20Cu–20Ni interlayer had a strength of ~ 700 MPa, but subsequent ageing increased this to 1400 MPa with ~ 10% reduction in area.

10.5.2 Aluminium alloys

Diffusion bonding of aluminium alloys has proved difficult because of the presence of a tenacious surface oxide film. The oxide film can be disrupted by large deformations and high temperatures in, for example roll bonding, but the introduction of soft interlayers in the form of dilute aluminium alloy cladding or foil inserts enable bonds to be made with smaller overall deformation. At present DB of aluminium alloys is primarily under consideration for sheet applications, especially with the development of superplastic aluminium alloys, and massive DB, used with titanium alloys, is practically unknown. The bonding conditions and properties of some aluminium alloys are shown in Table 10.3.

Table 10.3 Diffusion bonding parameters and bond strengths of metallic and intermetallic joints

Material	*Technique*	*Temperature (°C)*	*Pressure (MPa)*	*Time (h)*	*Joint strength (MPa)*
Similar metals					
Ti–6Al–4V	Solid phase	930	2–3	1.5	575 (shear) 990 (tensile)
Ti–6Al–4V (Cu and Ni interlayers)	Liquid phase	950	-	1–4	950 (tensile)
IMI 834 (nr α alloy)	Solid phase	990	5	0.5	1040 (tensile)
Al–Zn 7475 T6	Solid phase	516	0.7	4	331 (shear)
Al–Li 8090 T6	Solid phase	560	0.75	4	230 (shear)
Al–Li 8090 T6 (Cu interlayer)	Liquid phase	552	0.75	2	215 (shear)
Intermetallics					
Ti_3Al (Super α_2)	Solid phase	1050	2.1	2	-
Ti–38 mass % Al	Solid phase	1200	15	1.1	225 (tensile)
Metal matrix composites					
Al–Li 8090/20 wt% SiC particulate T6	Solid phase	560	1.5	4	100 (shear)
Al–Li 8090/20 wt% SiC particulate (8090 interlayer) T6	Solid phase	560	1.5	4	150 (shear)
Al–Li 8090/20 wt% SiC particulate (Cu interlayer) T6	Liquid phase	560	0.75	4	> 107 (shear)
Ti–6Al–4V/45 vol.% SiC cont. fibre-butt joint	Solid phase	900	10	3	700 (tensile)
Dissimilar materials					
Ti–6Al–4V/304 stainless steel (V and Cu interlayers)	Solid phase	850	10	1	450 (tensile)
Ti–6Al–4V/Stellite HS6 (Ta and Ni interlayers)	Solid phase	900	20	1	406 (shear) 798 (tensile)
Ti_3Al (Super α_2)/Ni alloy HS242 (Ta and Ni interlayer)	Solid phase	990	20	1	488 (shear)

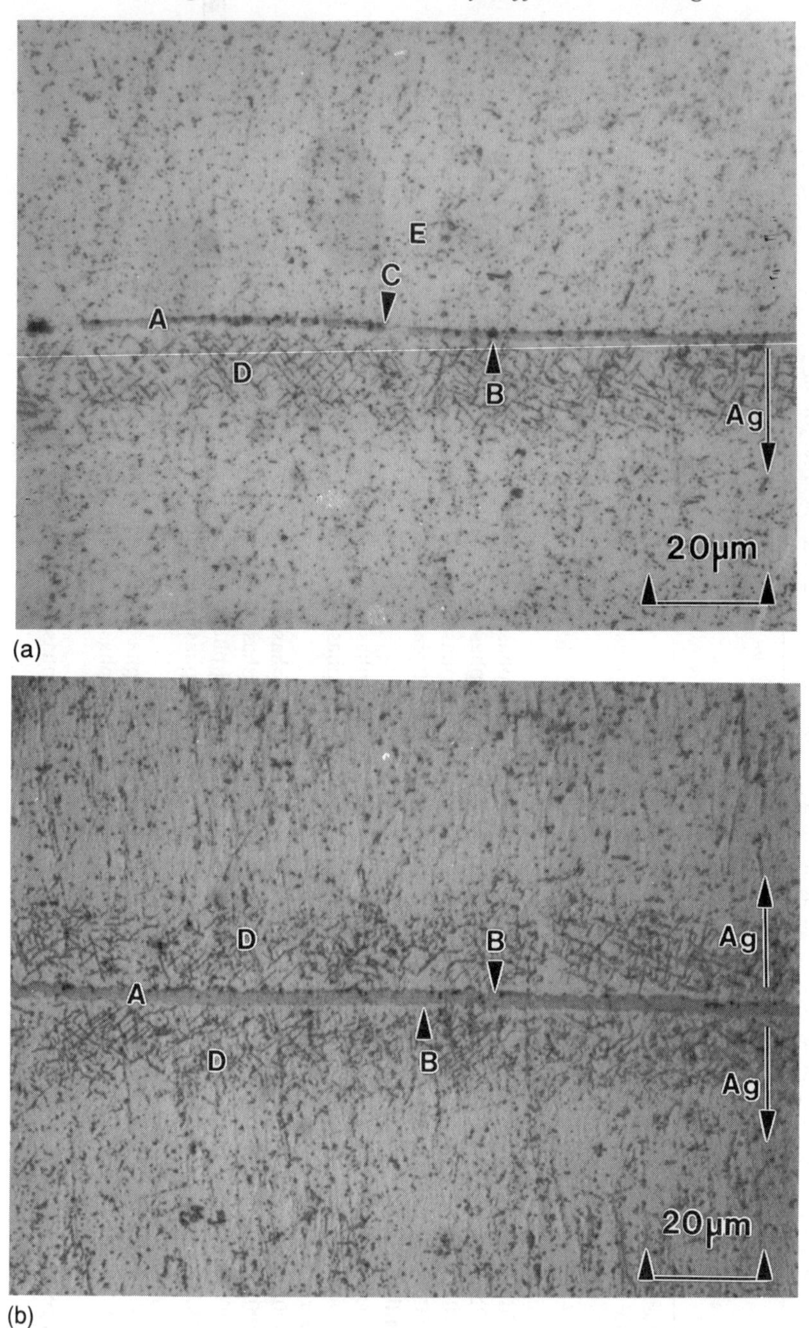

Figure 10.14 Optical micrographs showing effect of oxide films on silver diffusion [5]. A, silver interlayer; B, oxide free interface; C, oxide film; D, silver rich precipitates; E, region in which silver rich precipitates are absent.

The production of oxide free surface on aluminium alloys is possible using argon ion sputter cleaning, followed by coating with silver. Such surfaces are readily bonded above the dissociation temperature of Ag_2O (approximately 200 °C) and in this respect the DB resembles that for titanium alloys. The barrier effect of the oxide layer may be seen with a joint between silver coated aluminium alloy (7010) and uncoated aluminium alloy (Figure 10.14) [5]; on the coated side the silver has diffused away but diffusion is prevented at the aluminium oxide film. This technique could be applied to most aluminium alloys.

Al–Zn–Mg 7475 alloy sheet has been bonded successfully using 5052 Al–Mg alloy as an interlayer at 500 °C and 2.76 MPa pressure. The bond shear strength was found to improve with increase in deformation from 3% to 15% and with increase in bonding time to 60 minutes. Bond shear strengths of 172 MPa and 241 MPa have been reported for as-bonded and T6 heat treated condition. Shear strength has also been found to increase without interlayers with bonding pressure and time until 4 hours at 516 °C and 0.7 MPa gave parent metal strength. This was due to recrystallization at the bond interface (induced by prior peening) leading to a non-planar interface.

Diffusion bonding of aluminium–lithium alloys has been reviewed [5] and is particularly attractive because the alloys combine high specific

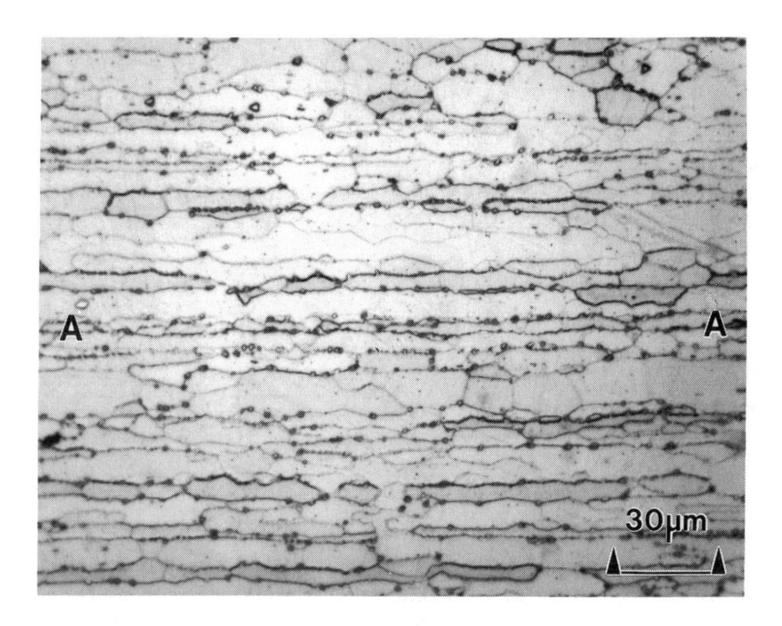

Figure 10.15 Section through a diffusion bonded joint in Al–Li 8090 alloy after solution heat treatment and age. Bond interface at A–A [7].

stiffness and strength with superplastic behaviour. A DB joint made in vacuum without interlayers is shown in Fig. 10.15. This joint exhibited parent metal compressive shear strength in the T6 condition (~ 230 MPa) and 90 ° room temperature peel strength of 54 N mm^{-1} (peak) and 18 N mm^{-1} (plateau); this can be compared with the corresponding

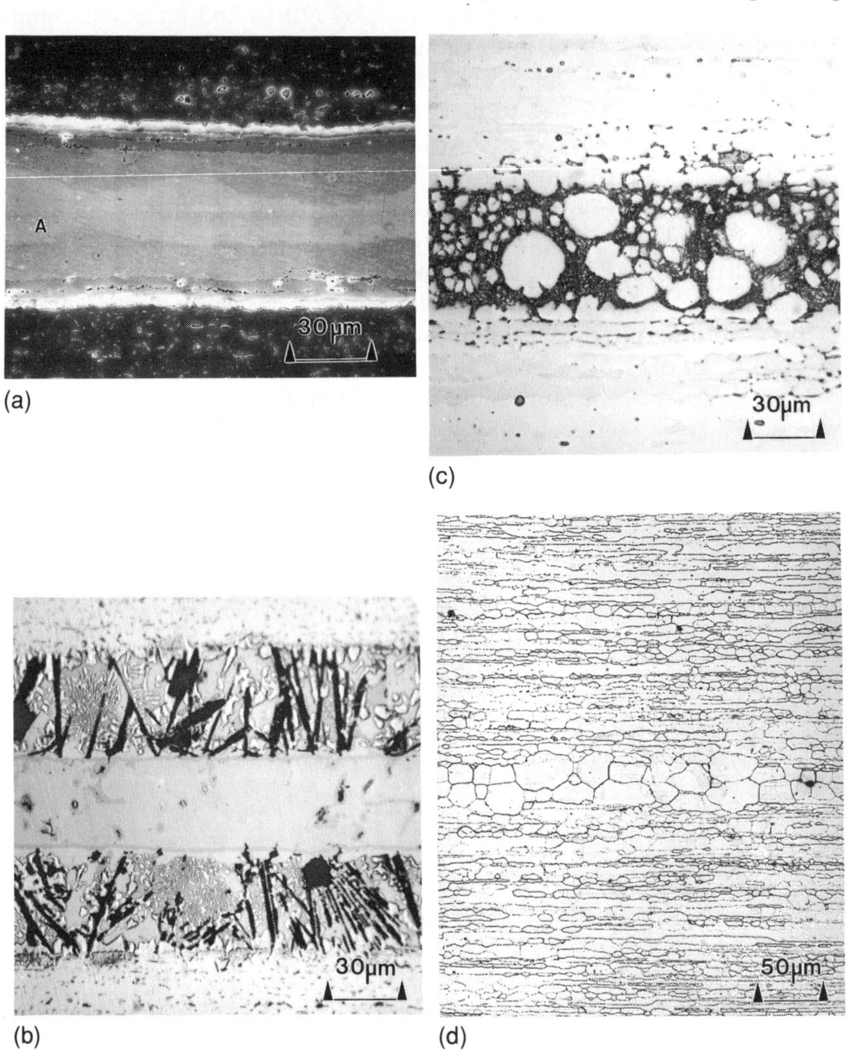

Figure 10.16 Micrographs of transient liquid phase (TLP) diffusion bond interface in Al–Li 8090 alloy: (a) solid state bonded at 475°C, showing copper layer at A; (b) TLP after 1 s at 530°C followed by water quench for 50 µm copper layer; (c) TLP after 9 s at 560°C followed by water quench for 10 µm copper interlayer; (d) TLP after 4 h at 560°C, solution heat treated (530°C for 20 min, water quench) and ageing (185°C for 5 h) for 10 µm copper layer [5].

strength values for adhesive bonded aluminium alloy joints of 30–40 MPa shear and 8 N mm^{-1} peel. Al–Li alloys may be bonded without interlayers since lithium compounds disrupt the surface oxide. However, because of the more rapid thermal oxidation of Al–Li alloys, particular care is required to avoid surface contamination during heating to the bonding temperature. DB with a liquid phase may be attractive and Al–Li 8090 alloy has been joined using a copper interlayer (Figure 10.16 (a)); with careful selection of the interlayer thickness the copper can be diffused away from the bond interface (Figure 10.16 (b–d)). In comparison with titanium alloys the DB of aluminium requires procedures of greater complexity and any deviation can lead to poor reproducibility of joint quality; however, the significant advantages for structural efficiency and weight savings when combined with SPF are very attractive.

10.6 DIFFUSION BONDING OF INTERMETALLICS

Intermetallics are ordered alloys with enhanced strength and oxidation resistance at operating temperatures between those of metals and ceramics (about 500 °C to 800 °C). Of particular interest are the low density alloys based on α_2-Ti_3Al and γ-TiAl. There is little quantitative data available yet but bonding conditions are shown in Table 10.3. The Ti_3Al alloys seem to diffusion bond readily at temperatures of 100–150 °C below the β transus to give parent metal microstructure, similar to conventional titanium alloys. It is also expected that parent metal properties should be easily obtained. However, higher pressures, compared to titanium alloys, are required for DB of the titanium aluminides due to the higher flow stresses of these intermetallics. The DB of the TiAl based alloys is more difficult than the Ti_3Al material since aluminium modified titanium oxides are formed and little interface migration may occur. DB joints with a planar interface had a tensile strength of ~ 40 MPa less than the parent metal at 800 °C and 1000 °C test temperatures. Plastic deformation of the surface prior to bonding can cause recrystallization at the bond interface and enhance bond strength [10].

10.7 DIFFUSION BONDING OF CERAMICS

Ceramics (Al_2O_3, SiC, Si_3N_4) combine low density with excellent corrosion, oxidation and wear resistance, and good strength retention above 1000 °C. Their disadvantages are low ductility and sensitivity to defects and notches at ambient temperatures. However, methods of toughening ceramics are being developed involving stress induced phase transformations and ductile metal–ceramic composites.

It is very difficult to obtain sufficient interface contact or diffusion for ceramic–ceramic bonding below about 1000 °C. However, many ceramics

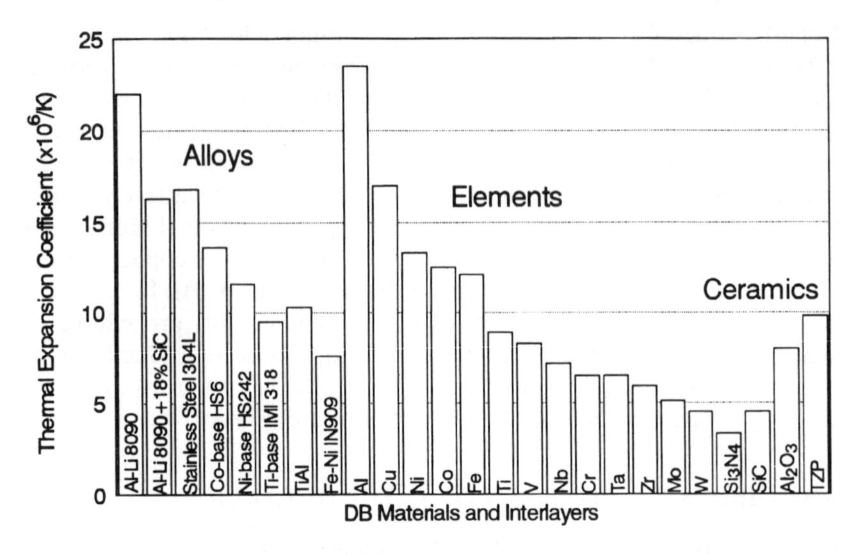

Figure 10.17 Thermal expansion coefficients of elements, metallic alloys and ceramics (TZP, tetragonal zirconia polycrystal).

can now be diffusion bonded either to themselves or to thicker metallic supports using reactive metal interlayers (containing Mg, Li, Zr, or Ti) and solid or liquid phase techniques, and this has been reviewed [11]. The joint strengths, which are greater than for conventional brazed or mechanically fastened joints, are therefore based on ceramic–metallic bonding. A major problem is presented by the mismatch in thermal expansion coefficients between ceramics and metals (Figure 10.17), which gives rise to high thermal stresses and cracking at the bond interface. An approximate estimate of the elastic stresses caused by thermal mismatch is given by:

$$\sigma_i = \frac{E_i E_j}{E_i + E_j} (\alpha_i - \alpha_j) \Delta T \tag{10.4}$$

where E = Young's modulus, α = thermal expansion coefficient, ΔT = temperature change from bonding temperature [11]. Clearly, the tensile stresses in the metal may exceed the yield stress, which could lead to a reduction in the stresses or to fracture, depending on the interface ductility. With solid state bonds multiple layers are used to decrease the stress gradient and a laminated microstructure is created at the interface (Figure 10.18) [12] For example Si_3N_4 has been bonded to Nimonic 80A with interfaces of the type Si_3N_4/Ni(0.5)/W(0.8)/Cu(0.5)/Nim80A, where the numbers in brackets denote the thickness in mm, to give strengths of about 125 MPa above 950 °C. Thin liquid phase interlayers can accelerate

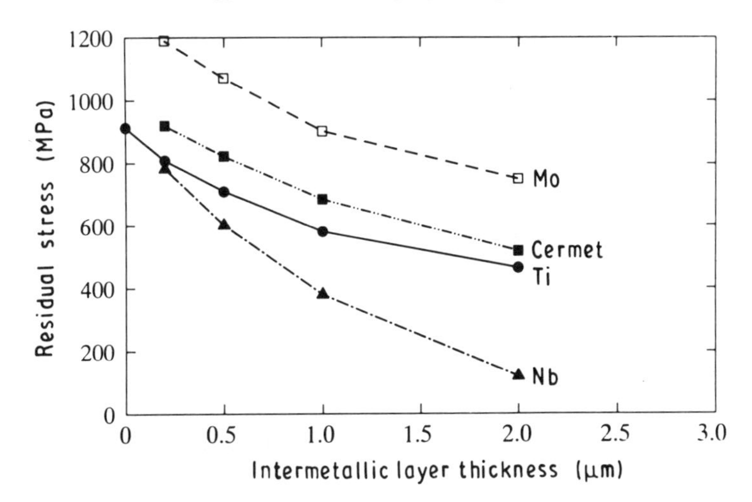

Figure 10.18 Effect of interlayer thickness on residual stresses of ferritic stainless steel (AISI 405)–alumina joints (1300°C, 100 MPa, 30 min) [12].

bonding, e.g. Al–Cu between Al_2O_3 at 1100 °C can produce joint strengths of 157 MPa, but metals reactive to the ceramic are necessary to ensure wetting. Some examples of the interlayers used, bonding conditions and joint strength are given in Table 10.4 [13]. In some bonding systems an oxygen partial pressure enhances bonding by producing metallic oxides which then react with the ceramic, e.g. Cu–alumina and Mo or Mn–alumina joints. Many reaction layers are brittle intermetallics and increase in thickness parabolically with time, whilst bond strength increases and then decreases; the layer thickness must therefore be controlled.

10.8 DIFFUSION BONDING OF COMPOSITES

Advanced continuous SiC fibre composites in titanium or titanium intermetallic matrices can lead to substantial improvements in unidirectional strength and stiffness at ambient and elevated temperatures while particulate ceramic reinforced alloys provide more isotropic properties. The processing of these materials is critical, since ceramic–ceramic contact can drastically reduce the strength. Although aluminium MMCs can be made by melt infiltration, reactive metals are restricted to solid state processing. Two process routes for continuous fibre titanium MMC are shown in Figure 10.19; the bonding conditions are similar to those for titanium. The coated fibre route prevents fibre contact, ensures uniform fibre distribution and allows high fibre volume fractions. Fibre surface enrichment in carbon or a TiB_2 coating reduces reaction at the fibre–matrix

Table 10.4. Summary of interlayers for reducing thermal expansion mismatch in ceramic/metal joints. (I) soft metal, (II) Composite, (III) laminate (soft metal/low expansion and hard metal) and (IV) crack layer

Group	*Ceramic/metal*	*Interlayer*	*Joining conditions*	*Strength (MPa)*
(I)	Al_2O_3/Type 321 steel	Al	873 K, 50 MPa, 30 min	70 (t)
	TZP/Type 316 steel	Cu	1273 K, 1 MPa, 4 h	52 (b)
	MgO/steel	Cu/Metal foam	1273–1473 K*	33 (t)
	Al_2O_3/steel	BA03	883 K, 10 MPa, 30 min	30 (t)
(II)	SiC/Al†	Cu–C fiber	1043 K (Al was joined at 823 K)*	–
	Al_2O_3/W ‡	Al_2O_3–W	2125 K, 8 h	–
	Al_2O_3/Fe ‡	Al_2O_3–Fe	1473 K, 3 GPa, 30 min	–
	TiN/Mo ‡	TiN/Mo	1573 K, 3 GPa, 30 min	80 (t)
	Al_2O_3/Fe ‡	FeO–Fe	1473 K, 29 MPa, 1 h	–
	TZP/W ‡	TZP–W	1673 K, 1 h in H_2	200–400 (b)
(III)	Al_2O_3/type 405 steel	Nb/Mo	1673 K, 100 MPa, 30 min	500 (b)
	Al_2O_3/type 316	Ti/Mo	1373 K, 9 MPa, 3 h	70 (t)
	Si_3N_4 steel			200 (b)
	SiC/steel	BA03/WC	883 K, 2 MPa, 1 h	150 (b)
	Sialon/steel			300 (b)
	Sialon/steel	Type 304/WC steel	1373 K, 5 MPa, 1 h	150 (b)
	Si_3N_4/Type 405 steel	Fe/W	1473 K, 10 MPa, 30 min	60 (t)
	SiC/super alloy	Ni/Kovar/Cu	1323 K, 54 MPa, 2 h	100 (t)
	Si_2N_3/super alloy	Ni/Kovar/Cu	1323 K, 54 MPa, 2 h	
	SiC/type 316 steel	Ti/Mo	1083 K, 0 min*	50 (s)
(IV)	Si_3N_4/type 405	Al/Invar (cracking in intermetallic compound)	1073 K, 0.15 MPa, 7 min*	60 (t)

* Brazing; the others are diffusion or eutectic joining.
† Soft metal with carbon fibre.
‡ Grading (cermet).
TZP: Tetragonal zirconia polycrystal.

interface. Since interface layers will be required to prevent excessive reaction or to minimize thermal stresses diffusion bonding could play a critical part

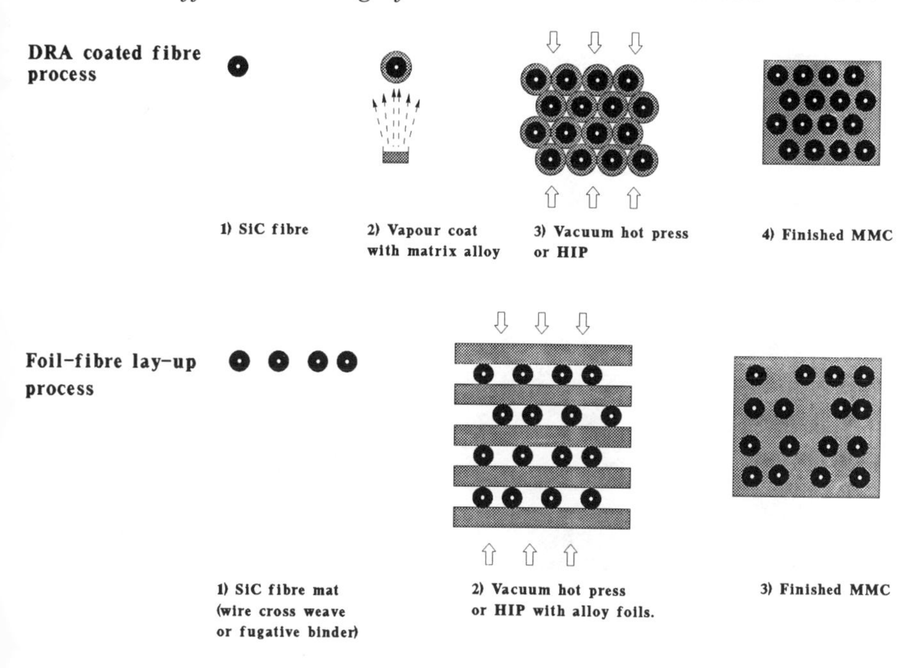

Figure 10.19 Solid state fabrication processes for metal matrix/SiC fibre composites [14].

in the production of advanced fibre composites because this process does not disrupt the interlayers [14]. Composite sheet is supplied as single or multiple plies and diffusion bonding may prove to be the most viable option for joining these materials to themselves or other materials.

Both solid state (Figure 10.20) and liquid phase diffusion bonding have been applied to aluminium particulate MMCs (Table 10.3) but higher bond strengths are obtained when particle–particle contact is avoided by the insertion of a matrix interlayer (Figure 10.21) [5].

10.9 DIFFUSION BONDING OF DISSIMILAR METALLIC MATERIALS

Often no single material can provide all the engineering requirements. The need to conserve expensive materials, to combine high strength with high toughness or to provide surfaces with corrosion or oxidation resistance, has led to the growth in interest in dissimilar metal joints. Diffusion bonding is particularly suited for such joints because it provides close control of the process variables. As for joints in ceramic–metal and metal–ceramic fibre composite joints, multiple interlayers are often necessary. However, the greater stress relaxation associated with

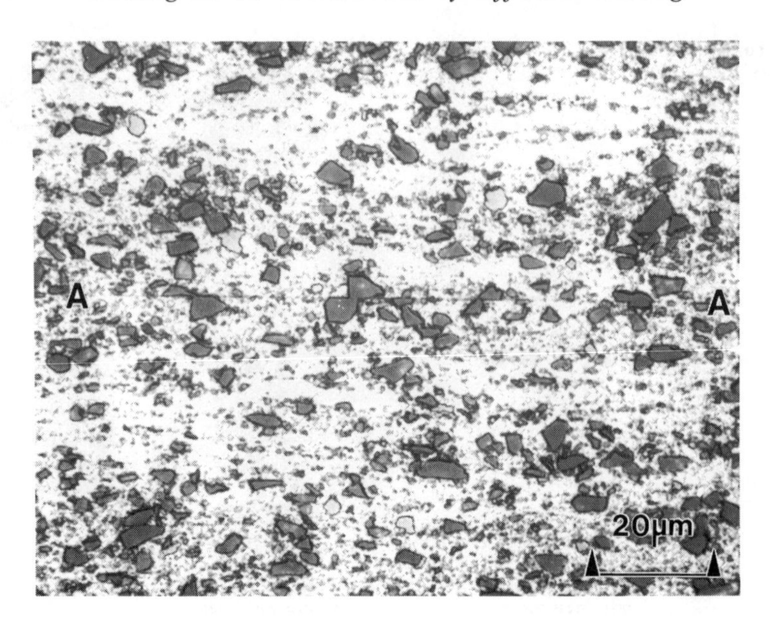

Figure 10.20 Solid state diffusion bond between Al–Li 8090 + 20wt% SiC particulate reinforced metal matrix composite (MMC) sheet with no interlayer (optical). Bond interface at A–A [5].

plastic deformation in metallic interlayers means that residual stresses due to thermal expansion mismatch may be less severe. In dissimilar metal joints care must also be taken to ensure chemical compatibility and minimum adverse reactions between the materials present in order to maximize the joint strength in service. The mechanical properties are particularly important, since dissimilar metal joints may be heavily loaded.

The interlayers should therefore ideally exhibit mutual solid solubility without intermetallic formation, a wide temperature range for bonding, and compatible thermal expansion and Young's modulus. An example of this type of joint is shown in Figure 10.22 [15]. The Ti–6Al–4V alloy has been diffusion bonded to the wear resistant cobalt base alloy Stellite HS6 using 25 μm thick tantalum and nickel interlayers. This joint had a tensile strength of ~ 800 MPa compared to ~ 600 MPa without interlayers, despite the formation of a Ni_3Ta intermetallic. The use of ductile interlayers may also enhance the toughness of joint interfaces. Many alloy combinations have been joined by diffusion bonding, including important commercial alloys. For example, titanium alloys have been bonded to stainless steels and nickel base alloys and steels have been bonded to cobalt base alloys, indicating the wide range of possible metal combinations [16,17].

Different alloys of the same base element may be diffusion bonded, for example a high temperature titanium alloy has been diffusion bonded to

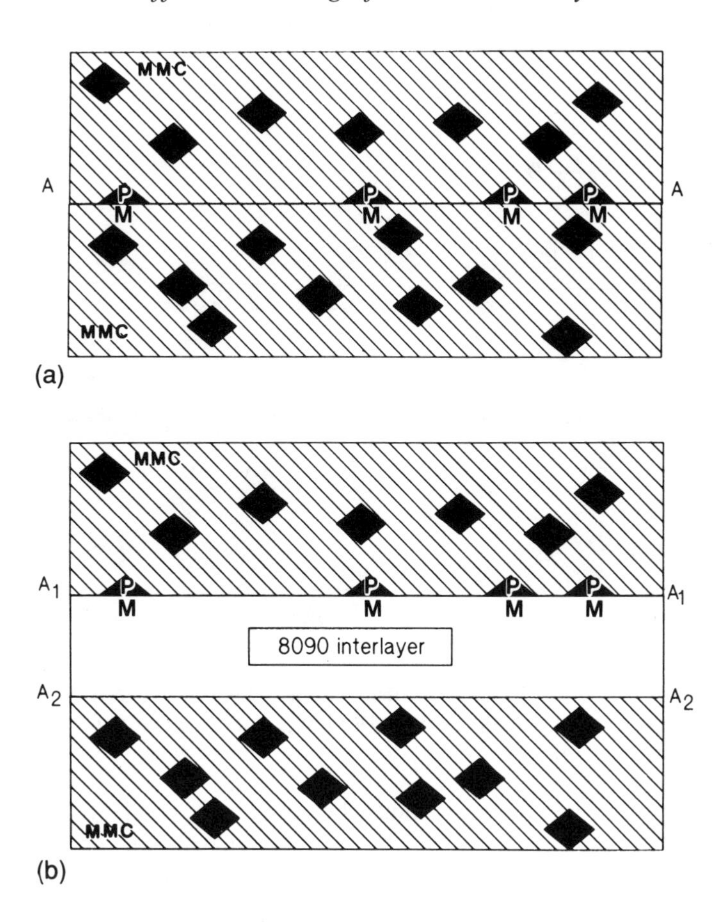

Figure 10.21 Schematic diagram of bond interfaces in MMC: P and M denote particle and matrix, respectively: (a) bond between MMC sheets; (b) Al–Li 8090 interlayer between MMC sheets [5].

a Ti_3Al base alloy. This is a relatively straightforward procedure for titanium alloys and steels using DB parameters for the lower temperature alloy of the couple. Future developments may exploit the potential of bulk macro- and micro-laminates produced by the bonding of stacked sheets or foils of dissimilar materials.

10.10 DIFFUSION BONDING OF METASTABLE ALLOYS

A wide range of alloys with unique microstructures are now being manufactured by rapid quenching from the melt or vapour phases. The alloys may be in the form of powders, ribbons or sheet. To preserve the

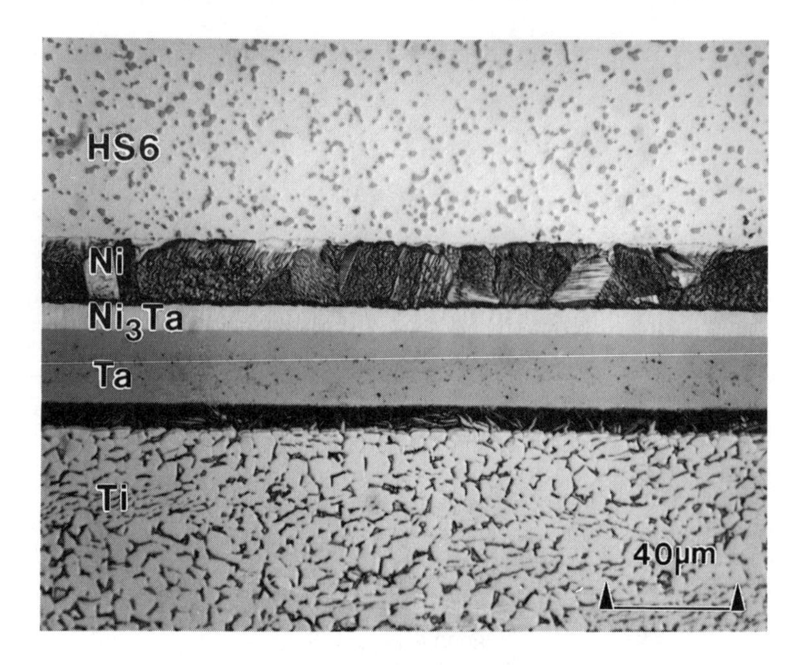

Figure 10.22 Microstructure of diffusion bond interface between wear resistant cobalt based Stellite HS6 and Ti–6Al–4V alloy, bonded at 900°C for 1 h and with 20 MPa (optical) [15].

microstructure the materials must be joined by solid state diffusion bonding. HIPping is ideal for the powder and ribbon and by careful design composites and dissimilar metal joints with complex shapes can be joined with minimum deformation.

10.11 MANUFACTURE OF COMPONENTS BY DIFFUSION BONDING TECHNIQUES

Diffusion bonds in components can be divided into those made by massive DB and those made between thin sheet; DB in thin sheet is most often associated with subsequent SPF and is dealt with below. Massive DB involves the joining of thick section machined parts under relatively high pressure (eg. 14 MPa, 2000 psi) applied by mechanical means, alternatively hot isostatic pressing (HIPping) may be used (section 13.6). Japanese workers have reported on a wide variety of Ti–6Al–4V alloy aircraft components diffusion bonded in vacuum (10^{-3} Pa) under a lower pressure of 2.94 MPa at 900 °C for ~ 2 hours. Massive diffusion bonded products tend to be substituted for complex machined or forged components to improve material utilization or cost and may therefore be heavily loaded.

The need for greater stiffness and strength at elevated temperatures has led to the most complex DB components to be proposed for turbine blades. These may involve thick and thin sections, dissimilar metals and metal–ceramic interfaces and HIP techniques may be used. The component must be sealed in an evacuated can for the pressure to be applied and after DB the can is removed unless it is an integral part of the component.

Diffusion bonding followed by superplastic forming has led to significant cost and weight savings for titanium alloy thin sheet structures. Ti–6Al–4V alloy sheet can be diffusion bonded by gas pressures of ~ 2 MPa (300 psi). The advantage over massive DB is that the pressure acts normal to the sheet surface whatever the shape being bonded and large area primary bonds with intimate contact are readily obtained. A pack

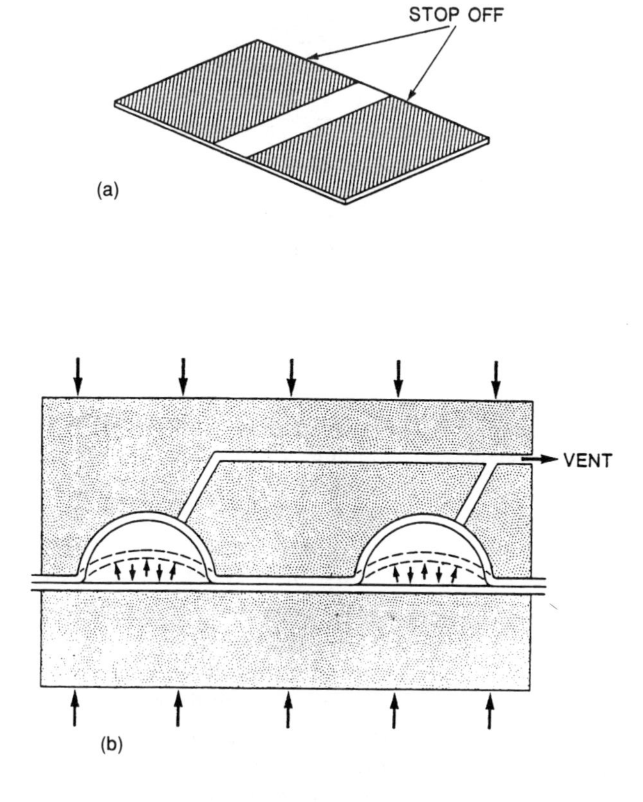

Figure 10.23 Schematic of superplastic forming of a 2 sheet structure using gas pressure [18]: (a) pack bonding using gas pressure; (b) SP forming of structure using gas pressure.

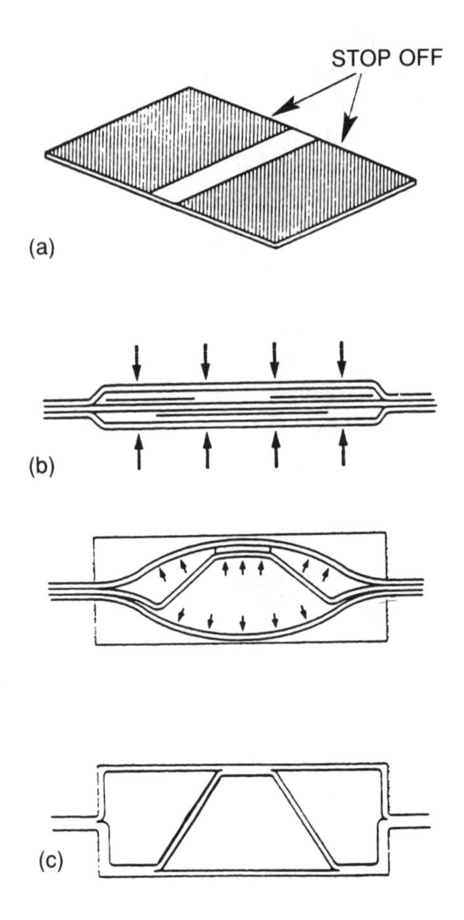

Figure 10.24 Schematic of superplastic forming of a 3 sheet structure [18]: (a) core sheet; (b) pack bonding; (c) forming.

bonding process can be combined with gas pressure superplastic forming (Figure 10.23) [18]. More complex 3 and 4 sheet structures may be made using stop-off to prevent bonding in predetermined areas and after bonding formed into shaped dies to produce honeycomb type structures (Figure 10.24) [18]. With Ti–6Al–4V alloy sheet some secondary bonds may be made after superplastic deformation. The surfaces for the secondary DB are exposed to the gaseous environment longer than the primary bonds with a greater risk of surface contamination; oxygen and water vapour must therefore be removed from the argon pressurizing gas. A 4 sheet structure in Ti–6Al–4V alloy is shown in Figure 10.25 and has been subjected to a ballistic projectile impact test. The damage is shown at A and detailed examination at B shows that fracture has not occurred in the diffusion bond interface but in the core sheet. The bond is

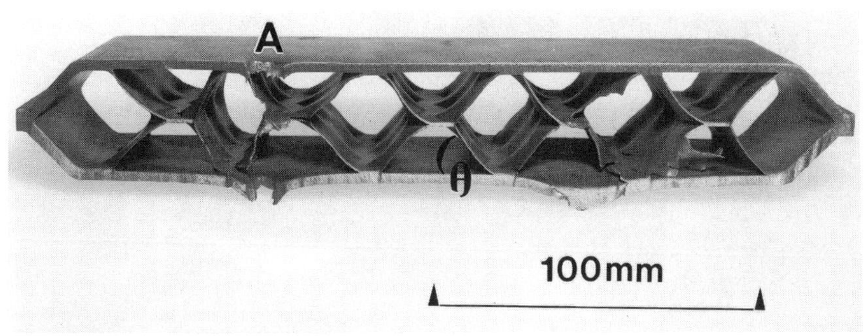

(a)

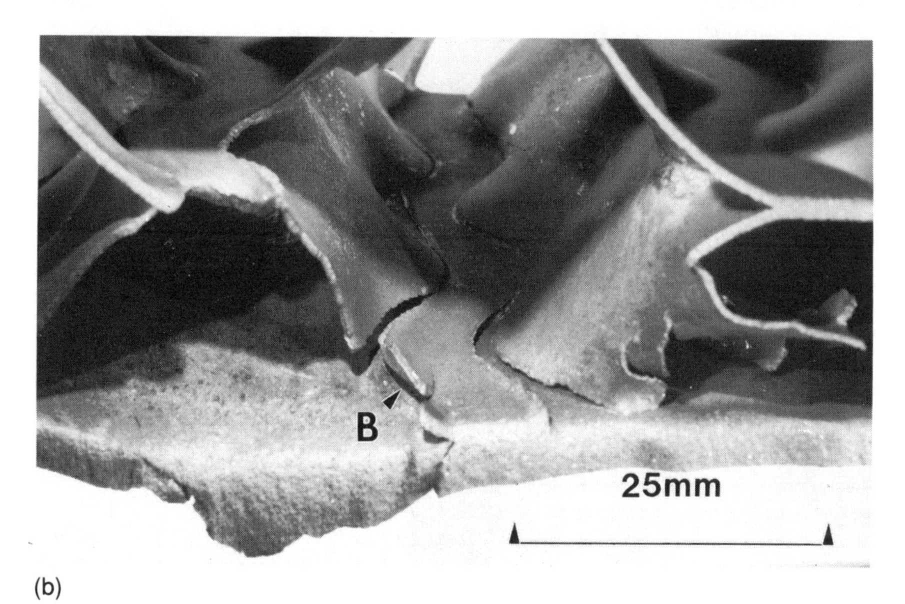

(b)

Figure 10.25 A SPF/DB 4 sheet structure in Ti–6Al–4V alloy: (a) ballistic projectile penetrated at A; (b) fracture in core sheet and unbroken bonded joint at B.

therefore not critical in this design. Several thousand titanium DB/SPF components are in service, for example as access panels on Airbus aircraft [19] and other civil and military applications [20,21]. A critical DB/SPF component is the fan blade from a civil gas turbine shown in Figure 10.26 [22].

During superplastic forming under gas pressure the bonded joint must resist peel; DB joints in titanium alloys have a high resistance to peel. However, the hot peel strength for aluminium alloy bonds may be insufficient without stiffening of the joint as shown in Figure 10.27 for

Figure 10.26 A civil gas turbine fan assembly with the fan blade fabricated by SPF/DB of a titanium alloy (courtesy of Rolls-Royce plc).

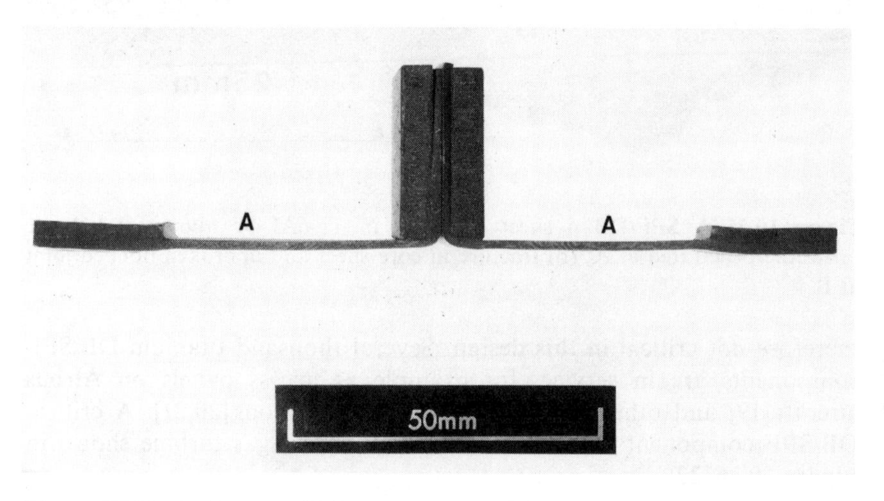

Figure 10.27 Liquid phase diffusion bonded 1.6 mm thickness Al–Li 8090 alloy sheet peel test piece with MMC stiffeners after testing at 530°C under superplastic conditions to 0.3–0.4 thickness strain in arms at A and absence of peel fracture [23].

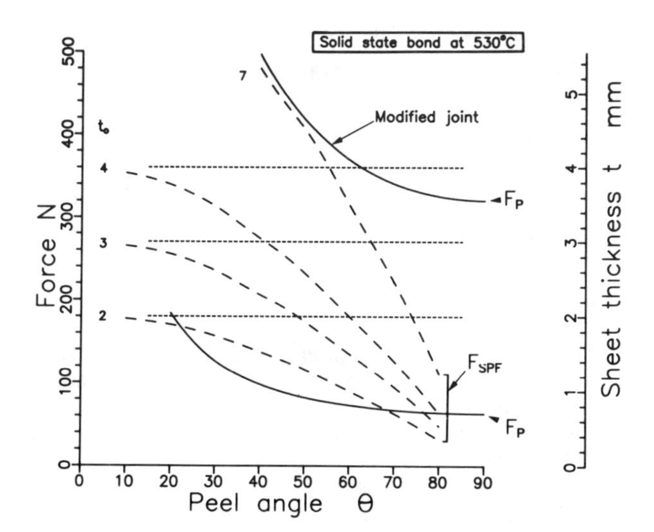

Figure 10.28 Peel force F_p and force for superplastic deformation in core sheet F_{spf} vs. peel angle θ for Al–Li 8090 alloy: t_0 is initial sheet thickness and t_i is sheet thickness during SPF [23].

90 ° peel test pieces [23]. The force for superplastic deformation of the core sheet F_{spf} decreases during forming (because the sheet cross-section decreases) according to the equation:

$$F_{spf} = \sigma_{spf} w_0 t_0 \cos\theta \tag{10.5}$$

where θ (in Figure 10.28)= angle between core and skin sheets, σ_{spf} = flow stress for SPF, and w_0 and t_0 are sheet width and thickness [23]. The force required to cause peel fracture $(F_a)_p$ also decreases with increasing θ. These forces are plotted in Figure 10.28 for 8090 Al–Li alloy. The curves show that for unstiffened joints peel fracture will occur when $t_0 > 2$ mm, whereas for stiffened joints t_0 can be up to about 7 mm. Since core sheets are usually < 3 mm, aluminium sheet structures with stiffened DB joints can be manufactured by DB/SPF.

10.12 CONCLUSIONS

The joining processes selected will depend on the materials, joint quality and strength requirements and costs. For demanding applications or for the more advanced materials, diffusion bonding offers many advantages and may be the only option for composites and metastable alloys. Future developments will include computer control of the bonding process, the use of sophisticated analytical techniques to characterise the joint microstructure and quality, and greater availability of large vacuum hot presses and HIP pressure vessels for manufacturing large components. More information on DB can be obtained from reference [24].

ACKNOWLEDGEMENTS

The authors would like to thank D. V. Dunford, H. S. Ubhi and C. M. Ward-Close for helpful discussions and the provision of figures.

REFERENCES

1. Schwartz, M. M. (1979) *Metals Joining Manual*, McGraw- Hill.
2. Schwartz, M. M. (1981) Diffusion welding. *Proc. conf. Welding Technology for the Aerospace Industry, Las Vegas Nevada 7–8, Oct. 1980*, pp. 1–40.
3. Owczarski, W. A. (1980) *Physical Metallurgy of Metal Joining* AIME, Pittsburgh PA, p. 166.
4. British Standard 1134, pt. 2, Aug. 1972.
5. Dunford, D. V. and Partridge, P. G. (1992) Overview diffusion bonding of Al–Li alloys. *Mats Sci. & Technol*, **5** (8), 385–98.
6. Wisbey, A. and Partridge, P. G. (1993) Diffusion bonding of the high temperature Ti alloy IMI 834. *Mats Sci. & Technol.*, **9**, pp. 441–4.
7. Partridge, P. G. (1989) Diffusion bonding of metals. AGARD Lecture Series 168 Superplasticity, pp. 5.1–5.29.
8. Williamson, J. R. (1981) Diffusion bonding in superplastic forming/diffusion bonding. *Proc. Conf. Welding Technology for the Aerospace Industry, Las Vegas Nevada 7–8, Oct. 1980*, 55–83.
9. Enjo, T., Ikeuchi, K., Akikawa, N. and Ito, M. (1980) Effect of superplasticity on the diffusion welding of Ti–6Al–4V alloys. *Proc. Conf. Titanium'80 Science and Technology, Kyoto, Japan 19–22 May 1980*, pp. 1097–1108.
10. Nakao, Y., Shinozaki, K. and Hamada, M. (1991) Diffusion bonding of intermetallic compound TiAl. *ISIJ International*, **10** (31), 1260–6.
11. Akselsen, O. M. (1992) Review diffusion bonding of ceramics. *J. Mats Sci.* (27), 569–79.
12. Suganuma, K., Okamoto, T. and Koizumi, M. (1986) *J. Amer. Ceram. Soc. Commun.* (67), 256.
13. Suganuma, K. (1990) *ISIJ International ceramic review*, **30**, pp. 1046–58.
14. Partridge, P. G. and Ward-Close, C. M. (1993) Processing of advanced continuous fibre composites: current practice and potential developments. *Int. Mats. Revs.*, **38**, pp. 1–24.
15. Wisbey, A., Partridge, P.G, and Thomson-Tur, M. A. (1993) The production of conventional Co base (Stellite HS6) wear resistant surfaces on Ti alloys. *Proc. Conf. 7th World Conference on Titanium, San Diego, Calif. USA, 28 June–2 July, 1992.*
16. Pearce, R. (ed.) (1987) *Proc. Conf. Diffusion Bonding*, SIS, Cranfield, Bedford.
17. Stephenson, D. J. (ed.) (1991) *Proc. Conf. Diffusion Bonding*, **2**, Elsevier Applied Science, London.
18. Stephen, D. and Swadling, S.J. (1986) Diffusion bonding in the manufacture of aircraft structure. *AGARD Proc. Conf. Advanced Joining of Aerospace Metallic Materials, Oberammergau, Germany, 11–13 Sept 1985*, pp. 7.1–7.17
19. Mansbridge, M. H. (1990) *Proc. 22nd Int. SAMPE Tech. Conf.*, p. 224.

20. Rolland, B. (1988) SPF–DB applications for military aircraft. *Proc. 6th World Conference on Titanium, Cannes, France, June 1988*, pp. 399–404.
21. Friedrich, H. E., Furlan, R. and Kullick, M. (1988) SPF/DB on the way to the production stage for Ti and Al applications within military and civil projects. *Proc. Conf. Superplasticity and Superplastic Forming, Washington, August, 1988*, pp. 649–64.
22. Rolls-Royce plc (1992) Private communication.
23. Dunford, D. V. and Partridge, P. G. (1992) Effect of joint stiffness on peel strength of diffusion bonded joints between Al–Li 8090 alloy sheet. *Mats. Sci. & Technol.* **8**, pp. 1131–40.
24. Kazakov, N. F. (1985) *Diffusion Bonding of Materials*, Pergamon Press, Oxford.

11

Adhesive bonding for aerospace applications

D. Driver

11.1 INTRODUCTION

Aviation and adhesive bonding have always been intimately connected – from the mythical flight of Daedalus and his son Icarus through to the more practical demonstrations of the Wright brothers and the supersonic capabilities of modern aircraft. While Icarus overextended the temperature capability of the hot-melt wax used to bond his feathered wings, the early aviators of World War I were more conservative in applying animal glues as size to improve the weather proofing and aerodynamic efficiency of their fabric-covered airframes and in the use of other natural glues based on milk extract and blood albumen to bond wooden fuselage frameworks and plywood panels.

By World War II synthetic adhesives were replacing natural glues, while wooden airframe structures were being superseded by metal (mainly aluminium). Only a few aircraft companies (in particular De Havillands in the UK and Fokker in Holland) developed the necessary expertise to adhesively bond metallic airframe components; instead, riveting of thin sheet structures became the favoured joining mechanism. This was in part because design/aeronautical engineers had a ready familiarity with mechanical fixing but were much less certain about the chemistry of adhesive bonding. In recent years, however, solid metal structures are giving way to stiff, lightweight laminate or honeycomb sandwich constructions in which adhesive bonding provides one of the few assembly options. Polymeric composite materials have also found increasing application, particularly in military and specialist aircraft. For such composite materials, bonding is not just a means of assembly, but is an integral part of the manufacturing route in which material and component are built up in the same processing operation.

11.2 BONDED WOODEN AIRCRAFT

Adhesives were first used extensively in primary structural aircraft applications, when early canvas-clad wooden fuselages with their complex internal struts and wire bracing were replaced by stronger, simpler, lighter structures in which the major loadings were borne by a stressed 'single skin' (French *monocoque*) construction in a similar manner to that whereby lobsters, snails, etc. withstand applied load on their shells rather than through a skeletal structure.

Louis Béchereau was one of the early exponents of monocoque construction, building a series of racing aircraft for Deperdussin which exploited adhesive bonding to achieve a smooth, tapered cylindrical fuselage. The aircraft bodies were built up from layers of wood veneers which were bonded together with casein glue over a mould which matched the fuselage shape. A Deperdussin seaplane racer won the first Schneider Trophy race at Monaco in 1913, while land versions established several world speed records. Unfortunately, Deperdussin's financial dealings were not as successful as his aircraft and in 1913 he was arrested for embezzling more than £1m.

The Germans took up casein-bonded stressed skin fuselage constructions in many of their World War I aircraft [1], while Fokker introduced a strong adhesively bonded cantilever wing construction [2]. Both these features, and the associated casein bonding, were subsequently combined in the Lockheed Vega of 1927. Such was the effectiveness of this American aircraft that during 1931 to 1933, under the control of either Wiley Post or Amelia Earhart, it established world records for solo transatlantic, transcontinental and round-the-world flights.

In the UK, De Havillands were the main exponents of bonded wooden aircraft. As early as 1913 they had designed a biplane that had a circular section monocoque fuselage which was 'a masterpiece of the cabinet maker's art and years ahead of its time' [3]. Part of that art lay in exploiting adhesive bonding to combine strong wooden joints with an effective streamlined shape. It was, however, the 'wooden wonder' of World War II – the Mosquito – for which De Havillands are rightly famous. As with many earlier monocoque constructions, the fuselage was built in two halves on a former and consisted of two skins of birch plywood between which was sandwiched balsa wood to give a strong, stiff, lightweight structure. All the wooden components – from the initial ply to the final two fuselage sections – were originally bonded using the traditional wood glue, casein; in later versions, however, more durable urea formaldehyde adhesives were introduced, with vinyl phenolic metal-to-wood bonding being used for naval versions which required folded wings for ease of storage on aircraft carriers. In all, some 7781 aircraft were built.

Although the Mosquito was undoubtedly the most famous wooden aircraft of World War II, it was not the only such craft. In Germany the Focke Wulf Ta 154 was developed using 57% wood:30% steel and with few strategic materials (such as aluminium) in the remaining 13%. This aircraft (unofficially dubbed the 'Moskito' by the Germans) made its first flight in July 1943. More than 250 aircraft per month were originally scheduled but in the event less than 25 were ever built, due to problems in bonding the wooden components. In 1929 the Goldschmidt company had patented a two-part, cold setting, acid hardening, phenolic film adhesive for bonding wood and in the mid-1930s began manufacturing the adhesive in commercial quantities in their Essen factory under the trade name Tego Film; it was Tego Film bonding which was intended for the Ta 154. However, the Essen factory was bombed by the Allies and new adhesive variants had to be tried. On the first two Ta 154 aircraft an excessively acid resin hardening agent was used which damaged the wood, causing both aircraft to disintegrate during high-speed trials [4]. The project was soon abandoned and attention concentrated on the Volksjäger (People's Fighter), a new lightweight fighter aircraft with superior performance over conventional fighters, thanks to the use of the new BMW 109 turbojet engine.

The Volksjäger, or Heinkel He 162 as it came to be known, had a semi-monocoque fuselage of duralumin but with nosewheel doors in plywood, together with shoulder-mounted wooden wings. The first prototype made its maiden flight in 1944, when a poorly bonded undercarriage door was torn off while the aircraft was flying at maximum speed. During a second demonstration flight, the wooden starboard wing disintegrated while the aircraft was making a low-level, high-speed pass. Investigations revealed that the acid in the phenolic bonding agent had again been insufficiently neutralized and had eaten into the wood. A new He 162 wing was developed which used a double skin structure extensively bonded with urea formaldehyde, but the redesign failed to go into production. Instead a new hardening agent was found for the phenolic resin and this modified adhesive [5] was applied to the He 162 and to the Messerschmitt 262 – the first turbojet aircraft to enter service (in autumn 1944) – as well as to the rocket-powered Me 163 Komet. By the end of World War II wooden aircraft had given way to metal and, with Germany barred from aircraft production, future use of bonding for aircraft structures was concentrated in Holland, the UK and USA.

Early bonded aircraft structures had tended to be based on trial and error. A more scientific marriage of adhesive bonding and aircraft construction began in the early 1930s when Dr Norman de Bruyne, a Cambridge physicist with an enthusiasm for flying, began developing synthetic adhesives which he tested on aircraft of his own design and construction. While de Bruyne's aircraft have now been lost, his adhesive formulations and his appreciation of the principles of bonding continue to be used to advantage within the aerospace industry.

11.3 PRINCIPLES OF BONDING

In de Bruyne's words [6], adhesives stick 'because of unsatisfied molecular forces emanating from a surface'. In their solid state materials remain bonded together due to relatively strong, short range (0.1–0.3 nm) primary chemical forces, together with weaker but somewhat more extended (up to 1 nm) secondary van der Waals forces. Within the solid such forces are in equilibrium, whereas at a free surface the balance is disturbed. If clean, atomically flat surfaces could be brought into intimate contact, then the interfacial forces would interact and reconstitute the bonding. Real surfaces are, however, neither clean nor atomically smooth. Typically lapped or polished surfaces contain millions of asperities, some 100–500 nm in height, while machined surfaces are appreciably rougher (3–6 μm or 6000 nm). Direct interaction between two abutting surfaces is therefore restricted to a few points of contact (typically less than 1% of the nominal overlapping surface area), while the remaining regions are separated by distances several orders of magnitude too great to achieve interatomic bonding. The only way to achieve close contouring of a surface is to introduce a liquid (i.e. wetting must occur). For effective load transfer that liquid must then transform into a solid. Adhesive bonding hence involves the two processes of **wetting** and **setting**.

11.3.1 Wetting

Whether a liquid spreads uniformly over a flat surface or remains as discrete droplets depends on the relative surface energies of the components. The equilibrium contact angle is achieved when:

$$\gamma_{SV} = \gamma_{SL} + \gamma_{LV}\cos\theta$$

where γ_{SV} is the surface free energy of the solid when in equilibrium with the liquid vapour, γ_{SL} is the surface free energy of the solid in contact with the liquid and γ_{LV} is the surface tension of the liquid in equilibrium with its vapour. For effective wetting $\theta \rightarrow 0\,°$ (i.e. $\cos\theta \rightarrow 1$). Typically γ_{SL} is small compared with γ_{LV} and hence (dropping the subscript v) for spontaneous spreading:

$$\gamma_L < \gamma_S$$

Table 11.1 provides a listing of surface energies (γ_S) for typical solid materials used in aerospace applications, together with that for water. In liquid form the aerospace materials have surface tensions (γ_L) which are

Table 11.1

	Surface energy (mJ^{-2})	
Steel	1000	Metals
Aluminium	500	
Water	72	
Wood	60	
Epoxy	47	Thermosets
Polyester	43	
Polypropylene	32	Thermoplastics
Polyethylene	31	
Silicones	25	
PTFE	20	

somewhat lower than their solid equivalents. Table 11.1 shows a clear distinction between metals with their high surface energies and the remainder. Although metals can be used as liquid adhesives (e.g. as in soldering and brazing), their ability to provide effective wetting is restricted to a limited number of metallic surfaces of little relevance to aircraft structures. It is only polymeric aerospace adhesives which will be considered here.

11.3.2 Setting

Having wet the surface, the adhesive must then transform to a loadbearing solid. This solidification process is achieved by one of three mechanisms:

- solvent loss by evaporation or diffusion
- cooling from above the melting point (hot melts)
- chemical reaction (polymerization)

Casein, the early wood glue, is one of the few aircraft adhesives which sets by water diffusion. Other common solvent-based adhesives are the rubber solutions which used to predominate in the early DIY markets.

Hot melts are thermoplastic resins which melt to a low-viscosity liquid but rapidly solidify again on cooling. Sealing wax is a common example. Although there are a limited number of thermoplastic aircraft adhesives, this type is more favoured in high throughput industries (such as footwear and packaging), where rapid bonding is necessary.

Most structural aerospace adhesives nowadays harden by polymerization – a chemical reaction in which individual small molecules (monomers) form long linear assemblages or cross-link into a complex three-dimensional interconnecting mass. Such polymerization can be triggered either by heat or by reacting with a hardener/catalyst.

11.4 AEROSPACE ADHESIVE TYPES

11.4.1 Casein wood glue

Casein glue is made by precipitating casein (the whey of Little Miss Muffet's curds and whey) from skimmed milk using sulphuric, hydrochloric or lactic acid. The casein is then dissolved in an aqueous alkaline solution which is often applied warm. As the solution cools (or its pH is changed) the viscosity can change rapidly and the liquid transforms into a semi-solid 'gel'. It is the ability of protein glues to penetrate into wood and effectively wet the surface, then gel quickly on cooling to a tacky phase for assembly, before finally setting into a solid, which has made them such attractive adhesives. The durability of casein glues is sensitive to the choice of alkaline solvent; sodium or ammonium hydroxide give relatively long life but poor water resistance, whereas calcium hydroxide provides good resistance to water but the glue gels irreversibly within an hour or two. All casein-glued joints will, however, weaken with prolonged exposure to moisture and therefore more permanent structural adhesives needed to be developed.

11.4.2 Synthetic wood glues – urea formaldehyde

Urea formaldehyde, the first thermosetting synthetic glue for aircraft components, was developed in the early 1940s. The resin was made by adding urea (NH_2–CO–NH_2) to an aqueous solution of formaldehyde (CH_2O) to form methylol (CH_2OH) ureas. The methylol groups are particularly reactive and under acidic conditions form long chain linear polymers:

$$HOCH_2 - [- NH - CO - NH - CH_2 -]_n - NH\ CO - NH - CH_2OH$$

In the presence of excess formaldehyde the polymer cross-links and sets to an infusible mass. Since the setting reaction results in the liberation of water, this adhesive is restricted to porous substrates such as wood, paper and cork. When correctly cured, urea formaldehyde has better water resistance and appreciably greater joint life than casein, though the rigid nature of the cross-linked adhesive caused difficulties with Mosquito aircraft in the Far East, due to strain incompatibility between the adhesive and the adjoining wood which swelled excessively in the humid climate.

11.4.3 Phenolic adhesives

For better weathering phenolic adhesives are preferred Phenol OH reacts with formaldehyde to produce methylol phenols OH CH_2OH in a similar manner to that for urea formaldehyde. Subsequent reactions are, however, more complicated. Excess formaldehyde and alkaline conditions cause water to split off and a resol structure to form:

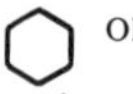

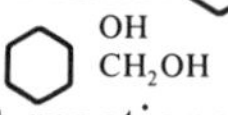

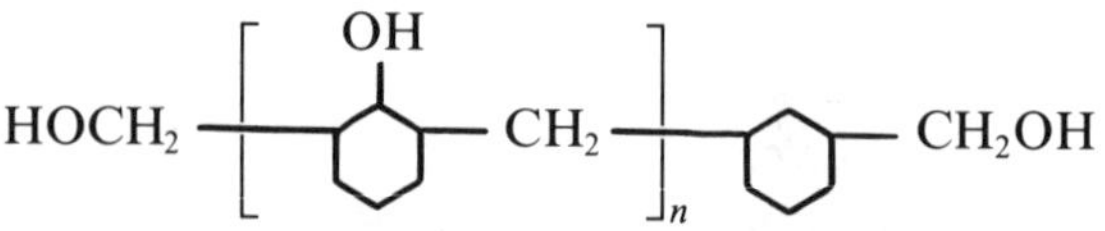

OH

$HOCH_2$ [ ] CH_2 [ ]$_n$ CH_2OH

On heating, the resol prepolymer becomes rubbery, due to molecular entanglement, before cross-linking more fully and curing to a thermosetting resin. Cross-linking takes place slowly at low temperatures but the addition of a strong acid to the resol resin can facilitate more rapid curing. This is the basis of two-part acidic hardening, cold-curing phenolic resins used for early aircraft assembly work. Care must be taken, however, in using just sufficient adhesive to achieve a timely cure; otherwise excess hardener can damage the joint, as with the German Ta 154 and He 162 aircraft. If excess phenol reacts with formaldehyde under acid conditions, then formation of methylol groups is slow and a stable novolak structure is formed:

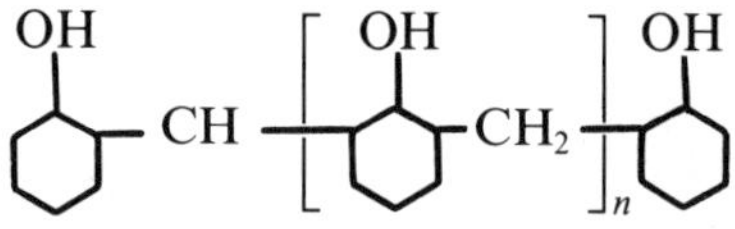

OH OH OH

CH [ ] CH_2 [ ]$_n$

On adding a cross-linking agent, the novolak is converted to the resol and can be heat cured. Cross-linking provides a strong but brittle bond and synthetic formaldehyde adhesives are therefore frequently blended with other polymers to improve toughness. Such modified phenolics formed the basis of the earliest metal-to-metal adhesives.

11.4.4 Vinyl phenolic adhesives

Some toughness can be introduced into phenol formaldehyde if the resol phenolic is co-cured with polyvinyl acetal. The flexibility arises because the methylol groups on the resol can only cross-link with the limited number of hydroxyl groups in the acetal polymer chain. Originally, the resol was applied directly to one of the surfaces to be joined and polyvinyl formal was then sprinkled over the sticky resin before closing the joint and applying heat and pressure to achieve the cure. This was the form in which vinyl adhesives were first developed by de Bruyne at Aero *Re*search *Dux*ford in 1943 and marketed as Redux. Recently Redux has also become available in the more easily handleable film form. Redux was the first synthetic structural adhesive for bonding metals and has continued to be used in Europe for over 50 years for aircraft applications, typical examples being given in the section on bonded metallic structures. The long service lives of many of these aircraft confirm the durability of Redux-bonded aluminium joints under a variety of adverse combinations of moisture, stress, fatigue, etc. A disadvantage of this system, however, is the slow cure, which requires the use of high temperatures and pressures in large-scale costly autoclave equipment.

11.4.5 Nitrile phenolic adhesives

Increased flexibility can be introduced into phenolic resin by the addition of nitrile rubber. Such adhesives are exceptional in requiring the more stable novolak (instead of resol with its reactive methylol groups) in order to maintain compatibility between the phenol and rubber constituents. One of the earliest examples of nitrile phenolic bonding was for the wing structure of the Chance Vought Cutlas swept-wing fighter aircraft (1948), where it was used to bond the aluminium skin to a balsa wood core. The low intrinsic resistance of balsa wood to moisture, fungus or fire meant, however, that it was soon replaced by aluminium honeycomb cored panels. Nitrile phenolic films, combined with riveting for additional structural integrity, were used to provide sealed containment for aviation fuel in the integral fuel tank/wing structures ('wet wings') of Convair F102 and F106 fighters, and in Convair 880 and 990 civil aircraft [1]; this innovative fuel tank technology has now become standard practice within much of the aircraft industry.

11.4.6 Epoxy-phenolic adhesives

As increased speeds led to aerodynamic heating of outer aircraft skins, so the need arose for structural adhesives capable of operating at temperatures in excess of 150 °C. The combination of high molecular weight epoxy resins with the resol phenolic yielded in the 1950s epoxy-phenolic adhesives with good strength retention up to about 250 °C but at the expense of flexibility. This adhesive has been used in the USA to bond aluminium alloy sandwich structures on the fuselage and wings of the Convair B58 Hustler (1958) mach 2 bomber.

11.4.7 Epoxy adhesives

Epoxy adhesives were initially developed in the 1940s and have since become established as one of the most versatile structural adhesive systems. They exist in a variety of forms (liquid, paste, film and foam) and are capable of bonding most materials (wood, metal, glass, ceramics, composites, rubber, concrete). Epoxy adhesives typically consist of two-part, chemically reactive systems which are cured either at room or elevated temperatures. Single-part, pre-mixed systems are also available which have to be heat-cured. The epoxy resins are thermosetting materials which contain the characteristic epoxy $-\overset{O}{CH-CH_2}$ group. Curing occurs typically through the use of an amine ($-NH_2$) hardener which opens up the epoxy ring and facilitates cross-linking. As with other highly cross-linked structures, the epoxies are intrinsically brittle but can be toughened with the addition of elastomers. The introduction of up to 15% by volume of liquid nitrile butadiene rubber (NBR) causes the rubber to precipitate out as small (~ 1 μm) spherical particles on curing (Figure 11.1), increasing the energy of fracture by a factor of 30–40. Epoxies have high strength up to ~ 200 °C, good solvent resistance and low shrinkage. Epoxy-based adhesives now dominate the aerospace industry, where they are used, mainly in film form, for honeycomb sandwich assembly, for structural bonding, for repairs and for assembly of aircraft interiors.

11.4.8 Acrylic adhesives

Acrylic adhesives were developed in the late 1960s from polymethyl methylacrylate (PMMA) chemistry. They were originally used not as an adhesive, but for aircraft windows and canopies, since the resin can be

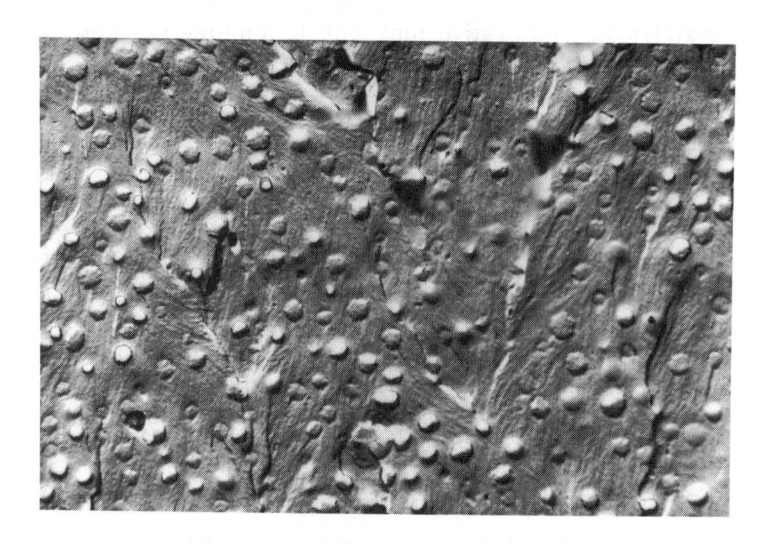

Figure 11.1 An electron micrograph of the internal structure of a typical toughened adhesive. Each spherical domain is approximately 1 μm in diameter. (from Lees [20])

polymerised in large volumes to generate low-density sheet with excellent light transparency and weathering characteristics. One advantage of acrylic adhesives (which differentiates them from the others described so far) is their rapid room-temperature curing rate. There is an initial incubation period during which time parts can be positioned, after which curing occurs within minutes, thereby avoiding long-term jigging. Such rapid room-temperature curing can be used to advantage in the aircraft industry for repair, though applications are limited to less stringent service conditions. As with epoxies, acrylic adhesives can be toughened by incorporating liquid rubber.

11.4.9 Anaerobic adhesives

Anaerobics are a sub-set of the acrylic adhesives and are based on the methylacrylate monomer, but also contain a mixture of initiators (peroxides), accelerators (amines), stabilisers (phenols) together with chelators which trap transition metals. This complex blend is introduced to prevent spontaneous free radical polymerisation. The most important inhibitor, however, is atmospheric oxygen since this readily reacts with free radicals to form more stable species. Once oxygen is removed, free radicals can be released (particularly at metallic surfaces) and the anaerobic is activated. The primary use of anaerobic adhesives is for thread-locking or to secure inserts, a particularly critical aerospace application being the retention of numerous bearings in their housings for the complex flight control linkage in Westland helicopters. Unlike most adhesives, which are permanent, anaerobics can be formulated with a wide range of shear strengths so that bonded bolts can often be released by application of a critical torque.

11.4.10 Cyanoacrylate adhesives

Cyanoacrylates are a further sub-group of acrylic adhesives which became commercially available in 1958 and have since been successfully marketed for DIY applications as 'super glues'. They are relatively expensive but can be cost-effective because only small quantities are required for almost instant bonding, thereby obviating expensive fixturing or production delays. The cyanoacrylate bond is relatively brittle and has found few aircraft applications except for non-structural uses such as display panels, nameplates, electronic interconnects, etc.

11.4.11 High temperature polyimide adhesives

A limitation of most structural adhesives is their inability to withstand temperatures over 200 °C for extended periods. Considerable attention has therefore been given to developing suitable polymer resins which could extend this temperature capability. To date polyimides have been the most commercially successful. Aromatic polyimides were first introduced in the early 1960s but became more fully exploited in the 1980s, when developments had significantly improved processability and handling characteristics. The polyimides are prepared by combining aromatic diamines with dianhydrides [7]. In thermosetting form they exist as low molecular weight chains, terminated at each end by cross-linking units. For the more amenable thermoplastic polyimides, flexible long chain polymers develop.

Polyimides are resistant to most chemicals (other than strong bases); they are abrasion, oxidation and fire resistant and are capable of continuous service temperatures of 250–350 °C. Their low density and resistance to high-energy radiation make them suitable for military and civil aerospace applications, where both the resin (as a matrix for glass and carbon fibre composite materials) and the adhesive (in film, paste or lacquer form) are used for repair and aircraft structural parts, space vehicles, missiles and electronic components.

11.4.12 Silicone adhesives

Silicone adhesives consist of a polymer chain which is based on silicon and oxygen:

$$\left[-O-\underset{\displaystyle R_2}{\overset{\displaystyle R_1}{\underset{|}{\overset{|}{Si}}}}- \right]_n$$

It is the characteristics of the R_1 and R_2 groups associated with the Si – O linkage which govern the behaviour of the polymer. If R_1 and R_2 are methyl groups (– CH_3), polydimethylsiloxane is formed. At low molecular weights (small *n*) the product is a release agent; at large *n* it is a grease.

If R_1 is an alkoxyl group (e.g. – O – CH_3) which can react with metal surfaces and R_2 is replaced with a reactive group (such as an epoxy), then at low *n* the product is a chemical coupling agent for epoxy adhesives (see surface priming). If the molecular weight is large and some R groups cross-link with other chains, then an elastomeric silicone adhesive is formed. The characteristic silicon–oxygen link is chemically more stable than the carbon–carbon bond, thereby giving silicone adhesives their higher temperature resistance. Silicone materials can also be formulated to provide resistance to mechanical vibration, thermal shock (including cryogenic temperatures), radiation and atomic oxygen. When combined with low levels of outgassing, these characteristics make the adhesive suitable for the harsh environment of space. The tiled heat shield on the space shuttle is bonded using a silicone adhesive. The tiles consist of 10% silica fibre, while the remaining volume is unfilled. These lightweight fibrous tiles are then coated on their outer surface with a thermal barrier layer of borate-silicate glass, which provides 95% of the thermal protection. In early flights, some tiles were ripped from the heat shield. Investigations have, however, confirmed that such failures occurred within the tile and were not caused by adhesive failure.

Adhesives based on carbon-carbon linkages oxidize to give gaseous volatiles. With silicone materials, oxidation can produce tenacious heat resistant, glassy refractory char which can withstand temperatures as high as 3300 °C for several minutes. Such silicone ablative coatings have been used for short-term protection of rocket nozzles and ramjet combustion chambers and are proposed for the nose cones of hypersonic vehicles.

11.5 SURFACE TREATMENTS

For effective bonding it has already been established that the adhesive must wet and contour the surface. If the surface contains micro-cavities into which the the adhesive can penetrate, then intuitively such features could provide sites for strong mechanical keying.

11.5.1 Wood surfaces

Wood, with its characteristic cellular structure (Figure 11.2), is a striking example where interpenetrating cavities are available to facilitate mechanical keying. When, however, wood is sectioned at an angle to the grain,

Figure 11.2 A 0.1 mm cube of wood showing the characteristic cellular structure (from Knight [21])

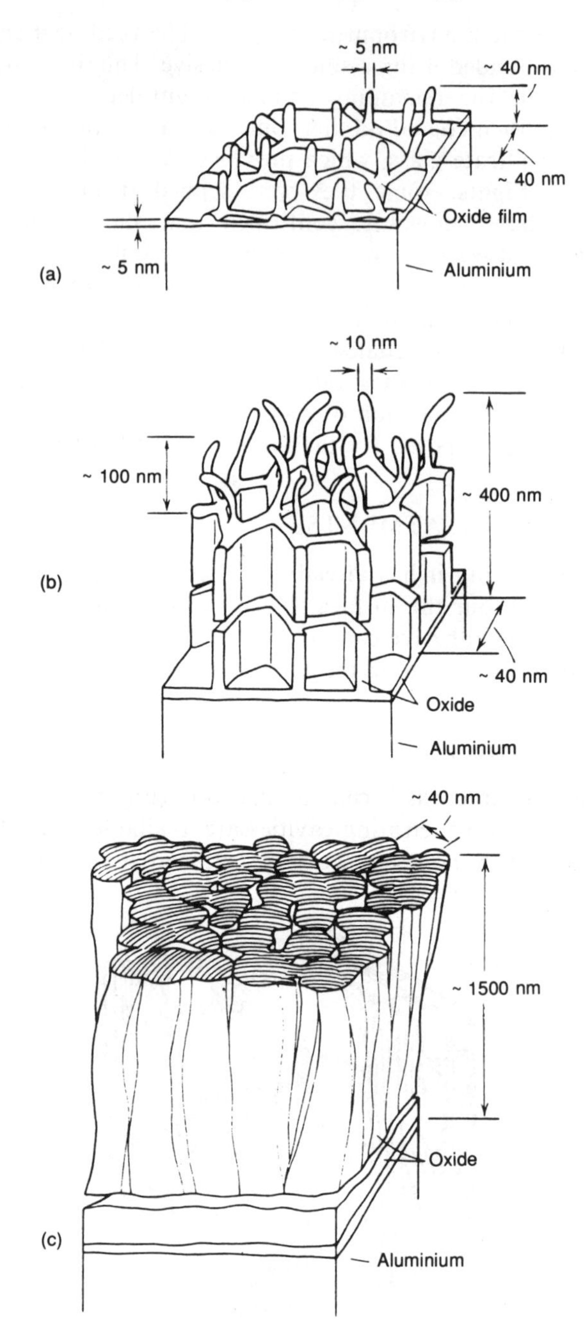

Figure 11.3 Surface morphology on aluminium following: (a) Forest Products Laboratory etch (FPL); (b) phosphoric acid anodizing (PAA); (c) chromic acid anodizing (CAA) (from Venables *et al.* [22])

the cells breaking the surface can be damaged by mechanical cutting or by the hot pressing operations used in plywood manufacture, thereby impairing surface penetration of the glue. To avoid this so-called 'case hardening effect' all plywood for aircraft applications was given a shot blasting treatment to reveal the open underlying cell structure.

Because the cellular cavities in wood can be extensive, an alternative problem arose with low viscosity adhesives. Adhesive could penetrate down cell cavities, starving the joint of sufficient adhesive to maintain a continuous bond line. These examples serve to illustrate the importance of optimizing surface morphology and of controlling the kinetics of adhesive flow during bonding.

11.5.2 Aluminium surfaces

Metals with high surface energies should be wet effectively by polymeric adhesives. Such high energy surfaces will, however, also have a strong affinity for water and possibly non-tenacious corrosion/oxidation products which could impair bonding. It is to improve the coherence of any oxide film, and to optimize the surface morphology for enhanced mechanical keying with the adhesive, that metallic aerospace materials are given sophisticated surface treatments [8].

Aluminium surfaces are prepared for adhesive bonding in the aerospace industry by using either chromic-sulphuric acid etch (the Forest Products Laboratory – FPL process) or phosphoric (PAA) or chromic acid (CAA) anodizing. The FPL process generates a cellular oxide structure with a regular array of protrusions (Figure 11.3(a)) which provides strong bonding through mechanical interlocking between the oxide and the well dispersed adhesive. A thin amorphous alumina film at the substrate interface also confers durability by providing a corrosion-resistant barrier.

The PAA oxide morphology is thicker and more well developed (Figure 11.3(b)), with deeper hexagonal cells and longer protrusions providing more effective mechanical interlocking. In contrast, the CAA treatment generates a thick barrier layer topped by tall columns of close-packed oxide (Figure 11.3(c)). This structure offers less interlocking potential than the earlier morphologies and hence for maximum bonding strength the PAA treatment is preferred.

11.5.3 Titanium surfaces

As with aluminium, anodizing of titanium prior to bonding can produce strong, durable joints, though the detailed surface oxide morphologies are different, due to the two-phase nature of many titanium alloys. With alloys such as Ti–6Al–4V, chromic acid anodizing (CAA) produces islands of aluminium-rich α phase surrounded by vanadium-rich β phase which contains a honeycomb structure of amorphous TiO_2 of similar cell diameter (~ 40 nm) and wall thickness (~ 5 nm) to that of anodized aluminium surfaces. The titanium oxide topography is, however, more even, with fewer protrusions than characterize the aluminium oxide structures. Below the honeycombed titanium oxide is a thin (20 nm), dense oxide barrier layer.

With the more recent sodium hydroxide anodizing (SHA) process, the macroscopic level of roughness is much greater than with CAA, with irregular oxide protrusions 1–2 μm high and wide separated by regions variously reported as containing a cellular structure of similar [9] or less pronounced [10] micro-roughness to that with CAA. An alkaline hydrogen peroxide etch (AHPE) has also been evaluated [11] but is not favoured because the hydrogen peroxide is unstable at the 65 °C bath temperature.

Above 200 °C, surface oxides on Ti–6Al–4V gradually dissolve in the alloy and therefore alternative surface treatments have to be applied. Plasma spraying is one such recent option. Powdered Ti–6Al–4V is injected into a plasma, where it melts and is projected at supersonic speeds onto a titanium substrate. Most molten particles 'splat cool' on striking the surface, though some cool prior to impact, thereby generating a mixed morphology of larger solidified nodules within a random distribution of agglomerated fine 'splat' particles with extended porosity. Recent experiments with a high-temperature (polyimide) adhesive have confirmed the improved durability of plasma sprayed and bonded Ti–6Al–4V substrates over their CAA treated equivalents [12].

11.5.4 Plastic/composite surfaces

Polymeric materials can have surface energies which are similar to the surface tension of typical adhesives and therefore some pretreatment is desirable. In some cases (such as the high-energy thermosets) a solvent wipe may be sufficient, but for critical structural aerospace applications degreasing coupled with a mechanical abrasion treatment is necessary. Moulded plastic components can also be contaminated with release agent, and cleaning is essential. For thermoplastic fibre-reinforced composites, solvent cleaning and abrasion are insufficient to achieve optimum joint strength [13]. Instead, the thermoplastic surfaces require Corona discharge treatments to increase surface energies.

Adhesive bonding of boron–epoxy or graphite–epoxy patches is now used for the repair of both metallic and composite aircraft structures. Repairs need to be carried out *in situ* and in applications where preparation and cure facilities are often limited. For composite/composite repairs, abrasion cleaning, to remove release agent, followed by light alumina grit blasting is effective. For composite-to-aluminium repairs, the Aeronautical Research Laboratories, Australia have undertaken extensive field trials [14]. Initially, alumina grit blasting was their sole surface treatment, followed by epoxy nitrile bonding and curing at 120 °C. More recently, however, it was found that epoxy bonded composite patches with excellent durability can be affixed to aluminium structures with a simple grit blasting and silane primary treatment – even when the aluminium cure temperature was as low as 50 °C.

11.5.5 Surface priming

To improve adhesion on glass, metals and ceramics, chemical bonding using silane coupling agents can be used.

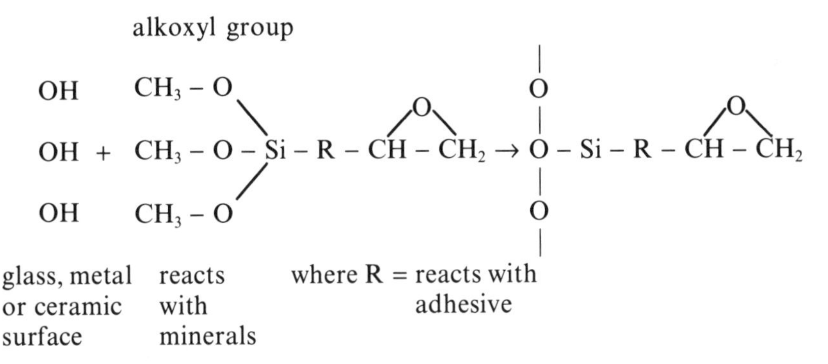

Alkoxyl groups in the silane are initially hydrolized to silanols (SiOH) which chemically bond to the surface, leaving, in this case, the characteristic epoxy $-CH\overset{O}{-}CH_2$ group to react with the adhesive. There is a variety of arguments [8] which suggests that the chemical coupling mechanism is too simplistic and alternative models have been proposed in which an absorbed polysiloxane film forms an open network on the adherend surface into which adhesive molecules can penetrate.

11.6 DESIGN OF BONDED JOINTS

Bonded joints must be designed to exploit the behaviour of the adhesive rather than adopting joint configurations used for mechanical fixing or welding. An adhesive bond is essentially a low-modulus polymer interlayer between two more rigid substrates. For effective load transfer, the bonded joint should have the following general characteristics:

- a large bond area
- a uniformly thin and continuous bond line
- loading in compression or shear
- minimum peel loading
- avoidance of stress concentrations.

While the requirement for a large bond area is justified, any assumption that the applied load is uniformly distributed across the bond overlap must be questioned. Only if the substrates were perfectly rigid and the adhesive elastic would this be true. In fact, the substrates also stretch elastically, the extent of that extension being dependent on the adjoining constraint. The adhesive must match this differential strain such that a stress concentration builds up at the edges of the joint which is given by:

$$\text{Stress concentration factor} = \frac{G_a l^2}{Eht}$$

where the common substrates have a stiffness E, thickness h and overlap l, while the adhesive layer is of thickness t and shear modulus G_a. The above factor demonstrates that decreasing the overlap area, or using an adhesive with a lower shear modulus, will reduce the stress concentration. Alternatively, stiffer, thicker adherends can be used. Increasing the thickness of the adhesive layer would also help to reduce the stress concentration.

Another critical characteristic of adhesives is their low peel and cleavage strength. As a simple guide, the relative ability of bonded joints to withstand compression, shear and peel varies in the ratio 1000 : 100 : 1. Ideally, the design should be modified to avoid peel, or alternatively local mechanical restraints (bolts, rivets or spot welds) can be introduced. The USSR has used weld bonding on the Antonov An24 turbo-prop aircraft to provide transverse strengthening through the tack welds, while shear loadings were carried by the continuous layer of adhesive. The assembly process involved spot welding through previously applied unreactive adhesive after which the adhesive was cured. Flexible (plasticized) adhesives were initially used, but spot welding tended to cause local burning of the plastisol which caused corrosion. Weld bonding was also assessed for assemblies on the Fairchild A10 and upper fuselage panels were chosen for production qualification. The original components were autoclave adhesively bonded, but computer-controlled weld bonding of the aluminium components was subsequently introduced on a cost-reduction basis.

The simple elastic model described above (originally due to Volkersen [15]) ignores the bending moments on a lap shear joint due to non-axial loading; these were subsequently considered by Goland and Reissner [16]. Additional refinements allow for adhesive plasticity, non-linear adhesive behaviour and fillet geometry, while more complex joint configurations have also been analysed [17]. Joint geometries of particular relevance to aerospace applications are illustrated in Figure 11.4. For thick sections, the stepped lap or scarf joints are preferred. When bonding composites to metal, the stepped lap joint is easier to lay up and it is this configuration which has been used to bond graphite-reinforced/

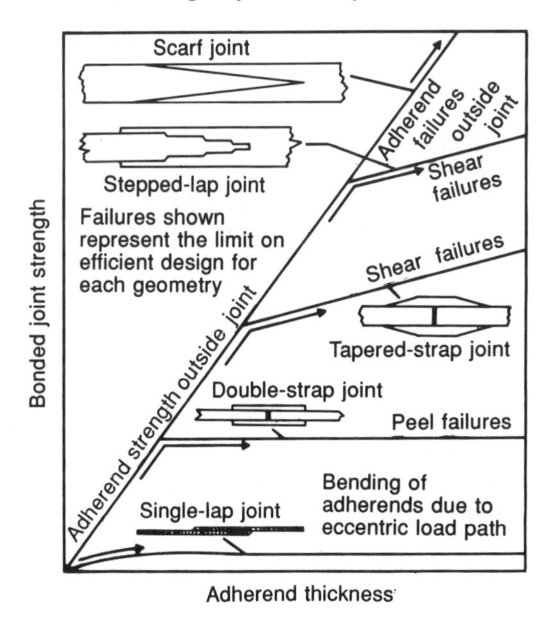

Figure 11.4 Typical joint geometries for different joint thicknesses (from Hart-Smith [23])

epoxy matrix composite wing skins to the titanium wing root of the McDonnell Douglas F18 fighter aircraft (Figure 11.5). Similar joints have also been used to bond boron–epoxy to titanium in the horizontal

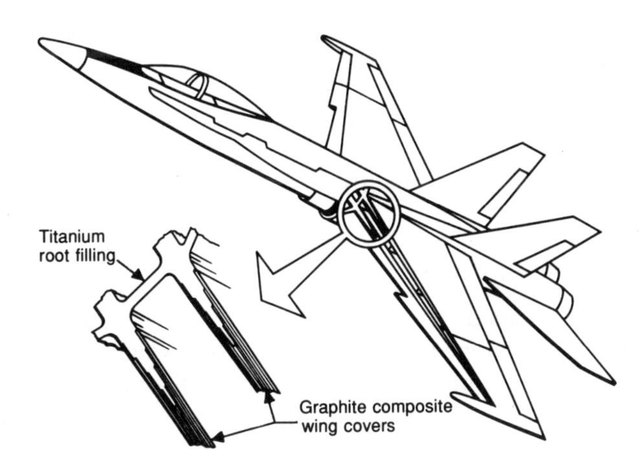

Figure 11.5 The composite wing to titanium fuselage bonded attachment in the F18 aircraft (after Krieger [24])

and vertical tailfins of the Grumman F14 and the McDonnell Douglas F15 aircraft.

11.6.1 Bonded metallic structures

The bonding of metallic aircraft components began in 1943 when the vinyl phenolic adhesive Redux was used to produce a wood-to-aluminium alloy composite spar for the wings of the De Havilland 103 Hornet aircraft. Subsequently, metal-to-metal Redux bonded stringers were introduced to strengthen the semi-monocoque Duralumin fuselage of the DH Dove (1948), with similar strengthening on the Heron (1950). Bonded stringers up to 25 feet in length were also used on double curvature surfaces in the DH Comet. The use of Redux bonding, particularly for doubler plate reinforcement around the windows of the Comet, was investigated as a possible cause of the structural failures in this aircraft in 1953. Metal fatigue was however subsequently identified as the failure cause and bonded doublers and stringers have continued to be used with considerable success through to the latest civil jet airliners such as the BAe 146, where Redux bonding of stringers to the top wing skin provides the largest example of a single-piece bonded component currently in use.

It was in Europe that commitment to adhesive bonding was initially most pronounced. In 1947 the Swedish aircraft industry began evaluating Redux and steadily increased its usage in their J29 (Tunnen), J32 (Lansen) and J35 (Draken) fighters. Most applications were for bonded laminates or for stiffened door panels, wings and fuselage structures using extruded J or Z stringers or rolled 'top hat' configurations (Ω). The J or Z sections gave optimum stiffening but were difficult to maintain in position during curing. The 'top hat' configuration was more stable but contained a closed section which could encourage corrosion. It has been the Dutch, however, who have continued to demonstrate the effectiveness of adhesive bonding. In 1950 Fokker introduced the F27, the first aircraft to be built primarily from bonded metallic components, and with the F28, F50 and F100 this trend has continued.

11.6.2 Honeycomb structures

While localized bonded reinforcement can increase the stiffness and strength of metal skinned structures, greater weight saving can be achieved with honeycomb materials in which an open cell structure is sandwiched between two thin layers of metal. Such aluminium honeycomb structures, originally developed by the Americans in the mid-1950s, offer great potential for adhesive bonding. The honeycomb can be made by either expansion of a bonded stack of foils or by bonding individual corrugated sheets. In the expansion route, thin sheets of metal have lines of phenolic or epoxy adhesive printed on to their surface; the sheets are stacked into a block, which is placed in a heated press and, when the adhesive cures, each sheet is bonded to the next in a localized region. When the stack is subsequently pulled apart, the unbonded regions separate to form a honeycomb block which is sectioned to the required thickness. If a regular hexagonal honeycomb array is to be maintained, it is critical that the adhesive printing is accurate.

Alternatively, sheets can be corrugated mechanically using suitably profiled rolls. Adhesive is then applied to the required surfaces and a honeycomb stack built up. The stacks are heated to cure and sectioned as before. To produce the final honeycomb sandwich, the core needs to be faced top and bottom with metal foil, which requires additional adhesive bonding between the facing sheets and the internal honeycomb. Early sandwich construction used vinyl phenolic adhesives for bonding. Since these adhesives give off volatiles during curing, bonding of the facing skins required that the honeycomb was perforated to provide adequate venting. Unfortunately, such ventilation also provided access for moisture which could corrode the honeycomb and degrade the bonds and therefore unperforated honeycomb and alternative adhesives are now preferred. Various modified epoxy and high-temperature polyamide epoxy adhesives have been used, but service experience has been mixed [1,11]. To achieve satisfactory performance it is necessary to choose a relatively corrosion-resistant aluminium honeycomb material, to optimise the skin and honeycomb surface treatments and to use a durable adhesive with effective wetting and filleting characteristics. Since the 1960s, epoxy bonded aluminium honeycomb-cored panelling has been widely adopted for aircraft components and typical applications include control surfaces on tail fins and windows, upper and lower main wing surfaces and similar regions where stiffening is required in thin sections.

11.6.3 Composite structures

In recent years honeycomb structures are being challenged by composites for many of the thin, stiffness-critical aircraft components. In the F14 and F15 the horizontal tail skin is in boron/epoxy, while in the F18 and the AV8B the same component is being made in a graphite/epoxy composite. Control surfaces on these aircraft, together with the Rockwell B1 bomber and the Boeing 757, are also either boron or graphite/epoxy. In the case of the F18, graphite/epoxy comprises 9% of the structure weight and 34% of the external surface [18], while the recent Beech Starship has a structure composed almost entirely of graphite/epoxy composite.

Structural polymeric foams and aramid honeycomb cores can variously be adhesively bonded to thin metallic skins or to continuous carbon/aramid fibre reinforced epoxy sheets to produce a wide range of sandwich constructions. Applications include flight control surfaces on Airbus and the McDonnell Douglas C17, while for the new Westland EH101 helicopter most material combinations are used for composite sandwich construction of the fuselage.

Since the earliest days of flight, adhesive bonding has been used to advantage in the construction, initially, of wooden and fabric airframes. As metallic structures predominated, synthetic adhesives contributed to lightweight assemblies by providing local stiffening and facilitating the construction of honeycomb/sandwich structures. Such contributions are not insignificant; in the Lockheed C-5A transporter 'plane, bonded honeycomb sandwich constructions total 24000 square feet in area [19], while for a Boeing 747 some 14000 square feet of adhesive is used [11]. As non-metallic composites find increasing aerospace applications, adhesive bonding is likely to become the major means of assembly. Not only will individual components be bonded, but adhesion provides the primary mechanism for combining the composite materials. The prospects for both aerospace and adhesive bonding appear promising.

REFERENCES

1. Albericci, P. (1983) Chapter 8, Aerospace applications pp. 317–50 in Kinloch A. J. (ed.) *Durability of Structural Adhesives*. Reprinted 1986. Elsevier Appl. Sci. Publ., Barking, Essex.
2. Weyl, A. R. (1985) *Fokker: The Creative Years*, pp. 197–203. Putnam, London. Reprinted 1987.
3. Jackson, A. J. (1978) *De Havilland Aircraft since 1909*, pp. 42–3. Putnam, London.
4. Smith, J. R. and Kay, A. (1978) *German Aircraft of the Second World War*, pp. 211–14. Putnam.
5. Smith, J. R. and Kay, A. (1978) *German Aircraft of the Second World War*, pp. 307–16, Putnam.
6. de Bruyne, N. A. (1957) *Ciba Technical Notes* no. 179. Reprinted 1963. Aero Research Ltd, Duxford, Cambridge.
7. Shaw, S. J. (1992) Polyimide adhesives, pp. 319–24 in Packham, D. E. (ed.) *Handbook of Adhesion*, Longman Scientific & Technical, Harlow, Essex.
8. Clearfield, H. M., Thomas, J. McNamara, D. K. and Davis, G. D. (1990) Surface preparation of metals pp. 259–75. *ASM Engineering Materials Handbook Vol. 3. Adhesives and Sealants.*
9. Kennedy, A. C., Kohler, R. and Poole, P. (1983) *Int. J. Adhes.*, **3**, 133.
10. Filbey, J. A., Wightman, J. P. and Progar, D. J. (1987) *J. Adhes.*, **20**, 283.
11. Poole, P. (1985) Adhesion problems in the aircraft industry. Chapter 9, pp. 258–84 in Brewis, D. M. and Briggs, D. (eds) *Industrial Adhesion Problems*. Orbital Press, Oxford.
12. Clearfield, H. M., Shaffer, D. K., Van Doren, S. L. and Ahearn, J. S. (1989) *J. Adhes.*, **29**, 81.
13. Kinloch, A. J., Kodakian, G. K. A. and Watts, J. F. (1992) *Phil. Trans. Roy. Soc. Lond.*, **A338**, 83.

14. Baker, A. A. (1987) Fibre composite repair of cracked metallic aircraft parts – practical and basic aspects. *Composites*, **18** (4), 293–308.
15. Volkersen, O. (1938) *Luftfahrtforschung*, **15**, 33.
16. Goland, M. and Reissner, E. (1944) *J. Appl. Mech*: Trans. ASME **66**, p. A17.
17. Adams, R. D. and Wake, W. C. (1984) Reprinted 1986. *Structural Adhesive Joints in Engineering*. Elsevier Appl. Sci. Publ., Barking, Essex.
18. Baker, A. A. (1988) Development and potential of advanced fibre composites for aerospace applications. *Materials Forum*, **11**, 217–31.
19. Hogemaier, D. J. (1990) End product non-destructive evaluation of adhesive-bonded metal joints, pp. 743–84, *ASME Engineering Materials Handbook Vol. 3, Adhesives & Sealants.*
20. Lees, W. A. (1984) *Adhesives in Engineering Design*, p. 58. The Design Council, London.
21. Knight, R. A. G. (1969) *The relative merits of mechanical and visual tests for assessing glue joints in wood*, p. 263, Adhesion – Fundamentals & Practice. Report on Int. Conf., Nottingham 20/22 Sept 1966. MoD. Publ. Maclaren & Sons, London (1969).
22. Venables, J. D., McNamara, D. K., Chen, J. M., Sun, T. S. and Hopping, J. L. (1979). *Appl. Surf. Sci.*, **3**, 88.
23. Hart-Smith, L. J. (1980) *Adhesive bonding of aircraft primary structures*, p. 3. Douglas Paper 6979 presented to Soc. Auto Eng. Inc. Aerospace Congress, Los Angeles 13–16 Oct (1980).
24. Krieger, R. B. (1986) *Stress analysis concepts for adhesive bonding of aircraft primary structure*. Report C174/86 IMechE.

12

Rapid solidification and powder technologies for aerospace

H. Jones

12.1 INTRODUCTION

Rapid solidification (RS) involves propagation of a solidification front at high velocity. This is most readily achieved by suitable treatment of a volume of melt. Suitable treatments include:

- dividing it up into a multitude of small droplets (atomization, emulsification or spray-forming) so that most of them can undercool deeply prior to solidification
- stabilizing a meltstream of small cross-section in contact with an effective heat sink (melt-spinning or thin-section continuous casting)
- rapid melting of a thin layer of material in good contact with an extensive heat sink, which may be the same or related material (electron or laser beam surface pulse or traverse melting).

In each case rapid solidification results from rapid extraction of the heat of transformation either directly by the external heat sink and/or internally by the undercooled melt (in which case the system rapidly reheats, i.e. recalesces during solidification). The large undercoolings developed amount to large departures from equilibrium, leading to formation of extended solid solutions and new non-equilibrium phases (crystalline, quasicrystalline or glassy) while the short freezing times give rise to size-refined and compositionally rather uniform microstructures as well as relatively high rates of throughput of material. The products of RS range from powder or flake particulate, through thin discontinuous or continuous ribbon or filament to thick spray deposits containing some trapped porosity. These products can sometimes be applied directly for aerospace applications as in the cases of finely divided light metal particulate used as the basis for space shuttle and satellite launch rocket fuel and signalling flares, and planar-flow-cast strip used for braze assembly of engine components. For most applications, however, they must be suitably bonded or consolidated into full size, fully dense sections or components. This may involve processes such as polymer bonding or liquid metal infiltration but most commonly involves powder metallurgy techniques such as hot pressing and/or hot working.

The purpose of the present chapter is to set out the principles of RS technology of interest to the aerospace industry, its suppliers and customers and to indicate some of the benefits that derive from processing via this route. Earlier reviews of the process technology have been published by Duwez [1], by the present author [2–4], by Savage and Froes [5] and by Anantharaman and Suryanarayana [6].

Its potential for aerospace applications has been assessed, with reference to aluminium alloys, by Staniek *et al.* [7], Millan [8,9], Sakata [10], Sakata and Langenbeck [11], Quist and Narayanan [12], Ronald *et al.* [13], Kuhlmann [14], Quist and Lewis [15], Jones [16], Billman and Graham [17], Yamauchi [18], Rainer and Ekvall [19], Gilman and Das [20], Gilman *et al.* [21], Frazier [22], Chellman *et al.* [23], Frazier *et al.* [24], Das *et al.* [25], Gilman [26], Eckvall *et al.* [27], and Chellman *et al.* [28], with reference to titanium alloys by Dulis *et al.* [29], Eylon *et al.* [30–32], Thompson [33], Moll *et al.* [34], Witt and Weaver [35], Froes *et al.* [36], Singer [37], Moll *et al.* [38,39] and Wasielewski *et al.* [40], and with reference to superalloys by Cox and van Reuth [41,42], Thompson [33,43], Bartos [44], Dreshfield and Miner [45], Patterson *et al.* [46], Tracey and Cutler [47], Wildgoose *et al.* [48], Lherbier and Kent [49] and Kent [50].

12.2 PRODUCTION TECHNOLOGIES

These will be discussed under the headings identified in section 12.1, i.e. droplet technologies, spinning technologies and surface melting technologies. The characteristics of representative examples are summarized in Table 12.1.

12.2.1 Droplet technologies

These make use of the fact that a pendant drop or free falling melt stream tends to break up into droplets as a result of surface tension. This process can be intensified, for example, by impingement of high velocity jets of a second fluid, by centrifugal force at the lip of a rotating cup or disc or by an applied electric field. The standard method uses high velocity gas or water jets (Figure 12.1). These form, propel and cool the droplets that may freeze completely in flight or may form individual splats or build up a thick deposit by impact on a suitable substrate. Powder or shot

Table 12.1 Characteristics of representative techniques for rapid solidification and their products

Technique	*Principal Operational Features*	*Characteristics of Products*
1. *Droplet Methods*		
Gas jet atomization	Impact of pressurized gas on to meltstream	Irregularly shaped or spherical powder particulate
Centrifugal atomization	Expulsion of melt film from periphery of a rotating disc or cup	Smooth spherical powder particulate
Splat quenching	Release of impacted droplets from a chill surface	Flakes down to 2 mm diameter by 1 μm thick
Spray forming	Deposition, spreading and coalescence of impacted droplets on to chilled or heated collector	Thick spray deposit which can have low porosity
2. *Spinning Methods*		
Free-jet chill-block melt-spinning	Impingement of cylindrical melt-stream on to a rotating chill	Continuous ribbon or flake ≃ 2 mm wide × 10 to 100 μm thick
Planar flow casting	Impingement of rectangular or multiple meltstream(s) at close proximity to rotating chill surface	Continuous strip up to 150 mm wide by 20 to 100 μm thick
Melt extraction/ melt overflow	Continuous pickup of melt by contact of a rotating disc with a source of melt	Continuous or short fibres of non-circular section or continuous wide strip
Rotating water bath process	Injection of a cylindrical meltstream into a contained rotating water annulus	Continuous wire circular in section or spherical powder particulate
3. *Surface Melting Methods*		
Electron or laser beam traverse remelting	Continuous rapid melting of material at the surface of a massive heat sink	A thin layer of rapidly solidified material at the surface
Nanosecond and picosecond pulse melting	Localized rapid melting of material at a surface	Localized rapid solidification at the surface around the point of treatment

For a more complete review of the characteristics of these and other related techniques see Jones [2–4], Savage and Froes [5] and Anantharaman and Suryanarayana [6].

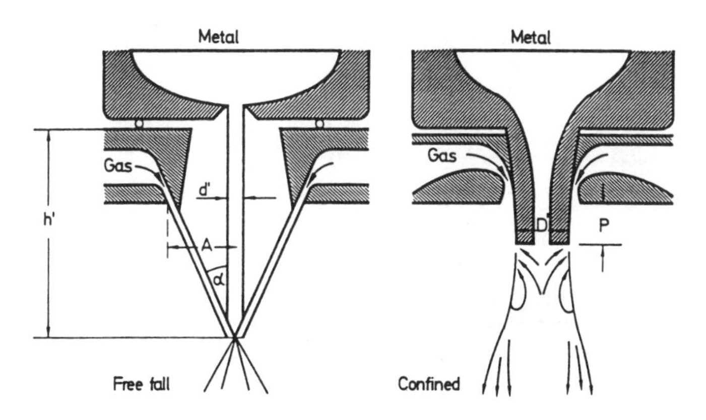

Figure 12.1 Principle of spray droplet formation by impingement of high velocity gas jets on to a free falling or emerging meltstream. (From Schmitt [51]).

particulate generated by such techniques displays a range of particle sizes and shapes even for a single set of operating conditions with increasing atomizing gas pressure, for example, increasing the yield of finer particle sizes and the more effective quench of water atomization leading to more irregular particle shapes. Although particles of more or less identical size and shape from the same powder sample can show quite different microstructures, smaller particles tend to cool more rapidly and/or undercool more prior to solidification so tend to solidify more rapidly (i.e. at higher front velocity). Splats formed from droplets of a given size tend to solidify even more rapidly because of more effective heat extraction from the larger surface area they offer, especially when at least one of their surfaces is in good contact with an efficient heat sink, such as a water-cooled rotating copper drum. A spray deposit can maintain the same microstructure as the equivalent splats provided that their solidification time is sufficiently less than the interval between deposition of successive splats at a given location on the substrate [52]. Atomizers range in size from laboratory units with capacities of less than 1 kg per run to commercial size facilities with capacities as large as 50 000 tonnes per year [53]. Powders destined for high performance applications tend to be atomized with inert gases or in vacuum to minimize formation of oxides or other potentially damaging inclusions.

12.2.2 Spinning technologies

These derive from the simplest system in which a single meltstream emerging from an orifice is stabilized by surface film formation or solidification before it can break up into droplets, to the most sophisticated

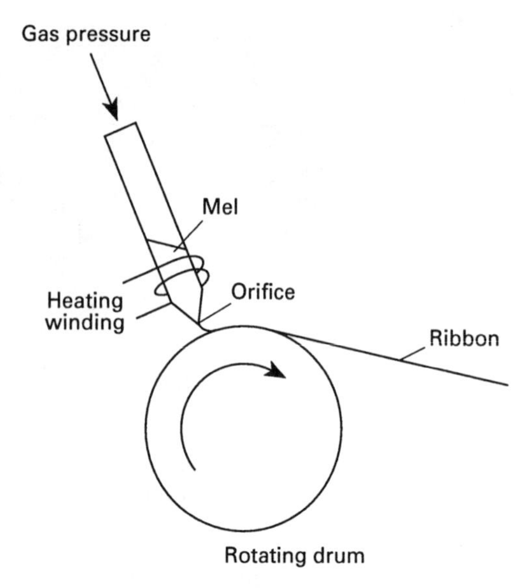

Figure 12.2 Principle of melt-spinning of continuous ribbon by impingement of a meltstream on to a rotating chill block. (From Overshott [54]).

in which one or more meltstreams are used to make wide or composite ribbons by impingement on single or twin chill roll surfaces. In standard free-jet chill-block melt-spinning (Figure 12.2) the meltstream forms a single ribbon typically 10–100 μm thick and a few millimetres in width by impingement on to a single chill roll rotating at a surface speed of some tens of metres per second. In planar flow casting the distance from the nozzle orifice to the chill surface is reduced to less than $\simeq 0.5$ mm to eliminate formation of the melt-pool and associated instabilities. Thin strip up to 500 mm in width can be cast in tonnage quantities by this method. Melt-extraction, in which the rotating chill roll forms the product by direct contact with the surface of a crucible of melt or the melted extremity of solid electrode, and its derivative, melt overflow, do not involve generating a meltstream.

Melt-extraction is well established as a cost-effective technique for producing tonnage quantities of staple fibres for reinforcement of concrete or refractories [55].

Continuous rapidly solidified filament with round section can be produced by the rotating water bath process in which a meltstream emerging from a cylindrical orifice is solidified directly on entry into a rotating annulus of water contained within the lip of a rotating water bath [56]. This process can also be used to generate powder particulate when the conditions result in break-up of the meltstream on entry into the water bath [57]. Free-jet chill-block melt-spinning and melt-extraction can be used to generate a flake product directly if continuity is interrupted by a series of regularly spaced notches on the chill-block. Such (aluminium) flake can be loaded into a polymer matrix to combine the conducting properties of a metal with the processing economies and flexibilities offered by plastic [58]. Melt-spun ribbon or planar flow cast sheet can be pulverized, for example, by means of a blade cutter mill [59,60], into platelets $\simeq 0.3$ mm across. These provide a more suitable feedstock for consolidation by powder metallurgy than continuous ribbon, for example, which exhibits a much lower packing density.

12.2.3 Surface melting technologies

These derive from spot or continuous welding techniques and differ from them only in that the depth melted is limited to ensure that the ensuing solidification will be sufficiently rapid. In its simplest form a single pulse or continuous traversing heat source is used to rapidly melt the surface of a block of material, the unmelted bulk acting as the heat sink during the subsequent rapid solidification (Figure 12.3). The resulting rapidly

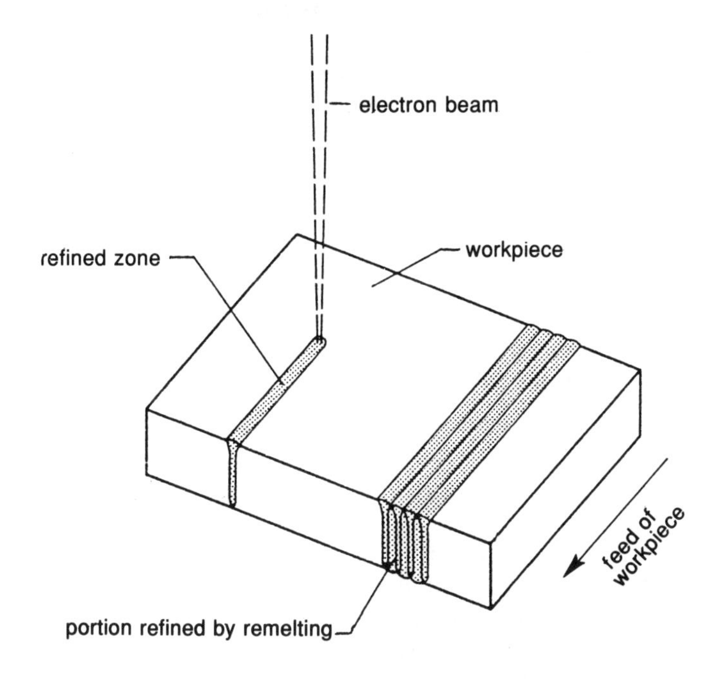

Figure 12.3 Principle of rapid solidification at the surface of a block of material following rapid local melting with traversing heat source. (From Lux and Hiller [61]).

solidified material has the same composition as the underlying parent material, although the rapid solidification may produce a quite different microstructure and much improved properties. A second possibility is to preplace or inject alloy or dispersoid additions at the surface so that they are incorporated into the melt zone to form a surface region of composition different from the underlying bulk. The third possibility is to melt a different material preplaced on the surface so that mixing with the underlying material is limited to the minimum required to produce effective bonding. All three variants offer the practical possibility of generating a more durable surface on an underlying material that is in all other respects entirely adequate for the application in view.

Both nanosecond and picosecond laser sources have been used to generate some quite spectacular nonequilibrium effects in surface melt zones as shallow as 0.1 μm in which cooling rates during solidification have been estimated to reach 10^{10} K/s [62] or more, and solidification times to be as short as 10^{-9} s [63]. Both traversing laser and electron beams have been used to treat entire surfaces via repeated incremental displacement of the beam by the width of the melt zone at the start of each new traverse [61]. Samples with crack-free relatively smooth treated surfaces can now be produced by appropriate control of the process parameters. The technique can also be used to develop coupling between the traversing beam and the solidification front so that the effects on resulting microstructure of systematic variations in front velocity up to $\simeq$ 1 m/*s* can be determined experimentally [64–66].

12.3 EFFECTS ON MICROSTRUCTURE

These divide naturally into those generated directly by rapid solidification and those deriving from subsequent processing.

12.3.1 Effects of rapid solidification

The presence of non-equilibrium phases in rapidly solidified microstructures implies that solidification has occurred at undercoolings and velocities sufficiently high for their formation to be kinetically favoured over the equilibrium phase or phases they displace. Metallic glass formation provides the simplest example, for which suitable alloying is employed, for example, to minimize the temperature interval between liquidus and glass formation temperatures within which crystallization can intervene before glass formation. Suppression of formation of primary intermetallic β phases in peritectic or hypereutectic composition ranges depends on their kinetics of nucleation and growth in competition with the terminal α-phase or a eutectic at temperatures below the extended α-liquidus or eutectic temperature. Minimization of the tem-

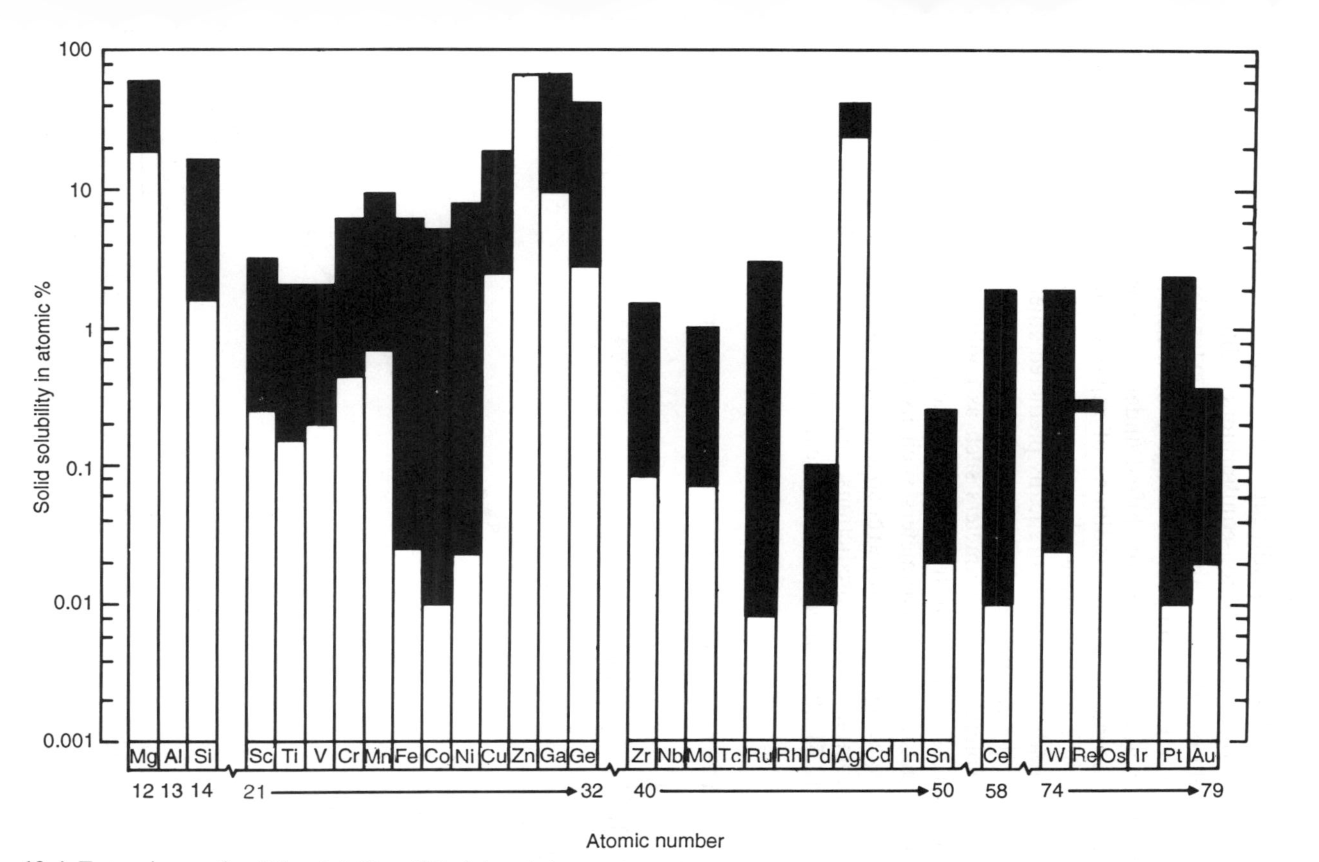

Figure 12.4 Extensions of solid solubility (filled bars) beyond equilibrium maximum values in aluminium as a result of rapid solidification, plotted as a function of atomic number of the solute. (From Midson and Jones [67]).

perature interval between the α-liquidus and the β-liquidus, eutectic temperature or liquidus of a non-equilibrium intermetallic again favours formation of the latter constituents. The composition of a phase resulting from growth under such non-equilibrium conditions can depart substantially from equilibrium values resulting in extensions of solid solubility (Figure 12.4). The composition range over which fully eutectic growth can occur is also extended or shifted under such conditions.

The outcome of given conditions of rapid solidification can be predicted on the basis of known limiting transformation temperatures and experimentally based or theoretically predicted kinetics of nucleation and growth of the competing phases and constituents. Observed relationships between solidification parameters such as cooling rate, local solidification time, solidification front velocity and temperature gradient and microstructural scale parameters such as grain size, dendrite cell size, primary spacing, secondary arm spacing and eutectic interphase spacing can be compared with predictions. Such relationships (Table 12.2 and Figure 12.5) allow predictions of scale of microstructure to be made

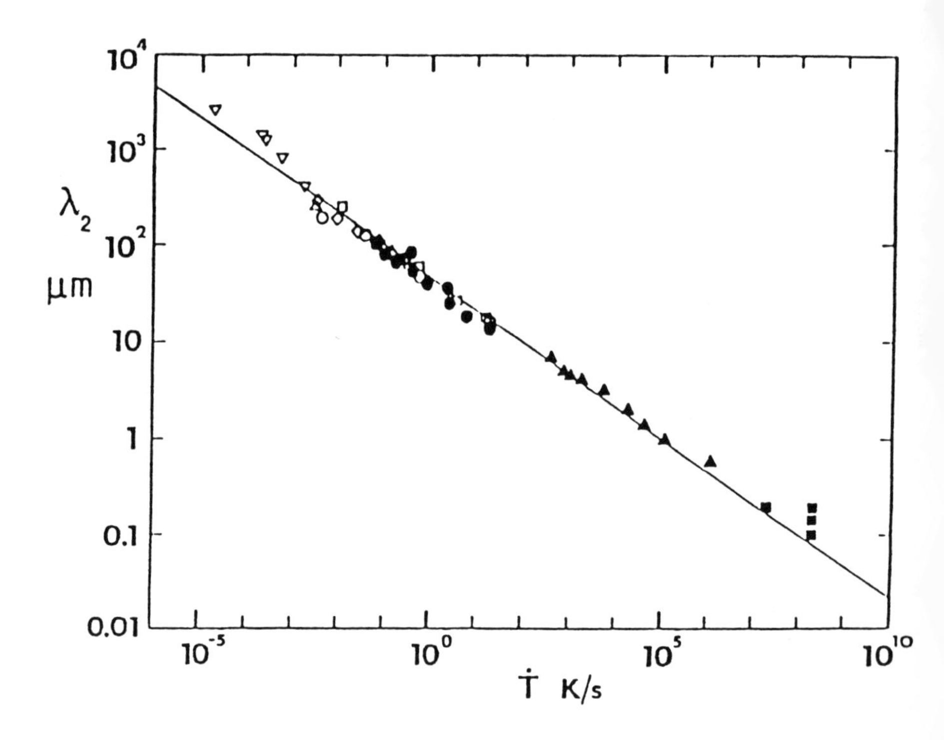

Figure 12.5 Dendrite cell size λ as a function of cooling rate $\dot{T}$ for Al–4 to 5 wt% Cu (open points) and Al–7 to 11 wt% Si (closed points). (From Jones [3]).

Table 12.2 Functional relationships between solidification parameters and scale of microstructure in rapid solidification

Scale parameter	*Functional relationship*	*Typical values and range of applicability*	*References*
Grain size λ_0	$\lambda_0 \dot{T}^{n} = k_0$	$K_0 = 1$ to 9 m $(K/s)^n$ with $n = 1$ for selected Fe, Ni and Ti based alloys for $3 \times 10^4 < \dot{T} < 5 \times 10^6$ K/s	[68]
Dendrite primary spacing λ_1	$\lambda_1 G^{1/2} V^{1/4} = K_1$	$K_1 = 30$ to 90 $\mu m^{3/4} K^{1/2} s^{-1/4}$ for αAl in Al–Cu and Al–Mg–Si and for Si in hypereutectic Al–Si, for $0.01 < V <$ 500 mm/s and $0.5 < G < 2000$ K/mm	e.g. [69–71]
Dendrite secondary arm spacing λ_2	$\lambda_2 \dot{T}^{n} = K_2$	$K_2 = 50\,\mu m (K/s)^n$ with $n = 1/3$ for Al–Cu(–Si) for 2 mm $< \lambda < 0.1\,\mu m$ and $10^{-5} < \dot{T} < 10^8$ K/s	[3,72]
Cellular spacing λ	$\lambda V^{1/2} = K_3$	$K_3 = 100$, 100 to 400 and $14 \pm 6\,\mu m^{3/2} s^{-1/2}$ for Ag–15 w/o Cu, Al–0.5 to 5 w/oMn and Al–8 w/oFe respectively, for $0.1 < V < 2400$ mm/s	[64, 66, 73]
Eutectic interphase spacing λ	$\lambda V^{1/2} = K_4$	$7 < K_4 < 50\,\mu m^{3/2} s^{-1/2}$ for Al–based. $K_4 = 9.4 \pm \mu m^{3/2} s^{-1/2}$ for Al–Al_2Cu for $8 < \lambda < 0.015\,\mu m$ and $0.002 < V$ 50 mm/*s*	[74]

provided that the conditions of rapid solidification are sufficiently well known or well characterized. Complications result from collisions between droplets at different stages of cooling and solidification in atomisation and spray forming processes, from variabilities in local contact with the chill surface and breakaway from it in spinning methods and from reheating of previously treated material from a subsequent adjacent traverse in surface melt traversing. Resulting microstructures are then a combination of the effects of rapid solidification and of subsequent heat treatment.

12.3.2 Effects of post-solidification treatments

Such treatments include **mechanical** ones such as cutting or pulverization of melt-spun ribbons or cold compaction of particulates, **thermal** ones such as vacuum or depurative degassing of particulate prior to hot consolidation, **thermomechanical** ones such as hot pressing, extrusion, rolling or forging of canned, cold or hot-pressed or spray-deposited material to complete closure of porosity and ensure metal to metal bonding and also possibly a final heat treatment. Appropriate control of

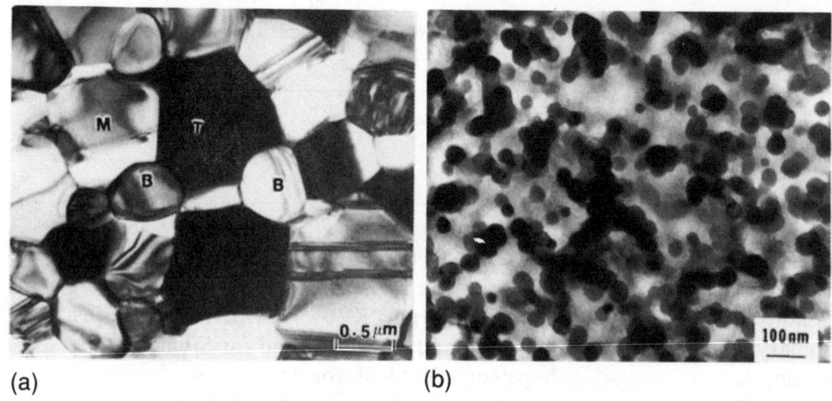

Figure 12.6 Examples of optimized microstructures produced from rapidly solidified feedstocks as a result of controlled thermomechanical treatment during or following consolidation: (a) submicron matrix grain size stabilized by high volume fraction of submicron boride particles in Ni–Mo–Cr–B resulting from controlled devitrification of a metallic glass precursor during consolidation. (From Das and Raybould [76]); (b) submicron silicide dispersion in rapidly solidified Al–Fe–V–Si as a result of subsequent thermomechanical processing to optimize fracture toughness and elevated temperature strength. (From Raybould *et al.* [77]).

such treatments is evidently necessary either to preserve the as-rapidly solidified microstructure essentially unchanged or, more commonly, to develop it into an optimum form which meets the particular requirements of the application. Dynamic compaction techniques have been developed (e.g. [75]) in which rapid deformation and melting is confined essentially to particle surfaces resulting in local welding with minimal temperature rise of the bulk of the material, so preserving its nonequilibrium structure essentially unchanged. Alternatively, hot consolidation can be carried out at temperatures low enough and over time intervals short enough to prohibit significant microstructural change. Where change is required this can be engineered as part of the consolidation cycle.

One example (Figure 12.6(a)) is the devitrification during consolidation of pulverized melt-spun glassy Ni-, Co- or Fe-base alloys to produce an ultrafine matrix grain size stabilized by a high volume fraction of submicron boride particles derived from the glass-forming addition [78], [79]. Another (Figure 12.6(b)) is the formation during consolidation of microcrystalline Al–Fe–V–Si alloy melt-spun particulate of a high volume fraction of submicron spherical silicide particles that are highly resistant to thermal coarsening, giving highly effective dispersion strengthening at elevated temperature combined with good corrosion resistance and adequate toughness [80]. The alternative Al–Cr–Zr–Mn system, which is susceptible to age-hardening following completion of consolidation, demonstrates that absence of an optimum thermal excursion to initiate precipitation during consolidation can actually be detrimental to final properties [81].

12.4 BENEFITS OF RAPID SOLIDIFICATION FOR AEROSPACE APPLICATIONS

Rapid solidification is a means both of optimizing processing of established materials and their products and of generating new materials and products derived from them. An example of a non-optimized process route is the production of brazing materials for gas turbine components by mechanical pulverization of cast material containing coarse and brittle second phases for application bonded with an organic binder in the form of transfer tape. This gives joints of variable integrity and strength due to incomplete accommodation of loss of binder during the brazing cycle and variability of local composition of the braze powder itself. Rapid solidification by melt-spinning (planar flow casting) allows production of these same or similar compositions in the form of continuous microcrystalline or metallic glass foils. These can be applied directly to produce more uniform and more reproducible brazed joints. They also

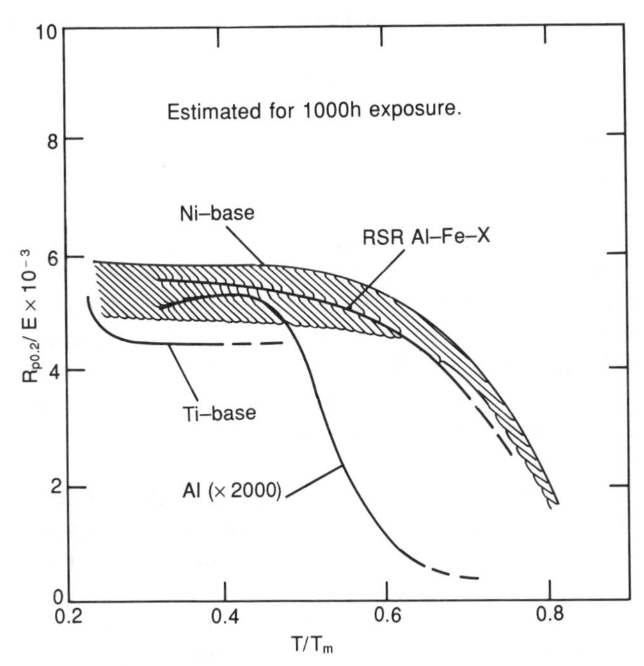

Figure 12.7 Normalized strength σ/E as a function of homologous temperature T/T_m for rapidly solidified aluminium–transition metal alloys compared with conventional aluminium, Ti- and Ni-based alloys. (From Singer and Couper [88]).

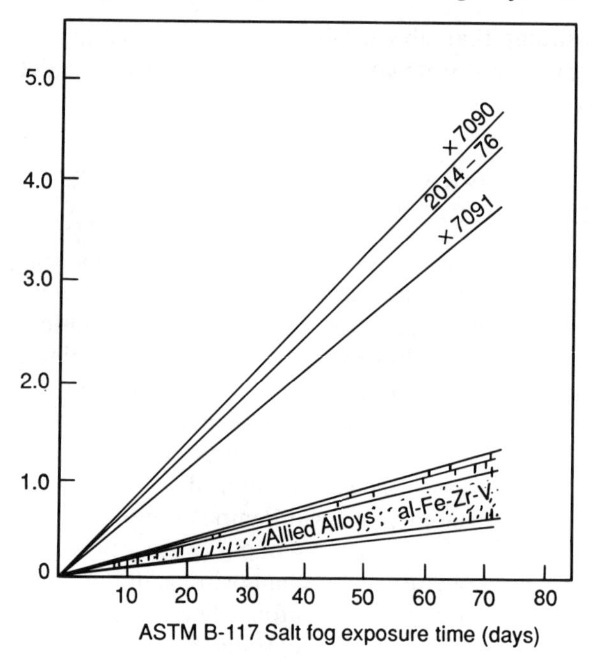

Figure 12.8 Saline environment weight loss as a function of exposure time for extrusions of rapidly solidified Al–Fe–V–Zr compared with P/M 7090, 7091 and wrought ingot 2014–T6. (From Adam and Lewis [89]).

have sufficient ductility to be formed to match part contours and to be cut or stamped into shape with minimal waste [82–85].

Examples of novel routes to finished products for aerospace include: manufacture of wafer turbine blades by precision bonding of stacked arrays of rapidly solidified thin sheet on to the surfaces of which cooling passages have been imprinted by photochemical etching or by other methods [86]: near net shape forming of high performance turbine discs from rapidly solidified prealloyed powder [49]: and laser cladding of turbine blades with hard surfacing alloy to provide optimum wear characteristics, freedom from heat affected zone cracking and consistent overall quality [87].

Examples of new materials for aerospace made possible by rapid solidification include novel aluminium-transition metal based alloys that combine (Figures 12.7 and 12.8) high strength up to 350 °C ($> 0.6T_m$) with good corrosion resistance [20, 21, 24–26]. Such materials offer both cost and weight savings when substituted for titanium alloys in the cooler end of the turbine compressor and for skins of future high speed transport aircraft. They also offer resistance to softening when substituted for 2014 aluminium alloy as an aircraft wheel subjected to temperature rises from the more effective braking provided by new carbon-carbon composite brake materials [25]. Other developments include the use of rapidly solidified aluminium alloy particulate as matrix materials for powder metallurgy mechanically alloyed or metal matrix composite materials containing respectively low and high volume fractions of fine and relatively coarse ceramic dispersoids [90, 91]. Spray co-deposition provides an alternative lower cost rapid solidification route for manufacturing such materials [92].

12.5 CONCLUSIONS

Rapid solidification and powder technologies offer numerous attractions for development of aerospace materials and their manufacture into components. These technologies are capable of generating higher performance from established materials as well as providing shorter or near-net-shape routes to final products. They also facilitate the processing of new monolithic or composite materials to meet entirely new performance requirements. Large-scale facilities are now operating at a number of sites to support the continuing development of materials, processes and products that depend on these versatile technologies.

REFERENCES

1. Duwez, P. (1968) In *Techniques of Metals Research*, Vol. 1, Part 1 (ed. R. F. Bunshah), Interscience, New York, pp. 347–58.
2. Jones, H. (1981) In *Treatise on Materials Science and Technology*, Vol. 20, (ed. H. Herman), Academic Press, New York, pp. 1–72.
3. Jones, H. (1982) *Rapid Solidification of Metals and Alloys*, The Institution of Metallurgists, London.
4. Jones, H. (1987) In *Enhanced Properties in Structural Materials by Rapid Solidification*, (eds F. H. Froes and S.J. Savage), ASM-I, Metals Park, Ohio, pp. 77–93.
5. Savage, S. J. and Froes, F. H. (1984) *J. Metals*, **36**(4), 20–33.
6. Anantharaman, T. R. and Suryanarayana, C. S. (1987) *Rapidly Solidified Metals: A Technological Overview*, Trans. Tech. Aedermansdorf, Ch 2.
7. Staniek, G., Wirth, G. and Bunk, W. (1980) *Aluminium*, **56**, 699–702.
8. Millan, P. P. (1982) In *High Strength P/M Al Alloys*, (eds M. J. Koczak and G. J. Hildeman), TMS-AIME, Warrendale, Pa, pp. 225–36.
9. Millan, P. P. (1983) *J. Metals*, **35**(3), 76–81.
10. Sakata, I. F. (1983) SAE Tech. Paper 831441.
11. Sakata, I. F. and Langenbeck, S. L. (1983) SAE Tech. Paper 831441.
12. Quist, W. E. and Narayanan, G. H. (1984) In *Materials and Processes: Continuing Innovations*, SAMPE, Azusa, Ca, pp. 374–91.
13. Ronald, T. M. F., Griffith, W. M. and Froes, F. H. (1984) In *P/M Aerospace Materials*, Vol. 1, MPR, Shrewsbury, Paper 25.
14. Kuhlmann, G. W. (1984) In *PM Aerospace Materials*, Vol. 1, MPR, Shrewsbury, Paper, 29.
15. Quist, W. E. and Lewis, R. E. (1986) In *Rapidly Solidified Powder Metallurgy Aluminium Alloys*, ASTM STP 890, pp. 7–38.

16. Jones, H. (1986) In *Materials in Aerospace*, Royal Aeronautical Society, London, pp. 97–121.
17. Billman, F. R. and Graham, R. H. (1987) In *Competitive Advances in Metals and Processes*, Vol. 1, SAMPE, Azusa, Ca, pp. 255–76.
18. Yamauchi, S. (1988) *Sumitomo Light Met. Tech. Rep.*, **29**(1), 69–81.
19. Rainer, R. A. and Eckvall, J. C. (1988) *J. Met.*, **40**(5), 16–18.
20. Gilman, P. S. and Das, S. K. (1988) In *PM Aerospace Materials*, MPR, Shrewsbury, Paper 27.
21. Gilman, P. S., Das, S. K., Raybould, D., Zedalis, M. S. and Peltier, J. M. (1988) In *Proc 2nd Int. SAMPE Metals and Metals Processing Conf.*, pp. 91–101.
22. Frazier, W. E. (1988) *Advanced Mater. Processes*, **134**(5), 42–6.
23. Chellman, D. J., Ekvall, J. C. and Rainen, R. A. (1988) *Met. Powd. Rep.*, **43**, 672–4.
24. Frazier, W. E., Lee, E. W., Donnellan, M. E. and Thompson, J. J. (1989) *J. Metals*, **41**(5), 22–26.
25. Das, S. K., Gilman, P. S., LaSalle, J. C. *et al.* (1990) In *P/M in Aerospace and Defense Technologies*, Vol. 1, MPIF, Princeton, NJ, pp. 77–82.
26. Gilman, P. S. (1990) *Metals and Materials*, **6**, 504–7.
27. Eckvall, J. C., Rainen, R. A., Chellman, D. J. *et al.* (1990) *J. Aircraft*, **27**, 836–43.
28. Chellman, D. J., Ekvall, J. C. and Flores, R. R. (1990) *Proc. Amsterdam Conf. on Light Alloys*, ASM- Internat., Metals Park, Ohio, pp. 277–88.
29. Dulis, E. J. *et al.* (1980) In *The 1980s: Payoff Decade for Advanced Materials*, SAMPE, Azusa, Ca, pp. 75–89.
30. Eylon, D., Field, M., Froes, F. H. and Eichelman, G. E. (1980) In *Materials 1980*, SAMPE, Azusa, Ca, pp. 356–7.
31. Eylon, D., Froes, F. H. and Parsons, L. D. (1983) *Met. Powd. Rep.*, **38**, 567–71.
32. Eylon, D. *et al.* (1985) Mater. Eng., **101**(2), 35–7.
33. Thompson, E.R. (1982) *Ann. Rev. Mater Sci.*, **12**, 213–242.
34. Moll, J. H., Peterson, V. C. and Dulis, E. J. (1983) In *Powder Metallurgy: Applications, Limitations and Advantages*, (ed. E. Klar), ASM, Metals Park, Ohio, pp. 247–98.
35. Witt, RH and Weaver, I. G. (1984) In *Titanium Net Shape Technologies*, TMS-AIME, Warrendale, Pa, pp. 29–38.
36. Froes, F. H., Eylon, D., Kosinshi, E. J. and Parsons, L. (1984) In *PM Aerospace Materials*, Vol. 2, MPR, Shrewsbury, Paper 40.
37. Singer, R. F. (1984) In *P/M Aerospace Materials*, Vol. 1, MPR, Shrewsbury, Paper 3.
38. Moll, J. H., Yolton, C. F. and Chandhok, V. K. (1988) In *PM Aerospace Materials '87*, MPR Shrewsbury, Paper 19.
39. Moll, J. H., Yolton, C. F. and Chandhok, V. K. (1988) *Industrial Heating*, **55**(5), 24–30.
40. Wasielewski, G. E. and Johnson, A. M. (1988) In *PM Aerospace Materials '87*, MPR, Shrewsbury, Paper 18.
41. Cox, A. R. and van Reuth, E. C. (1978) In *Rapidly Quenched Metals III*, Vol. 2 (ed. B Cantor), The Metals Society, London, pp. 225–31.
42. Cox, A. R. and van Reuth, E. C. (1980) *Metals Techno.*, **7**, 238–43.
    Cox, A. R., Moore, J. B. and van Reuth, E. C. (1979) *Proc. Agard Conf. on Advanced Fabrication Processes, Florence, 1978*, Paper 12.
43. Thompson, F. A. (1978) *Met. Prod. Rep.*, **33**, 162–5.

44. Bartos, J. L. (1980) In *PM in Defence Technology*, MPIF, Princeton, NJ, pp. 81–113.
45. Dreshfield, R. L. and Miner, R. V. (1980) *Met. Powd. Rep.*, **35**, 516–20.
46. Patterson, R. J., Cox, A. R. and van Reuth, E. C. (1980) *J. Metals*, **32**(9), 34–9.
47. Tracey, V. A. and Cutler, C. P. (1981) *Powd. Met.*, **24**, 32–40.
48. Wildgoose, P. *et al.* (1981) *Powd. Met.*, **24**, 75–86.
49. Lherbier, L. W. and Kent, W. B. (1990) *Int. J. Powd. Met.*, **26**, 131–7.
50. Kent, W. B. (1990) In *P/M in Aerospace and Defense Technologies*, MPIF, Princeton, NJ, pp. 141–5.
51. Schmitt, H. (1979) *Powder Met. Int.*, **11**, 17–21.
52. Oguchi, M., Inoue, A., Yamaguchi, H. and Masumoto, T. (1990) *Mater. Trans. JIM*, **31**, 1005–10.
53. Anon (1989) *Adv. Mater. Proc.*, **135**(6), 12.
54. Overshott, K. J. (1979) *Electronics and Power*, **25**, 347–50.
55. Edgington, J. (1977) In *Fibre-reinforced Materials*, Institution of Civil Engineers, London, pp. 129–40.
56. Ohnaka, J. (1985) *Int. J. Rapid Solidification*, **1**, 219–36.
57. Raman, R. V., Patel, A. N. and Carbonara, R. S. (1982) *Progr. in Powder Met.*, **38**, 99–105.
58. Holbrook, A. L. (1986) *Progr. in Powd. Met.*, **41**, 679–84, *Int. J. Powder Met.*, **22**, 39–45.
59. Gélinas, C., Angers, R. and Pelletier, S. (1988) *Mater. Lett.*, **6**, 359–61.
60. Pelletier, S., Gélinas, C. and Angers, R. (1990) *Int. J. Powd. Met.*, **26**, 51–54.
61. Lux, B. and Hiller, W. (1971) *Praktische Metallogr.*, **8**, 218–25.
62. Von Allmen, M., Huber, E., Blatter, A. and Affolter, K. (1984) *Internat. J. Rapid Solidification*, **1**, 15–25.
63. Spaepen, F. (1987) In *Undercooled Alloy Phases*, (eds E. W. Collings and C. C. Koch), TMS, Warrendale, Pa, pp. 187–205.
64. Boettinger, W.J. Shechtman, D., Schaefer, R. J. and Biancaniello, F. S. (1984) *Met. Trans. A*, **15**, 55–66.
65. Zimmermann, M., Carrard, M. and Kurz, W. (1989) *Acta Met.*, **37**, 3305–13.
66. Gremaud, M., Carrard, M. and Kurz, W. (1990) *Acta Met. Mater.*, **38**, 2587–99.
67. Midson, S. P. and Jones, H. (1982) In *Rapidly Quenched Metals*, (eds T. Masumoto and K. Suzuki), The Japan Institute of Metallurgy, Sendai, Vol. 2, pp. 1539–44.
68. Jones, H. (1991) *Mater. Sci. Eng.* A133, pp. 33–9.
69. Young, K. P. and Kirkwood, D. H. (1975) *Met. Trans.*, **6A**, 197–205.
70. McCartney, D. G. and Hunt, J. D. (1981) *Acta Met.*, **29**, 1851–63.
71. Moir, S. A. and Jones, H. (1991) *J. Cryst. Growth*, **113**, pp. 77–82.
72. Jones, H. (1984) *Mater. Sci. Eng.*, **65**, 145–56; *J. Mater. Sci.* **19**, 1043–76.
73. Juarez, J. A. Jones, H. and Kurz, W. (1988) *Mater. Sci. Eng.*, **98**, 201–5.
74. Juarez, J. A. and Jones, H. (1987) *Acta Met.*, **35**, 499–507.
75. Morris, D. G. (1981) *Metal Science*, **15**, 116–24.
76. Das, S. K. and Raybould, D. (1985) In *Rapidly Quenched Metals*, (eds S. Steeb and H. Warlimont), Elsevier, Amsterdam, pp. 1787–90.
77. Raybould, D., Zedalis, M., Das, S. K. and Udvardy, S. (1988) In *Rapidly Solidified Materials*, (eds P. W. Lee and R. Carbonara), American Society for Metals, pp. 59–66.
78. Ray, R. (1982) *Met. Progr.*, **121**(7), 29–31.
79. Raybould, D. (1984) *Met. Powd. Rep.*, **39**, 282–6.

80. Skinner, D. J., Bye, R. L., Raybould, D. and Brown, A. M. (1986) *Scripta Met.* **20**, 867–72.
81. Adkins, N. J. E. and Tsakiropoulos, P., (1990) *Proc. P/M 90*, Vol. 2, The Institute of Metals, London, pp. 302–6.
82. de Cristofaro, N. and Henschel, C. (1978) *Welding J.*, **57**(7), 33–8.
83. de Cristofaro, N. and Datta, A. (1985a) In *Rapidly Solidified Crystalline Alloys* (eds S. K. D. Das *et al.*), TMS, Warrendale, Pa, pp. 263–82.
84. de Cristofaro, N. and Bose, D. (1985b) In *Rapidly Solidified Materials*, (eds P. W. Lee and R. Carbonara), American Society for Metals, Metals Park, Ohio, pp. 415–24.
85. Rabinkin, A. (1989) *Welding J.*, **68**(10), 39–46.
86. Anderson, R. E., Cox, A. R., Tillman, T. D. and van Reuth, E. C. (1980) In *Rapid Solidification Processing: Principles and Technologies II*, Claitor's, Baton Rouge, La, pp. 416–28.
87. MacIntyre, R. M. (1986) In *Laser Surface Treatment of Metals*, Martinus Nijhoff, Dordrecht, pp. 545–549.
88. Singer, R. F. and Couper, M. J. (1986) In *Advanced Materials R & D for Transport, Light Metals 1985* (eds R. J. H. Wanhill, W. J. G. Bunk and J. G. Wurm), Les Edns de Physique, Les Ulis, France, pp. 139–50.
89. Adam, C. M. and Lewis, R. E. (1985) In *Rapidly Solidified Crystalline Alloys* (eds S. K. Das *et al.*), TMS, Warrendale, Pa, pp. 157–83.
90. Ovecoglu, M. L. and Nix, W. D. (1986) *Int. J. Powd. Met.*, **22**, 17–30; *High Strength Powder Metallurgy Aluminium Alloys II*, TMS, Warrendale, Pa, pp. 225–42.
91. Mellanby, I. J. (1990) *Met. Powd. Rep.*, **45**, 689–91.
92. Willis, T. C. (1988) *Metals and Materials*, **4**, 485–8.

13

Hot isostatic processing

B.A. Rickinson and S. Andrews

13.1 INTRODUCTION

Hot isostatic processing (HIPping) involves the application of high gas pressure at an elevated temperature to components in order to remove internal pores and voids. HIPping technology was developed during the 1950s, initially as a means of diffusion bonding nuclear reactor components and removing porosity in hard metals (tungsten carbide). However, commercial use is now mainly focused on the densification of high-performance castings and consolidation of metal powders, the aerospace industry being only one of a wide range of industries to utilize the process.

13.1.1 Equipment

A typical HIP vessel design can be seen in Figure 13.1. High standards of design must be adhered to in the construction of a HIP vessel to ensure operational safety when handling high temperatures and pressures. A HIP unit is a sophisticated pressure vessel with a furnace incorporated in the design. There are various designs of vessel: forged monolithic vessels or wire wound vessels with an external yoke closure or threaded end closure. With either type the material to be processed may be loaded from the top or bottom according to customer requirements.

Forged vessels are usually made of ductile low-alloy steel. This material must be so tough, or the design stress so low, that a crack cannot propagate during normal plant operation. Wire wound vessels consist of a high strength alloy steel wire wrapped around a similar steel liner so that the liner is always in a state of compression. In these vessels the critical crack size for fast fracture must be so large that the liner would be perforated before the critical size could be attained. This is called the leak-before-break criterion.

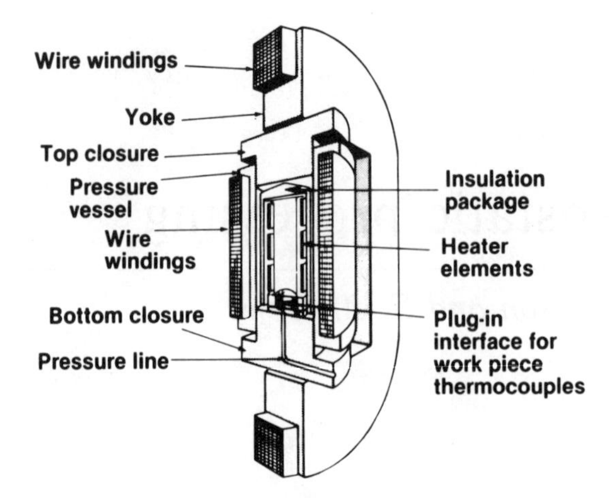

Figure 13.1 Cut-away view of an ABB Quintus HIP unit. Components to be processed are placed within the central core and enclosed within the heater-insulation package. The wire-wound design extends the useful operating life. (Courtesy of ABB Industry Ltd.)

Property requirements for vessel materials are: tensile strength, fatigue strength (fatigue occurring because of temperature and pressure cycling), and, most important, fracture toughness because the elastic strain energy within the vessel may help to propagate a crack.

If the wire wound vessel were to fail the windings would absorb a considerable amount of the energy released, but, as with all HIP vessels, it is preferable to locate the vessel in a concrete pit to provide an additional safety factor. Non-destructive testing of the vessel must be conducted at regular intervals.

The use of argon as a processing gas may develop an asphyxiation hazard should a gas leak occur. It is therefore necessary to monitor oxygen levels in the vicinity of the HIP vessel.

HIP vessels with working volumes in excess of 1 m diameter and 3 m height have now been produced (Figure 13.2), thereby providing the opportunity to achieve considerable economies of scale and making HIPping viable for lower value components. Smaller units are also useful for the economic processing of smaller batches, and are also advantageous when incompatible processing conditions restrict the combination of work to develop a large load.

13.1.2 Furnace

The furnace located within the insulation package of the vessel, is normally made from molybdenum or graphite and is heated electrically.

Figure 13.2 The megaHIP plant at HIP Ltd, Chesterfield, the largest available for sub-contract use in Western Europe at the time of writing.

Other materials can be used, i.e. kanthal, tungsten, platinum, etc. and furnace types can be of radiation or convection design.

13.1.3 Gas

In most HIP systems the pressurizing gas is argon because it is chemically inert and insoluble in most materials. It is also important that the gas be of high purity to avoid the possibility of contamination. Nitrogen, argon–oxygen mixtures or helium can also be used as the pressurizing medium for certain applications, e.g. the HIPping of ceramics.

13.1.4 The HIP cycle

HIPping can best be described as a batch process, i.e. it is carried out in discrete cycles. Table 13.1 details the stages in a typical HIP cycle. Table 13.2 shows typical HIP temperatures and pressures for a range of materials.

Table 13.1 The hot isostatic processing cycle

- Preparation
  Components are placed on trays, or within furnace furniture, and are suitably separated for expansion effects or batch control.
- Loading
  Individual trays are assembled into supporting furniture and the completed charge lowered into the vessel. The vessel is closed.
- Purging
  The pressure vessel is evacuated and purged with the pressurizing gas.
- Equalization
  High purity argon gas is admitted to the vessel until vessel pressure has risen to storage pressure.
- Compression
  Gas is pumped via a compressor from the storage system to the vessel until the required start pressure is reached.
- Heating
  The required HIPping temperature is achieved and sustained for between 1–4 hours depending on the application.
- Cooling
  At the end of the sustain period, the furnace is allowed to cool.
- Charge removal
  The charge is removed from the pressure vessel.

Table 13.2 Typical HIPping temperatures and pressures

Alloy type	*HIP temperature(°C)*
Superalloys	1180–1230
Steels	1100–1180
Nimonic	1100–1150
Titanium	900–950
Copper	750–950
Aluminium	480–520
Magnesium	300–450
Zinc	250–350

13.1.5 Quality controls

Temperature monitoring and control is achieved by employing thermocouples within the furnace elements and distributed throughout the workload. Temperature tolerance is usually specified to within + 15/– 10 °C of the nominal value, although in practice, closer control is normally possible. High purity argon gas is used and impurity levels are monitored during the cycle. Concentrations of oxygen, oxides of carbon, and hydrocarbons are all measured to ensure that gas purity remains within the customer specified limits.

Pressure is measured both by transducers and also by calibrated analogue pressure gauges, all traceable to National Standards. Process data are recorded during the cycle and full traceability of records is maintained to Civil Aviation Authority and Ministry of Defence requirements.

13.2 REMOVAL OF POROSITY

Porosity may be caused by shrinkage occurring as the material cools through the liquid-to-solid transition, by gas becoming trapped within the material, or by the generation of gas by reaction. During the HIP cycle material moves to fill the voids by plastic yielding. Beyond this stage creep and diffusion mechanisms ensure that complete closure of pores and removal of defects occurs: the boundaries of the defect are not simply pushed together to form a planar crack.

A component will not normally distort, because pressure is applied isostatically, i.e. equally in all directions. Limited dimensional change only occurs if porosity levels are very high, are localized, or the overall size of the component to be processed is very large.

13.2.1 Surface connected porosity

HIPping cannot remove surface connected porosity. This is because the gas impinges on the porosity as though it were an extension of the surface of the component. However, this problem can be overcome by coating or encapsulating the component, thus, in effect, making the surface connected porosity internal. This may then be removed by HIPping, and the coating or encapsulating material is subsequently machined off. Encapsulation techniques are explained in section 13.6.

13.3 BENEFITS OF HIP

The elimination of internal porosity results in a number of benefits:

- **Enhanced component integrity** Reject and re-work levels can be reduced, minimizing both work-in-progress and delivery times.
- **Improved tensile properties** Tensile strength and elongation are increased and scatter of results is reduced. The effect varies with alloy type and casting technique but typical improvements are a 5–10% increase in tensile stength and 50–100% increase in elongation. Since there is little change in grain size during HIP treatment, any change to yield strength is minimal.
- **Enhanced fatigue properties** The improved fatigue performance of HIPped castings can rival that of forgings, especially for steels.

Typically a 3–10 fold increase in fatigue life results from HIP of a cast structure, together with an improvement in stress to fracture.

- **Improved machined surface finish** The removal of porosity is useful for the improvement of seal surfaces or where coatings are to be applied. Such improvements in mechanical properties allow castings to be considered for applications which have traditionally been reserved for forged components.

13.4 APPLICATIONS OF HIP

13.4.1 Nickel alloys

The precision casting of nickel-based superalloys offers a near-net-shape production route which is used extensively in the aerospace industry for the production of blades and vanes for gas turbines. As an example, the blades in Figure 13.3 are produced by such a route.

Such components are specified in nickel or cobalt superalloy to provide high temperature creep resistance and oxidation resistance where tolerances must be maintained to within a few microns even at operating temperatures in the range 1000–1200 °C under high centripetal stress. Forging routes are not practical other than for the simplest solid. For this reason, most turbine blades are now made by vacuum investment casting to a near-net shape, with HIPping used extensively to remove residual porosity.

HIPping at a temperature within the range 1160–1220 °C at a pressure of 103 MPa and for 4 hours reduces the cored microstructure and refines eutectic phases. HIPping at higher temperatures leads to almost complete homogenization.

The main benefits of HIPping nickel alloys are the enhancement of fatigue strength and tensile ductility, reduction in the scatter in creep life and decrease in foundry scrap and inspection costs.

HIPping can also be used to rejuvenate used turbine blades. During operation under creep conditions, the structure slowly coarsens and micron sized voids gradually begin to form by mechanisms such as grain boundary sliding. Creep voiding substantially increases the potential for early failure of the blade. Given the high initial cost of a turbine blade, there is a large incentive for an operator to attempt to prolong the life of blades, within limits, by the elimination of creep voids by HIPping.

13.4.2 Steel

Investment cast steel parts suffer similar microcavitation effects arising from shrinkage during the liquid/solid transition. This form of porosity defect contributes to a marked reduction in properties when compared to those of forged products.

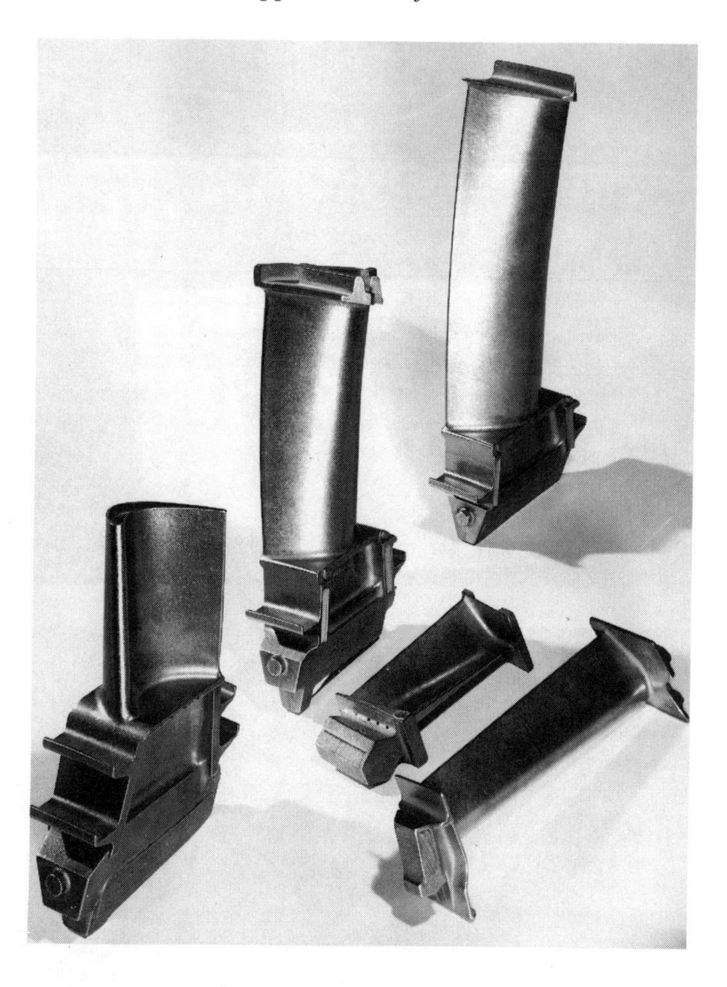

Figure 13.3 Cast superalloy turbine components (courtesy of A.E. Turbine Components Ltd).

The correct selection of HIP parameters develops a balance of improved properties which are particularly appropriate where improved performance of the component is the ultimate objective. Such applications as hydraulic thrust reverser components, valve parts and reciprocating parts for automotive use fall into this category.

In recent years the growth of applications for steel investment castings in aerospace has also increased the use of HIP. Indeed, the improvement of ferrous investment castings to compete with forgings for aerospace applications is now a major market. Using a preferred route of vacuum melting and vacuum investment casting followed by HIPping, internal

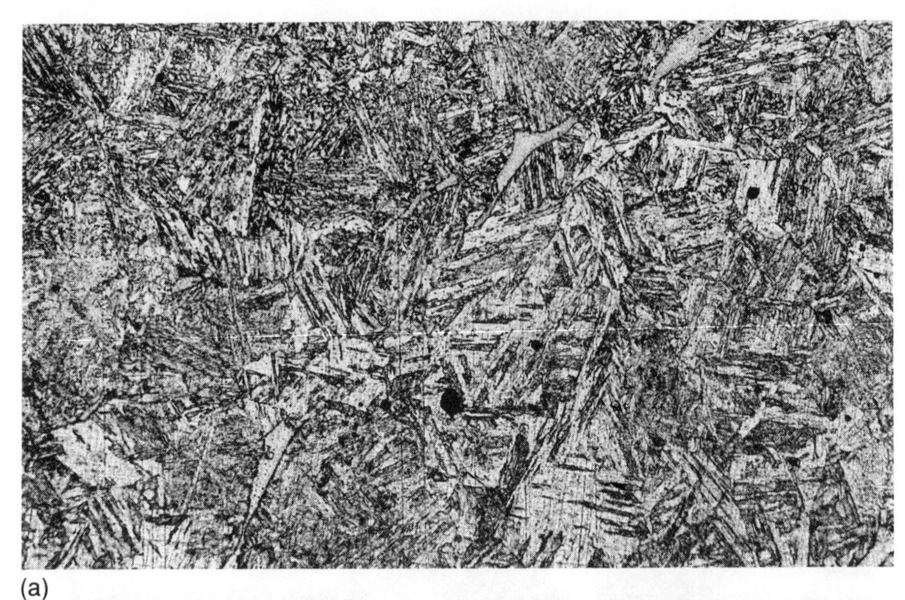
(a)

(b)

Figure 13.4 Micrographs of 17.4 PH steel: (a) as-cast condition × 200; (b) cast and HIPped condition × 200.

porosity is eliminated and properties are improved. Figure 13.4 shows the microstructure of 17.4 PH steel in the as-cast and cast-and-HIPped conditions.

The fatigue life of investment casting made of materials such as maraging steels and low alloy steels can be doubled, with the endurance ratio rising from 0.2 to 0.4. This is as good as, or better than, the best wrought or forged components which typically have ratios of 0.36–0.42.

Proof stress, ultimate tensile strength and elongation are improved by HIP treatment to match properties of wrought material. It should be noted that data scatter is reduced by comparison to a rolled/forged part, since the cast part is relatively isotropic with little texture or preferred orientation.

13.4.3 Aluminium

Cast aluminium alloys have, and are expected to continue to play, an important role in the aerospace industry. Strength has been increased to match that of components typically machined from solid, and product integrity and reproducibility have also been improved. This is largely due to the application of hot isostatic pressing to such parts as, for example, the aircraft steward seat casting shown in Figure 13.5, which is made in 507–1306 aluminium alloy. The decision to HIP the steward seat was taken because the end-user required a Class 1 radiographic standard in all parts of the casting on a consistent basis with existing tooling.

As with other alloys the benefits of HIPping aluminium are: the removal of porosity, and improvement in mechanical properties, of

Figure 13.5 Aluminium alloy aircraft steward seat casting (courtesy of Kent Aerospace Castings Ltd).

Table 13.3 Typical property improvements after HIPping of investment and sand-cast aluminium alloys

Alloy Type			
LM 16	(Al–5% Si–1.25 Cu–0.5 Mg)		
	(all material in heat-treated condition)		
	TS (MPa)	0.2 PS (MPa)	El (%)
As-cast	230	206	1.6
HIPped	266	205	3.3
L99	(Al–7% Si–0.4 Mg)		
	(all material in heat-treated condition)		
	TS (MPa)	0.2 PS (MPa)	El (%)
As-cast	258	211	1.9
HIPped	275	215	4.0
Fatigue comparison L99			
	Stress (Mpa)	Life (cycles)	
As-cast	137	2×10^5	
	103	1×10^6	
HIPped	137	6×10^5	
	103	3×10^6	
L155	(Al–4% Cu–1% Si)		
	(all material in heat-treated condition)		
	TS (MPa)	0.2% PS	El (%)
As-cast	310	221	5.5
HIPped	351	229	8.0

X-ray standard and of machined surfaces, plus reduction of property scatter. Table 13.3 gives a range of test results indicating the improvement of properties after HIPping in investment and sand-cast aluminium alloys. It is important to note that after HIP treatment and removal of porosity from cast alloy products remaining fracture initiation sites are large intermetallics, inclusions or surface-connected porosity since these represent the largest defects within the structure.

Considerable research data is available in the literature to demonstrate the positive benefits from HIP treatment of cast alloys. New processes, including Densal®, are available to remove porosity and enhance properties at low cost. In addition, the installation of larger HIP systems, bringing economies of scale, has made the HIPping of lower value materials such as aluminium more viable.

13.4.4 Titanium

The HIPping of titanium and its alloys is now well established, particularly for both rotating engine and static frame parts destined for the aerospace industry. Special techniques are required to overcome casting problems of titanium alloys which relate to low fluidity and high chemical activity of the liquid metal. Although internal microporosity is minimized by the careful control of foundry practice and the imposition of centrifugal force, there is still a need, particularly in regard to aerospace use, to ensure the absence of internal cavities. HIPping is applied to 100% of alloy castings for aerospace use to maximize properties and casting reproducibility. It is necessary to weld any surface penetrating defects prior to HIP treatment.

Titanium alloys are typically processed at a temperature of 920 °C and a pressure of approximately 100 MPa. Because of the metal's high chemical activity the argon used must be of very high purity. The HIP temperature is close to, but below the β transus temperature, to avoid excessive grain coursening. Since oxygen is fully soluble in titanium at HIP processing temperatures and pressures this factor ensures that any removed voids are free from oxide inclusions after the HIP cycle.

Low and high cycle fatigue and stress rupture properties are substantially enhanced by HIP, as shown in Figures 13.6 and 13.7, and the development of a precision investment cast and HIP manufacturing route, as opposed to machining from solid, achieves significant cost savings. The removal of feeding gates and grinding are significant yield and processing costs in the making of titanium castings. For large structural parts the complex gating systems, which consume large amounts of titanium alloy, can be minimized if HIPping densification is specified. Effectively some microshrinkage can be tolerated since this will be removed during the HIP processing. The homogenization of the cast

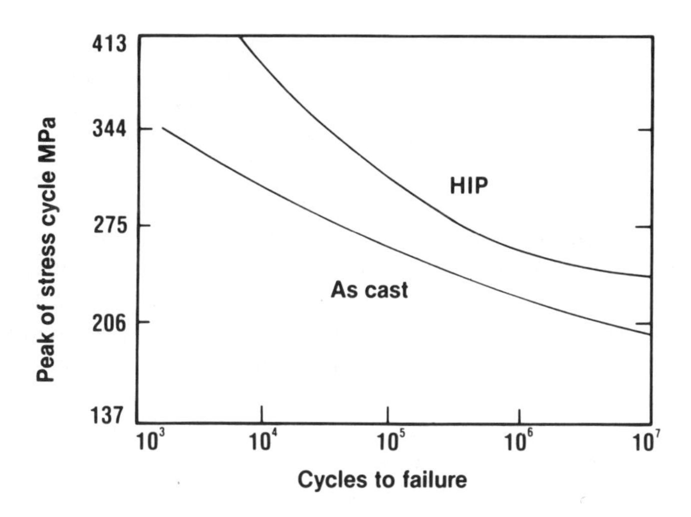

Figure 13.6 Effect of HIPping on fatigue properties of Ti–6 wt% Al–4 wt% V at 316°C.

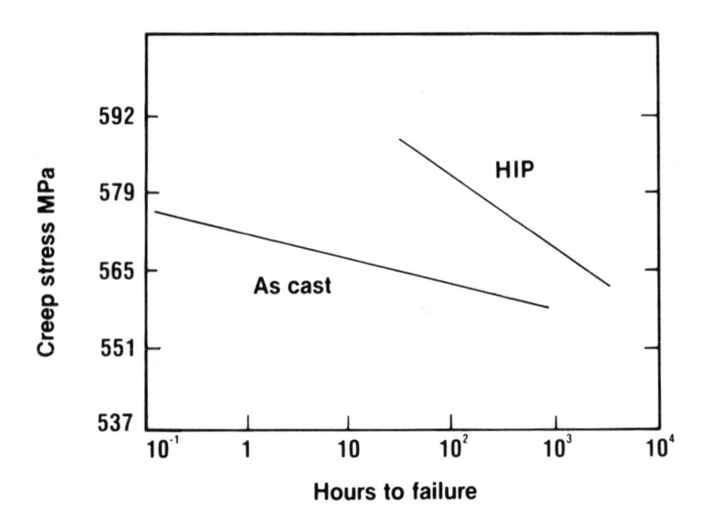

Figure 13.7 Influence of HIPping on creep life of Ti–6 wt% Al–4 wt% V at 399°C. (Figures 13.6 and 13.7 from MCIC Report, November 1987, MCIC 77–34, *Hot Isostatic Processing*, p. 63, Figure 35, prepared by Batelle Memorial Institute.)

structure also results in improved corrosion properties and more uniform response to chemical machining.

13.5 POWDER PRODUCTS

The utilization of HIPping was given considerable impetus by the development of gas atomized high quality pre-alloyed powders. Such powders, being spherical, free flowing and extremely clean, can offer good property mixes in the consolidated state (Chapter 12). The use of high gas pressure by HIP was found to be an extremely controllable means for such particle bonding/densification.

Powder for consolidation by HIPping is encapsulated in a gas-tight envelope which readily deforms at the processing temperature, thereby transmitting the pressure. The prerequisites for the envelope are that it must be strong, gas tight, inert and plastic under the applied temperature and pressure conditions and compatible with the material to be pressed in order to minimize surface contamination by the envelope material. It must also be readily removable. As mentioned in section 13.2.1, encapsulation techniques are also used for surface sealing where porosity is surface connected, and for interface bonding.

The powder capsule may be made of sheet metal, glass or ceramic. Sheets of mild steel, stainless steel or nickel alloy are most often used, and these are removed by machining or chemical milling. Glass may be used when sheet metal containers do not have the required properties, e.g. for HIPping ceramics at temperatures of 1500–2000 °C, and ceramic solid material is used when support for a large mass of loose metal powder is required.

Evacuation of the capsule prior to sealing must be rigorous or residual gas will dissolve into the component or chemically combine with the powder to reduce the cleanliness of the compacted product. It is often necessary to evacuate such capsules at temperatures in excess of 350 °C to ensure freedom from adsorbed water vapour. The volume of the encapsulated powder will reduce by as much as 30–40% during processing as a result of densification. Such shrinkage should be isotropic, although care must be taken to ensure an even packing density to achieve this effect.

Powder methods are most often used when a cast and forge route is difficult or wasteful, when the combination of properties required cannot be readily achieved by other methods, or when a powder route is cheaper than other routes, e.g. for high melting point metals. HIPping is used on encapsulated powders to produce billets for further shaping, semi-finished shapes for closed die forging and near-net shapes for final machining.

Current commercial powder applications are for toolsteels, stainless and high speed steels and nickel or cobalt based superalloys. There is also developing interest in relation to other materials such as ceramics and metal matrix composites.

13.5.1 Toolsteels

Toolsteels are used for tool and die purposes involving metal cutting, shaping and forming. The alloy must be capable of high hardness, abrasion resistance, toughness and dimensional stability. Production of such high speed toolsteel billets from encapsulated powders was one of the first major commercial applications of HIP equipment. The size of such billets is only limited by HIP unit size. Gas atomized powders are normally used so that each particle has nominally the composition of the melt and a fine carbide structure.

HIPping is carried out within the temperature range 1050–1160 °C and at a pressure of 100 MPa. The time at temperature and pressure is as short as possible to avoid unnecessary coarsening of the microstructure but sufficient to ensure full density at the billet centre. Consolidation of the powder gives a product with a uniform structure of fine carbides. This results in the following advantages:

- refined structure regardless of size
- improved hot workability
- improved grindability

- rapid hardening response due to fine carbides
- fine grain size in the final product
- improved toughness and isotropic properties
- more predictable dimensional changes during heat treatment
- minimal hot working required due to the fine as-consolidated structure

Most tool steel powder consolidated by HIPping is shaped to conveniently sized billets which are then hot forged and converted to bar stock. Such stock is made into cutting tools and die tools which have shapes too complex to be achieved by HIPping directly to a near-net shape. Tools produced by a powder metallurgy route are often more expensive but more cost effective than those produced by conventional methods. Such cost benefits arise because HIPped products demonstrate improved performance at higher cutting speeds, are less prone to tool chipping and have reduced regrinding costs under severe operating conditions.

13.5.2 Superalloy powder

A powder route is an alternative route for nickel-based superalloys, particularly for components such as compressor discs. Atomized superalloy powder is HIPped direct to a near-net shape. The advantages are that segregation is avoided and the size and distribution of carbides and intermetallics are controlled. For example, the combination of dissimilar alloys for the hub and rim section of a wheel prepared by a HIP powder route, combines toughness and enhanced creep properties at centre and edge respectively. Costs are reduced by following a powder metallurgy route because material is conserved; however, shape changes are more difficult to predict.

13.5.3 Ceramic powders

The use of HIPping for the consolidation of ceramic powders is a small but growing application, particularly because the production of net or near-net shapes can reduce costly and difficult machining operations. A number of processing routes have been developed, which in the main employ a form of glass encapsulation to transmit HIPping gas pressure. The enclosure of injection-moulded ceramic parts within a multi-layer glass capsule has, with subsequent HIP processing, developed a standard of product which is suitable for high integrity rotating parts subject to constant or cyclic thermal stress. Structural ceramic materials such as silicon carbide and silicon nitride are HIPped at temperatures between 1750 and 2200 °C and pressures up to 200 MPa to produce fuel-efficient and lightweight turbine components for civilian and military use. The high melting point of these materials is their main advantage relative to metal alloys such as the superalloys, but this also imposes demanding processing conditions, both for HIPping and for any other manufacturing routes.

13.5.4 Metal matrix composites

Particulate MMCs may be made by powder routes, and HIPping can play a valuable role within such a route to maximize properties. In the simplest case, both metallic and non-metallic powders are carefully blended in the appropriate volume fractions. Control of particle size and mixing method is necessary to ensure a uniform microstructure of the final part. Powders may be pre-compacted by conventional uniaxial pressing or HIPping techniques before the 'green' handleable part is inserted within a sheet metal capsule. During processing at temperature and pressure the softer metal matrix is extruded between the harder ceramic particles to promote compact densification. Matrix materials including aluminium, titanium, copper and ferrous alloys are currently under development. Solid-state bonding helps to avoid the adverse reactions which may occur between the particles and the matrix at elevated temperatures in liquid–metal routes. HIPping may be equally appropriate to the densification of MMCs produced by liquid–metal infiltration routes and stir-cast routes.

13.6 DIFFUSION BONDING

As a direct result of the requirements of the aerospace industry considerable resources have been devoted to the development of HIP diffusion bonding in recent years. During the HIP diffusion bonding process two surfaces are held in contact at a high temperature. The bonding pressure applied causes the surface asperities to move close enough together for short range interatomic forces to operate. Although extensive macroscopic plastic deformation does not occur, localized, or microplastic flow does take place at points where surface asperities come into contact. The pressures at the points of contact are high because the contact areas are small and locally the yield stress can thus be exceeded. Once the surfaces are in contact, inter-diffusion across the interface can start to occur and this is the origin of the bond. The mechanisms of bonding are discussed in more detail in chapter 10. An interlayer is sometimes placed between the surfaces to be bonded. Made of a thin layer of metallic material, it can prevent the formation of brittle compounds and alleviate stresses due to thermal expansion coefficient mismatch.

HIP diffusion bonding exhibits certain advantages over traditional joining methods such as welding. Pressure is applied isostatically during

HIP bonding, and therefore a coherent metallurgical bond can be achieved across a large interfacial area with many re-entrant angles. The surface preparation need not be as stringent with HIP bonding because some plastic deformation occurs at the interface. In addition, complex cylindrical bonds can be achieved using the HIPping route, and powders and porous bodies can be simultaneously densified and bonded to a substrate. There is also evidence to suggest that in the case of HIP bonding of a metal and ceramic, the metal will flow into surface defects and fill the undulations on the surface of the ceramic to a greater extent than in conventional diffusion bonding.

Encapsulation of the component prior to processing is necessary in order to isolate the interface from the isostatic gas medium. Methods of encapsulation for HIP diffusion bonding fall into three categories:

- Direct welding of the circumference of the contact area between two parts. This is the simplest method but is not always applicable.
- Application of a sleeve around the contact area between the two parts. This is only applicable for simple geometries, since the sleeve must be welded to the workpiece so that it is leak-proof.
- Encapsulation of part or all of the workpiece. This method eliminates welding problems and almost any geometry can be produced. This method also allows diffusion of a layer of powder material to a solid part.

Figure 13.8 details the steps taken in HIP diffusion bonding.

Diffusion bonding is normally enhanced by increasing temperature and pressure. However, the time at the sustain temperature should be a minimum, both for cost considerations and to avoid the formation of brittle intermetallics, excessive grain growth and secondary recrystallization.

Diffusion bonding is also used for the production of powder parts and powder-clad parts for actuator systems to develop inherently good corrosion resistance materials with improved fatigue strength, which are both weight and cost competitive.

Bar and near-net shapes may be produced from stellite-type materials using a diffusion bonding route. Such components are difficult to fabricate using conventional techniques.

13.7 OTHER APPLICATIONS

13.7.1 Surface engineering

Alloys used in the hot sections of gas turbines such as those based on nickel or cobalt have been developed on the basis of their elevated-temperature mechanical properties. At high temperatures, aluminium or

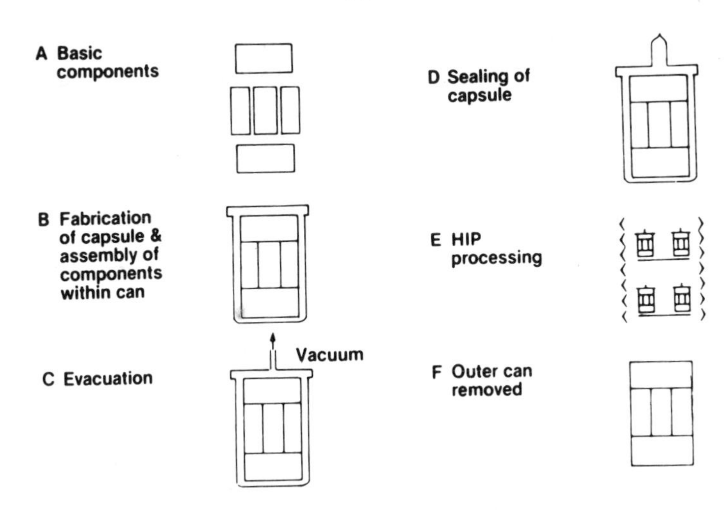

Figure 13.8 Steps in HIP diffusion bonding.

chromium oxides will be formed from aluminium or chromium present in the alloy. While the composition of some alloys will permit the formation of protective oxides in this way, spalling may result from thermal cycling, which will lead to a rapid depletion of the aluminium or chromium content of the alloy. Their life can be extended by providing additional resistance in the form of a discrete protective coating.

Aluminide diffusion type coatings are widely used on high pressure gas turbine blades. Traditionally these coatings have been produced by processes such as pack and slurry cementation or hot dipping. Recent work has shown, however, that an ion vapour deposition (IVD) and HIPping route can produce coatings which are equivalent to those formed by pack techniques but which have certain advantages.

The coatings are formed by placing the component in an IVD chamber, electron beam melting and ionizing the aluminium, and depositing a uniform layer of the aluminium on to the component. Subsequent HIP treatment diffuses in the aluminium to form nickel aluminide. This process route develops a coating as an integral part of the surface and sub-surface structure rather than as a discrete attached layer. Other advantages of the IVD/HIP route include the ease of masking components during IVD simply by using aluminium foil or tape, and a high degree of reproducibility and uniformity of coating thickness compared with conventional process techniques, especially on larger components. It is also feasible to modify the coating to produce more complex aluminides, e.g. the incorporation of platinum to increase the coating life at higher temperatures is possible.

It has also been shown that a nitride-based diffusion coating on titanium alloys can be developed using a modified HIP atmosphere. This is used to improve the wear resistance of components.

13.7.2 Infra-red materials

Zinc sulphide is an advanced polycrystalline material used for infra-red transmission applications. It is produced by the chemical vapour deposition (CVD) process and has several features. It combines high hardness with low cost, has uniform optical properties and exhibits excellent high temperature performance. However, in its un-HIPped state it scatters light in the visible spectrum owing to growth imperfections and hydride impurity. HIP is used to remove microporosity and cure the deficiencies.

Vacuum cast calcium aluminate is another infra-red transparent glass used in aerospace applications. The material is transparent at both visible and infra-red wavelengths, has high mechanical strength and a wide spectral range. HIPping removes pores and improves the optical quality of the material.

Such materials are suitable for infra-red windows on high speed aircraft, military surveillance equipment and other disposable aerospace applications.

INDEX

Page numbers appearing in **bold** refer to figures and page numbers appearing in *italic* refer to tables.